Water Stress in Crop Plants and Its Management

Editors

Prabhat Kumar Srivastava
Assistant Professor of Botany
KS Saket PG College, Ayodhya - 224123, UP, India

Parul Parihar
Assistant Professor of Botany
Department of Biosciences and Biotechnology
Banasthali Vidyapith, Niwai - 304022, Rajasthan, India

Richa Upadhyay
Assistant Professor of Botany
Mihir Bhoj PG College, Dadri, GB Nagar - 203207, UP, India

CRC Press
Taylor & Francis Group
Boca Raton London New York

CRC Press is an imprint of the
Taylor & Francis Group, an **informa** business

Cover illustration courtesy of Dr. Yashwant Singh

First edition published 2025
by CRC Press
2385 NW Executive Center Drive, Suite 320, Boca Raton FL 33431

and by CRC Press
4 Park Square, Milton Park, Abingdon, Oxon, OX14 4RN

CRC Press is an imprint of Taylor & Francis Group, LLC

Library of Congress Cataloging-in-Publication Data (applied for)

ISBN: 978-1-032-56123-3 (hbk)
ISBN: 978-1-032-56124-0 (pbk)
ISBN: 978-1-003-43398-9 (ebk)

DOI: 10.1201/9781003433989

Typeset in Times New Roman
by Prime Publishing Services

Preface

Water stress is a condition when a plant transpires more than the water it takes in from the soil. Water stress is caused by drought, inadequate rainfall, high salinity, extreme temperatures and higher light intensities. Water stress has become a growing concern due to global warming and climate change. Water stress causes several negative effects on plants viz. reduced growth, development and photosynthetic activity and hence productivity. Thus water stress poses a severe threat to food security amid a global population rise. More reactive oxygen species (ROS) are produced in response to water stress leading to oxidative stress, disruption in the electron transport chains and metabolic abnormalities. Plants face mineral deficiency under water deficit and become more susceptible to pests and diseases. Water stress not only negatively affects the morphological, anatomical, physiological, biochemical, hormonal, cytological, reproductive and embryological parameters but also a considerable spectrum of adaptations has been identified if the water stress lasts for a longer period. Plants deploy enzymatic and non-enzymatic antioxidant defense mechanisms to mitigate stress-induced oxidative damage. Plants can also adapt to water stress by osmotic adjustments, osmoprotection and stomatal closure favoring the plant survival by maintaining cellular turgor pressure. Plants respond to water stress by activating a signaling cascade involving metabolites, free radical scavengers, antioxidants, unsaturated fatty acids and phytohormones.

Innovative approaches have shown possibilities for reducing damage, such as rhizobacterial inoculation, mycorrhizal fungi and nitric oxide treatment. Genetic engineering and focusing on enhancing key enzymes strengthens plant resilience against water scarcity. Exogenous application of compounds like glutathione, proline and polyamines can increase the plant's ability to cope with water stress. Further, to develop high-yielding water stress-tolerant cultivars for changing climates integrating information across molecular pathways using genomics, proteomics, metabolomics and transcriptomics influenced by microarrays and RNA-sequencing data is imperative. The role of miRNAs has also been featured utilizing CRISPR/Cas9 technology.

The climate change severity needs a deep understanding of the consequences of water stress. This book is the result of the strenuous effort of many scholars working across the world, especially in India and Kenya. The editors have every hope that this concerted effort would enrich the current knowledge about water stress in plants and will open newer paths for future research endeavors in exploring drought-tolerant crop varieties with potential agronomic value, ecosystem restoration and land rehabilitation. The editors thankfully acknowledge all the contributors to this book. The editors gratefully acknowledge the CRC Press, Taylor & Francis Group, Florida, The USA, which made the book in its present form.

Prabhat Kumar Srivastava
Parul Parihar
Richa Upadhyay

Contents

1 | An Introduction to Water Stress in Plants

Dwaipayan Sinha[1*] and Pramod Kumar Tandon[2]

[1] Department of Botany, Government General Degree College, Mohanpur, Paschim Medinipur, India

[2] Department of Botany, Lucknow University, Lucknow, Uttar Pradesh, India

Water stress is a growing concern due to changes in climatic conditions and global warming. Climate change has engulfed much of the globe, leading to aridity and drought that affect human livelihoods. Water scarcity has impacted plant growth and productivity. This poses a threat to food security amid a booming global population. This chapter provides an overview of the global water stress situation. Efforts have been made to highlight the impact of water stress on plants and their growth.

1. Introduction

The term "water stress (WS)" refers to the inability, or perceived inability, of human populations and ecosystems to meet their demand for freshwater. Compared to "water scarcity," the term "water stress" is indeed broader and more inclusive. Water shortage occurs when the requirement for water surpasses the amount of water available in a certain area. It takes into account several material aspects that are associated with water resources. These include water access, quality, and abundance. Water accessibility is usually defined by the availability of infrastructure as well as the cost of water, among other things. The consumption and extraction of water both provide valuable data that can illuminate the current level of water stress being experienced (CEO WATER MANDATE SECRETARIAT, 2017a). A geographic region might experience water shortage when there is an imbalance between the availability and demand for water. An important challenge that the world's poorest countries face is a water supply that is either limited or inadequate. However,

*Corresponding author: dwaipayansinha@hotmail.com

the people of these nations are not the sole individuals who must shoulder the strain of water scarcity (Harhay, 2011). According to UNICEF, a severe scarcity of potable water impacts a minimum of four billion people globally, or around two-thirds of the entire world population, for a minimum of one month in a given year. Poor water supply in their home nations affects over two billion people. It is conceivable that by 2025, fifty percent of the global population would reside in areas with a water shortage. Reports indicate that by 2030, a severe water shortage could necessitate the relocation of as many as 700 million people due to water scarcity issues. Around the world, one in four children will reside in areas with a serious water shortage by the year 2040 (Water Scarcity | UNICEF, n.d.). Climate change is causing increasingly severe droughts and floods, already impacting people's access to water. One of the key causes of this issue is rising global temperatures. Climate change impacts the water cycle, altering the timing, location and intensity of rainfall. It also results in weather events worsening over time. In the coming years, rising global temperatures will lead to greater water evaporation, resulting in higher atmospheric water vapor levels and more frequent heavy, and intense rainstorms (National Geographic). Temperature increases cause disruptions in rainfall patterns and the whole water cycle, affecting water scarcity and water-associated risks. Water and climate change are closely related. Climate change is widely acknowledged to have complex effects on water resources. The majority of the impacts of climate alteration are related to water, such as irregular rainfall patterns, shrinking ice sheets, sea level rise, flooding, and droughts (United Nations, n.d). Due to changes in climate and precipitation patterns, the drought spells are likely to increase making the earth drier than before (US Geological Survey, n.d). Water stress has a deleterious impact on various aspects of plant physiology, including photosynthetic ability. If the stress is prolonged, plant growth and output suffer considerably (Osakabe et al., 2014). Drought is defined not only by water scarcity or water stress (WS) but also by rising temperatures, all of which affect water accessibility and result in scarcity. Because of global climate change, the intensity and duration of hot, dry days will increase significantly. The likelihood of drought and temperature stress occurring simultaneously is also expected to rise, which will affect several agronomic parameters, including mass and other development characters, phenology, and features that affect crop output. Additionally, important physiological characteristics such as water content, stability of membranes, photosynthesis, and associated characteristics including quantity of pigment, stomatal conductance, and chlorophyll fluorescence will all be severely affected (Bhardwaj et al., 2022). All these have resulted in a decrease in the yield of crop plants (Lesk et al., 2016; Zhang et al., 2018). This chapter is an attempt to broadly overview water stress (WS) in plants. Efforts have been made to highlight the causes and different patterns of water stress (WS) in plants.

2. Types and Quantum of Water Stress

Water stress is defined as the condition that arises when the demand for water, or its extraction significantly exceeds the available renewable water resources. The World Resources Institute calculates water stress by assessing the proportion of total yearly water extractions to total renewable water supplies in a given area. This is utilized as a benchmark WS. According to this fraction (% of extractions to renewable resources), the following categories of WS are demarcated:

<10% = low stress
10-20% = low-to-medium stress
20-40% = medium-to-high stress
40-80% = high stress
>80% = extremely high stress (Ritchie & Roser, 2017).

Figure 1 represents the water stress by countries.

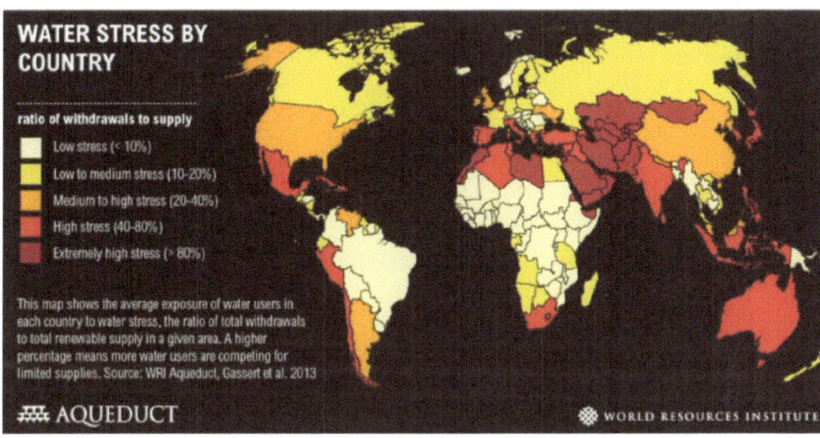

Figure 1: Water stress by countries (Kuzma Samantha et al., 2023d). (World Resources Institute)

According to the United Nations, water scarcity affects around 700 million people across 43 nations. A nation is considered to be experiencing water stress when its total renewable freshwater resources per person per year range from 1,000-1,700 cubic meters (Chibani Achref, 2022). By 2025, when the number of people living in places with absolute water scarcity reaches 1.8 billion, over two-thirds of the global race may be susceptible to water stress. According to current environmental change projections, between 75 and 250 million people in Africa are expected to live in areas with extreme WS by 2030. There would be a population shift of 24 million to 700 million people because of water constraints in dry and semi-dry areas. There are more countries experiencing water scarcity in Sub-Saharan Africa than in any other region (United Nations, n.d). The global need for water is predicted to increase in the years to come. The population shifts to urban areas will increase the demand for services. A growing middle class might demand more water for agriculture and power plants. The World Research Institute examined and ranked future water stress in 167 nations using a combination of climate simulations and social situations for the period between 2020 to 2040. In 2040, it is predicted that 33 countries would experience severe water scarcity (Maddocks et al., 2015) (Figure 2). Figure 3 pictorially represents the water-stressed countries in 2040.

RANK	NAME	ALL SECTORS
1	Bahrain	5.00
1	Kuwait	5.00
1	Qatar	5.00
1	San Marino	5.00
1	Singapore	5.00
1	United Arab Emirates	5.00
1	Palestine	5.00
8	Israel	5.00
9	Saudi Arabia	4.99
10	Oman	4.97
11	Lebanon	4.97
12	Kyrgyzstan	4.93
13	Iran	4.91
14	Jordan	4.86
15	Libya	4.77
16	Yemen	4.74
17	Macedonia	4.70
18	Azerbaijan	4.69
19	Morocco	4.68
20	Kazakhstan	4.66
21	Iraq	4.66
22	Armenia	4.60
23	Pakistan	4.48
24	Chile	4.45
25	Syria	4.44
26	Turkmenistan	4.30
27	Turkey	4.27
28	Greece	4.23
29	Uzbekistan	4.19
30	Algeria	4.17
31	Afghanistan	4.12
32	Spain	4.07
33	Tunisia	4.06

http://ow.ly/RiWop

WORLD RESOURCES INSTITUTE

Figure 2: Top water stress countries in the year 2040 (Maddocks et al., 2015). (World Resources Institute)

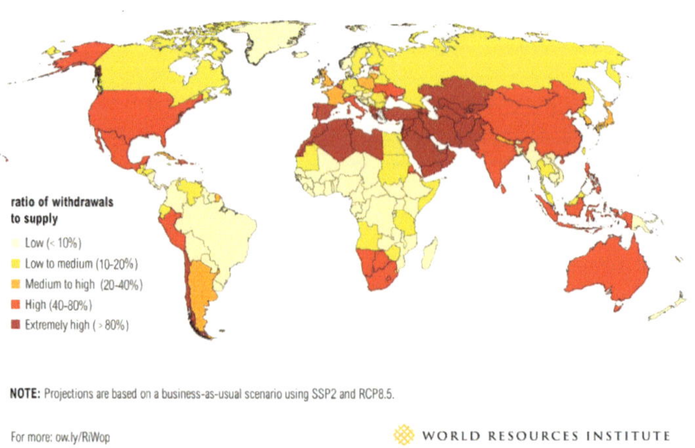

Figure 3: Pictorial representation of water stress by countries in the year 2040 (Maddocks et al., 2015). (World Resources Institute)

As evident from Figure 2, 33 countries could face severe water shortages by the year 2040. As many as nine nations in the Middle East will be amongst the 33 most water-strained by 2040. The region relies significantly on groundwater and desalinated sea water, making it globally the lowest water-protected. The region will also confront unprecedented water-related issues shortly (Maddocks et al., 2015).

The Arabian region faces a severe water shortage because it boasts 5% of the global populace yet just 1% of the globe's freshwater reserves. About a dozen Arab countries have critical water shortages, according to UN assessments. Sustainable water resources supply less than 500 m^3 of water per person per year. While the Middle East's agricultural sector contributes less than 10 percent to the region's GDP, it uses 85 percent of the region's fresh water. Increasing water use and the loss of freshwater supplies can be traced back to a few key causes.

- Superior living environments.
- Urbanization and industrial production growth.
- Rapid population increase.
- Failure to implement water-saving practices.
- The inability to desalinate seawater due to a lack of energy.
- Mass internal migration as a result of regional disputes.
- Disinterested citizens.

About 70% of Lebanon and Jordan's freshwater comes from perennial rivers. On the other hand, surface water and seasonal rivers are crucial for Oman, Saudi Arabia, Syria, UAE, and Yemen. In the other nations of the region, groundwater supplies only a third of the water required. Water scarcity on both the national and regional levels is creating disputes due to the unsustainable rate at which groundwater is being extracted. Approximately 80% of Yemen's internal conflicts are caused by the country's increasing population and inadequate water management (Abumoghli Iyad, 2021). North Africa is one of the world's most water-stressed areas. According to the World Resources Institute, Libya is suffering from acute WS, Tunisia, Algeria, and Morocco are suffering from extreme WS, while Mauritania is suffering from moderate to elevated WS. Rapid population growth brought on by various factors, such as rising income levels and expanding urbanization, has made water shortages worse in North Africa and around the world. Datasets of the year 2018 from the World Bank show that most of the countries in North Africa are water-scarce, with less than 1,000 cubic meters of renewable fresh water per person (Chibani Achref, 2022).

India is considered one of the most water-stressed countries in the world, as it holds only about 4% of the global water supply while supporting 18% of the global population. According to a recent study conducted by the NITI Aayog, a considerable portion of Indians endure extreme WS. This problem is exacerbated by India's reliance on a progressively erratic monsoon to meet its water needs. While the country is experiencing increasing floods and droughts, environmental change is anticipated to exacerbate the country's need for water supplies (The

World Bank, 2023). The Falkenmark Index estimates that 820 million Indians, spread over twelve river basins, have per capita water availability near to or below 1000 cubic meters, the official cutoff for water scarcity. In India's rural areas, 82% of the families lack individual piped water supplies, and 163 million people lack access to clean water near their homes. India's surface water is contaminated to an extent of 70%. India's average per capita water availability is projected to decrease to 1341 m^3 by 2025 and further to 1140 m^3 by 2050. These figures approach the official water shortage threshold. To address the expected water supply shortfall by 2030, an estimated expenditure of INR 20,000 crores is required (NITI Ayog, 2019).

Central Asia (CA) is facing one of the world's most severe water crisis, which has significantly impeded the region's progress towards achieving Sustainable Development Goals (SDGs). According to reports, California's water situation is becoming worse. The "severe WS" group includes Turkmenistan and Uzbekistan, both of which are experiencing an increase in Water Stress Index (WSI) levels. Tajikistan is designated as a " High WS" country. Both Kazakhstan and Kyrgyzstan show "moderate WS". Additionally, in all the CA countries, a smaller percentage of the countryside populace has safe drinking water compared to the urban population (Wang et al., 2022).

Specialists warn that Egypt may approach a situation of "complete" water scarcity within the next couple of years. Nearly all Egyptians receive their drinking water from the Nile River, which is where about 90% of the country's people reside. According to the United Nations, the country has an annual shortage of water and would most likely face water scarcity by 2025. Rising sea levels, according to water management experts, are creating saltwater breaches that not only impair water sources but also destroy agricultural soil (Elasfar Dara, 2023).

The scenario of Tunisia concerning water scarcity is also grave. The average temperature of Tunisia is anticipated to rise by 3.8 degrees Celsius by 2050, while yearly precipitation is expected to fall by at least 4%. As the nation experiences a severe drought for the fourth consecutive year, Tunisia has implemented water restrictions. The state water distribution corporation, Sonede, regularly shuts off the mains water flow every night between 9 p.m. and 4 a.m. The agriculture ministry has now made it illegal to use water for car cleaning, agriculture, and spraying parks and other areas. According to reports, reservoirs around the nation are about 30% under capacity. The Sidi Salem reservoir, which supplies water to Tunis and the rest of the country's north, is only approximately 16% filled (Cordall, 2023).

By 2030, Morocco anticipates reaching "absolute water scarcity" (under 500 cubic meters per person in a year). While Morocco has not seen the large-scale turmoil or domestic upheavals that some of its North African neighbors have, it has seen "thirsty protests" caused by water scarcity in the south (Hill Thomas M & Pimentel Martin, 2022). Morocco is presently considered to be in a state of "structural WS" because the amount of renewable water resources per person decreased from 2,560 m^3 to around 620 m^3 per person per year between 1960 and 2020. The capacity of the Kingdom's overall water storage increased tenfold during this time because of the construction of more than 120 sizable dams. However, over a majority of the last decade, the real amount of water held in the major dams of the nation has decreased. When the most recent drought struck the Kingdom, the general water level was only about 33%, posing a threat to the security of the water supply in certain Moroccan river basins and forcing the government to take several emergency measures (Benjelloun Meryam, 2022).

3. Worldwide Drought Situation

A drought is a prolonged dry period that can occur in the natural climatic cycle. It is a slow-moving disaster characterized by an absence of precipitation, which leads to water scarcity. Droughts are the most serious hazard to cattle and crops in nearly every region of the globe, impacting approximately 55 million individuals each year. Droughts threaten people's livelihoods, increase the risk of sickness and death, and cause mass migration. Water scarcity affects 40% of the world's population, and by 2030, 700 million individuals may be forced to flee owing to drought (Benjelloun Meryam, 2022). Throughout most of South Asia, which includes western India and southern and Central Pakistan, severe droughts began in the early 2000s. The South Asian regions have historically been among the world's most drought prone. In the last fifty years, at least one drought has been observed in Afghanistan, India, Pakistan, and Sri Lanka, while droughts are also a regular occurrence in Bangladesh and Nepal. Due to a severe drought in the districts of Tharpakrakar and Mirpur Khas in 2012, Pakistan declared an emergency and many people had to be relocated. Extreme drought in Cambodia had an impact on late-season and longer-duration genotypes (Miyan, 2015).

India experiences drought in areas with both high and infrequent rainfall. Approximately 68% of the nation is vulnerable to droughts of varying severity. 35 percent of the regions are at risk for drought when they receive between 750 to 1125 millimeters of rain, and 33 percent are at risk for chronic drought when they receive less than 750 millimeters. It is also not unusual for the Himalayan region to experience water shortages. India has already experienced severe droughts that have had a long-lasting effect on the country's water resources, agriculture, economy, and way of life (Rawat et al., 2022). The Gangetic plains and southern India saw substantial groundwater depletion as a result of the 2015–2016 drought, which also had an impact on nearly 330 million people in ten states. Drought and scarcity of water have damaged agriculture in many sections of the country, especially Maharashtra and the Gangetic Plain. Additionally, in the summer of 2016, water storage in 91 reservoirs across India was at its lowest point (about 15% of normal levels) (Mishra et al., 2016).

During the droughts of 2004, 2009, and 2012, there was a significant lack of terrestrial water storage, with the climax occurring in the Ganga basin in 2009. Increasing extraction of groundwater in the wake of soil water shortage and increasing demand for drinking water are identified as the primary contributing variables. Increased groundwater demand since 2008 may be related to regional warming (Goldin, 2016).

The dry and semiarid region of Pakistan experiences regular droughts as a result of shifting rainfall patterns. Drought-related disasters are posing major dangers to farming communities' ability to make a living, as well as to global food shortages, environmental migration, and sustainable development (Ashraf et al., 2021). The status of the predominantly arid nation has been worse as the per capita water accessibility has decreased from 5,060 cubic meters per year in 1951 to only 908 cubic meters in 2022. Rising temperatures and a lack of water in South Punjab and Sindh's already water-stressed regions have a serious negative impact on cattle. The Indus Delta has decreased by 92%. The Indus Basin's health is being restored by the UN and the Ministry of Climate Change's Living Indus initiative (NDTV WORLD, 2022). Due to its dry climate, high exposure, and poor adaptation capacity, the province of Balochistan in Pakistan is one of the most susceptible to droughts (Jamro et al., 2020). SPI was employed in a study to describe Balochistan's droughts from 1975 to 2010. The findings show that farmers' perceptions of drought and climatic variability are consistent with the conclusions drawn from secondary climatic data and information. The Mann-Kendall test results suggest that there are negative trends in precipitation for over 70 percent of Balochistan's stations, ranging from 137.2 to 283.4 mm across 36 years. These patterns have the potential to influence drought prevention, regional governance of water resources, and agricultural decision-making (Ashraf & Routray, 2015).

The Middle East's drought condition is rapidly worsening. In 2021, Syria endured the most severe drought in more than 70 years affecting millions of people's access to drinking water, electricity production, and agricultural water. As a result of the water shortage, wheat production plummeted from 2.8 million tonnes in 2020 to barely 1.05 million tonnes in 2021. Wheat prices have now reached record highs as a result, further taxing already tight household budgets (Relief Web, 2022a). Since 1991, Lebanon has gone through four severe drought seasons, each with a precipitation deficit of at least 40%. These droughts affected all of Lebanon's significant agroecological zones, with coastal plains, mountains, and inland regions all suffering crop and income losses. These dry years were difficult for the nation's water budget because of a combination of quick snowmelt, drying springs, and a late start to the autumn season. Many water users in Lebanon were alerted by the exceptionally severe drought of 2014, highlighting the country's increasing vulnerability to drought. The amount of water in rivers and aquifers decreased since the country only received half of the 812 mm of rainfall it typically receives annually (USAID & IWMI). Jordan has had repeated drought events throughout the last many decades, which has had a negative influence on the nation's social development and economic progress. Drought was more common in the 1970s, 1980s, and 1990s; the worst drought to be documented in the past 32 years happened in 1999-2000 (WMO, 2014). Jordan's annual rainfall fluctuates significantly from year to year. The primary constraints restricting agricultural production are the predominance of arid conditions and variable rainfall distribution. According to the UN, droughts occur 2.43 times every 10 years, and their incidence is on the rise (Climate Change Knowledge Portal, n.d). The eastern Mediterranean basin has experienced unusually scarce rainfall since the autumn of 2020, which has exacerbated the drought in Iraq. Water flowing into Iraq has declined from a peak of almost 80 billion cubic meters per year in the 1970s to less than 50 cubic meters in May 2021 (Relief Web, 2023).

Water scarcity and drought are also prevalent on the African continent. Throughout the past century, Algeria had numerous droughts. The first was observed in 1944, and they were common in the 1980s and 1990s. This phenomenon had a particularly negative impact on the country's northwest and central Plains, leading to a sharp decline in crop yields and water availability as well as severe crop destruction. The Chellif Plain (136%) saw the most significant water shortage stages in 1981 and 1989, and the Ghriss, Mitidja, and Sidi Bel Abbes plains (136%) experienced them in 1990 (Meddi et al., 2014). On a list of 184 countries, Algeria comes in at number 18 for drought exposure. Drought affects 3,763,800 people, or 10% of the population, according to estimates. In June 2003, Algeria had a record heat wave with temperatures over 40°C for 20 straight days, which is thought to have contributed to 40 fatalities. In a warming world, more of these occurrences are expected (Climate Change Knowledge Portal). According to a recent assessment by the International Federation of Red Cross and Red Crescent Societies, Mauritania is experiencing its most severe drought conditions since 2012 and fares worse than all Sahel nations that have faced rainfall deficits over the past decade. At least 34 of the 48 regions are impacted, especially the southernmost ones. Following a dismal 2021 rainy season that was recorded in 81% of Mauritania's monitoring stations, 1.4 million people are now at risk of food insecurity. The difficulties have compounded due to the continuing drought and the conflict in the region (Nature News, 2022). Due to frequent dry spells and low rainfall in 2021, Mali faced the worst shortage of rain in five years, putting 1.9 million people at risk of acute food insecurity throughout the nation, primarily in the districts of Kayes, Gao, Mopti, Segou, and Timbuktu (Relief Web, 2022b). Around 30 million people reside in the Lake Chad region, which has long struggled with the effects of desertification and drought. The lake has decreased by 90% since the 1960s as a result of climate change and resource exploitation (United Nations Convention to Combat Desertification, 2023). The current drought, which is affecting Somalia, Kenya, and Ethiopia, is maybe the worst in 40 years, and it is predicted to worsen (United Nations, 2022).

Additionally, since 2019, central-southern South America has consistently experienced severe drought conditions. One of the worst and longest droughts in recent memory is currently affecting the area. Crop output, the economy, and ecosystems are already feeling the detrimental effects. This enormous-scale drought was triggered by a series of smaller, geographically separated sub-events that were all influenced by a severe lack of precipitation. By the end of March 2023, severe vegetative stress had begun to develop throughout Uruguay, northern Argentina, and southern Patagonia due to a lack of precipitation mixed with warmer-than-normal temperatures. Crop yields have fallen as a result. The soybean harvest in Argentina is anticipated to be the smallest since 1988/89 in 2023, with production projected to be 44% lower than the average of the previous five years. The projected Argentinian GDP for 2023 has already decreased by 3 percentage points as a result of the drought (European Commission, 2023). According to the National Meteorological Service of the National Water Commission (Conagua), runoff, lack of rainfall, and drought have an impact on roughly 80% of Mexico's territory. According to the research, only 20.49 percent of the country's geographical area is unaffected, while 79.51 percent falls into the category of abnormally dry to incomparable drought (teleSUR, 2023).

4. Impact of Water Stress on Plant Growth and Development

Water-deficit stress is defined as a condition in which plant water potential and turgor are reduced to a level that disrupts normal activities (Shao et al., 2008). Water stress is characterized as a modest water loss that produces stomatal closure and restricts gas exchange (Abdalla et al., 2022). Desiccation is a much higher loss of water that may damage metabolic activity and cell structure, eventually leading to the cessation of enzyme-catalyzing activities (Moore et al., 2006; Yathisha et al., 2020; Van Der Walt et al., 2022). Water stress is defined as a reduction in water content, turgor (Coussement et al., 2021), leaf water potential (Osakabe et al., 2014), wilting (Wang, 2014), and a decrease in cell expansion and growth (Coussement et al., 2021). Severe WS can reduce photosynthesis (Wang et al., 2018), and lead to metabolic malfunction (Ghannoum, 2008; Hu et al., 2020). Water stress can affect Cdc2 kinase activity by delaying its passage from the G1 to the S phase or from the G2 to the M phase, leading to differences in the number of nuclei in the G1 and G2 phases (Schuppler et al., 1998). This inhibits the cell division process.

Drought stress caused a considerable drop in chlorophyll concentration in an investigation on sugarcane. Drought stress-induced plasmolysis was further detected in the cells with the migration of chloroplasts to the center of the cell. The shape of the chloroplast changed from long oval to round (Zhang et al., 2015). In maize, polyethylene glycol-induced drought stress led to a swollen thylakoid membrane accompanied by its disorganization. Furthermore, the intrathyllakoid space expanded, and there was a reduction in the length-to-width proportion and the size of chloroplasts (Shao et al., 2016). In *Xanthoceras sorbifolium*, chloroplasts encircled and gathered more starch granules and plastids under moderate drought stress. Under extreme drought stress, chloroplasts expanded dramatically, their outlines became wavy, thylakoids were disorganized, lamellar gaps widened, starch particles deteriorated, and the number and volume of plastoglobules rose significantly (Hu et al., 2023).

Water stress also has a profound impact on the respiratory processes of plants. In a soybean experiment, extreme WS triggered a significant shift of electrons from the cytochrome to the alternate pathway. Electron partitioning rose from 10% to 12% under well-watered or mild water-stress conditions to over 40% under severe WS. As a result, during acute WS, the predicted rate of mitochondrial ATP synthesis was reduced by 32% (Ribas-Carbo et al., 2005). Another study found that dryness combined with high temperatures caused mitochondrial disorganization. The mitochondria were either devoid of cristae or had a small number of cristae. This was supported by a high level of electron transparency in the matrix. Moreover, the mitochondria were of larger size in the stressed plants (Grigorova et al., 2012).

When plants are stressed due to a shortage of water, their photosynthesis slows, resulting in slower development and a greater incidence of early senescence. Water scarcity has a direct effect on the physiology, morphology, and molecular makeup of plants (Imadi et al., 2016). Water stress is bad for plant growth and development because it prevents stomata from opening and reduces internal water transport, which prevents access to the resources needed for photosynthesis (Sun et al., 2020). Water scarcity reduces the water potential of the soil, which in turn reduces the number of leaves generated by each plant as well as the size, longevity, and number of individual leaves (Shao et al., 2008). According to one study, increased water shortage resulted in a considerable fall in growth indices and biomass of peanuts. In addition, there was a reduction in the height of the plant and its leaf area index (Emara et al., 2023). Another study indicated that WS resulted in a reduction in the leaf area of *Polygala myrtifolia* and *Viburnum tinus*. Moreover, the dry mass was also reduced in both plants (Tribulato et al., 2019). In wheat, WS resulted in a decrease in the height of the plant, length of the spike, number of spikelets per spike, and grains per spike (Mirbahar et al., 2009). In another study, moderate and severe WS resulted in a decrease in the weight of spikes, grains, and 1000 grain weights. In addition, there is a reduction in fresh weight in plants subjected to severe stress. The dry weights were also reduced in water-stressed plants (Zhao et al., 2020). In maize, WS resulted in leaf area index and plant height (Laskari et al., 2022). In another study, drought stress reduced the area of the leaf, the height of the canopy, the number of kernels per spike, and the unit weight of the kernel during various stages of plant growth (Song et al., 2019). In soybean, the drought stress prevented an upsurge in the height of the soybean plant and the area of its leaves. This inhibition became more pronounced as the intensity of drought stress, as well as its duration and frequency, continued to grow (Dong et al., 2019). Drought stress greatly hampered the buildup of biomass in the root, stem, pod wall, and seed in another study on soybean, and the quantity of pods and seeds also dropped. Except for the root biomass distribution ratio, the distribution ratio of dry matter in seed and pod walls was greatly reduced by drought stress, whereas the distribution ratio of dry matter in roots and stems was significantly enhanced (Du et al., 2020).

5. Similarities and Variations Among Water, Drought, and Salt Stress

Salinity and drought are two key abiotic variables that limit food crop productivity globally. Drought and salinity reactions in plants are typically equivalent (Chaves et al., 2009). Salinity stress is commonly generated by irrigation with poor-quality water, and it shares many characteristics with drought stress, particularly in the initial phases of the stress response. This is due to the osmotic effect of these stressors, which inhibits water absorption via the root system (Navarro et al., 2013). For instance, the osmotic effect, which marks the beginning of salinity stress, resembles drought stress quite a bit. High salinity and water scarcity both produce osmotic stress in plants, which respond by synthesizing and accumulating suitable solutes or osmolytes to assist with cellular osmotic adjustments (Kumar et al., 2017). Plant roots have efficient systems for detecting low water potential, which can result from a genuine lack of water in the environment due to insufficient precipitation, or the presence of excessive salt ions in the existing water. Plants are unable to absorb enough water for proper growth and development in such circumstances, resulting in an induction of prevalent stress-related signal transduction pathways, which include higher levels of ABA (Ikegami et al., 2009). When water potential falls, osmotic changes in the root system help to preserve some turgidity, and the water potential gradient is reestablished for water intake. These modifications are responsible for root development in conditions of low water potential. Under conditions of salinity or drought, the stimulation of these stress-related signal transduction pathways results in significantly reduced shoot growth, particularly leaf growth. The drop in cellular oxygen and carbon dioxide (CO_2) levels caused by partial stomatal closure is a primary contributor to growth slowdown. Plant cells produce a range of reactive oxygen species (ROS) in response to salt stress and drought stress (Cruz De Carvalho, 2008; Kesawat et al., 2023). ROS levels are high in both the apoplast and the cytoplasm. Drought and salt cause increased ROS levels, which disturb normal cellular metabolism and redox equilibrium. Additionally, the structure of plant cell walls is modified as a result of ROS, which either inhibits cell growth by tightening the cell wall or promotes cell growth by loosening the wall structure (Uddin et al., 2016).

Early salinity and drought reactions are quite similar because both create WS, which delays growth, limits stomatal aperture, and results in plant nutrient deficits (such K^+ and Ca^{2+}). Long-term salt exposure, on the other hand, generates ionic stress in plants, which hinders photosynthesis and induces leaf senescence, both of which harm growth (Ma et al., 2020). Continuous root elongation happens in plants during long-term drought stress, which may be explained by the requirement for the plant to seek groundwater (Brunner et al., 2015). Long-term salinity stress might cause heavier roots to store more chloride. Excessive ion accumulation, particularly Na^{2+}, has a deleterious impact on photosynthetic components, lowering enzyme performance and pigment synthesis. These challenging conditions decrease the rate of CO_2 absorption, and the excess light that the plant does not utilize may lead to the development of ROS, resulting in oxidative stress. Several species with salt exclusion systems prevent salt from accessing plant cells or reduce the amount of salt inside the cytoplasm by compartmentalizing it in vacuoles Drought and Salinity Stress Responses and Microbe-Induced Tolerance in Plants (Khan et al., 2019; Ma et al., 2020).

6. Conclusion

Water stress refers to the inability, or perceived incapacity, of humans and ecosystems to meet the demand for freshwater. Water scarcity is currently gripping a large portion of the planet, spurred by a shift in weather patterns and a temperature rise. This has severely impacted the human race and the plants which form an integral part of the food. Consequently, food security has become a matter of grave concern due to diminishing yields from food crops. Furthermore, salinity and heavy metal stress are intricately linked to water constraints. As the water becomes scarce in the soil and adjoining milieu, the metal ions or salt molecules become much more concentrated thereby posing a direct threat to plants and animals with plants possibly the first line to succumb to the stress due to their immovable condition. Moreover, plants are first in the line in the process of biomagnification of these deleterious heavy metals and salts. Thus, the environmental change-induced WS has made its impact on various domains of life and the ecosystem. It seems that there are no immediate remedial measures to overcome the water scarcity as the damage to the environment has already been done and its detrimental manifestation. However, a holistic approach is required to overcome this crisis. This involves several platforms which includes social, environmental, and scientific arenas. Special efforts are required to expand forest cover to ensure both adequate and well-distributed rainfall. Additionally, there is a need to promote social awareness regarding water conservation among the masses. In this regard, proper planning from the policymakers and implementation by the administration holds the key. Thus, taking a holistic approach to conserving water is crucial for long-term sustainability. Initiating efforts at the grassroots level, starting with individual families can create a ripple effect that extends globally. This would hopefully be beneficial to the booming global population which is currently facing acute WS.

References

Abdalla M, Ahmed MA, Cai G, Wankmüller F, Schwartz N, Litig O, et al. 2022. Stomatal closure during water deficit is controlled by below-ground hydraulics. *Ann Bot,* **129:** 161–170.

Abumoghli Iyad. 2021. Available at: https://www.ecomena.org/water-arab/.

Ashraf M, Arshad A, Patel PM, Khan A, Qamar H, Siti-Sundari R, et al. 2021. Quantifying climate-induced drought risk to livelihood and mitigation actions in Balochistan. *Nat Hazards (Dordr),* **109:** 2127–2151.

Ashraf M & Routray JK. 2015. Spatio-temporal characteristics of precipitation and drought in Balochistan Province, Pakistan. *Natural Hazards,* **77:** 229–254.

Benjelloun Meryam. 2022. Available at: https://www.worldbank.org/en/news/press-release/2022/07/20/moroccan-economy-slows-in-wake-of-drought-and-commodity-price-rises.

Bhardwaj A, Devi P, Chaudhary S, Rani A, Jha UC, Kumar S, et al. 2022. 'Omics' approaches in developing combined drought and heat tolerance in food crops. *Plant Cell Rep,* **41:** 699–739.

Brunner I, Herzog C, Dawes MA, Arend M & Sperisen C. 2015. How tree roots respond to drought. *Front Plant Sci,* **6:** 1–16.

CEO WATER MANDATE SECRETARIAT. 2017a. Available at: https://ceowatermandate.org/posts/water-scarcity-water-stress-water-risk-actually-mean/.

Chaves MM, Flexas J & Pinheiro C. 2009. Photosynthesis under drought and salt stress: Regulation mechanisms from whole plant to cell. *Ann Bot,* **103:** 551–560.

Chibani Achref. 2022. Available at: https://arabcenterdc.org/resource/confronting-water-scarcity-in-north-africa/.

Climate Change Knowledge Portal Available at: https://climateknowledgeportal.worldbank.org/.

Climate Change Knowledge Portal Available at: https://climateknowledgeportal.worldbank.org/country/algeria/vulnerability.

Cordall SS. 2023. Water ban in drought-stricken Tunisia adds to growing crisis. *The Guardian.*

Coussement JR, Villers SLY, Nelissen H, Inzé D & Steppe K. 2021. Turgor-time controls grass leaf elongation rate and duration under drought stress. *Plant Cell Environ,* **44:** 1361–1378.

Cruz De Carvalho MH. 2008. Drought stress and reactive oxygen species: Production, scavenging and signaling. *Plant Signal Behav,* **3:** 156–165.

Dong S, Jiang Y, Dong Y, Wang L, Wang W, Ma Z, et al. 2019. A study on soybean responses to drought stress and rehydration. *Saudi J Biol Sci,* **26:** 2006–2017.

Droughts and Climate Change. USGS Web Page. Available from: https://www.usgs.gov/science/science-explorer/climate/droughts-and-climate-change.

Du Y, Zhao Q, Chen L, Yao X & Xie F. 2020. Effect of drought stress at reproductive stages on growth and nitrogen metabolism in soybean. *Agronomy,* **10:** 302.

Elasfar Dara. 2023. Available at: https://abcnews.go.com/US/alternative-egypt-worries-climate-change-dam-project-threaten/story?id=98481819.

Emara Eman IR, Moursy MAM & Hamed LMM. 2023. Response of certain peanut (Arachis hypogea L.) varieties to water regime using different irrigation systems in new reclaimed areas. *Journal of the Saudi Society of Agricultural Sciences,* **22:** 245–260.

European Commission. 2023. Available at: https://joint-research-centre.ec.europa.eu/jrc-news-and-updates/drought-conditions-threaten-economy-and-ecosystems-south-america-2023-05-22_en.

Ghannoum O. 2008. C4 photosynthesis and water stress. *Ann Bot,* **103:** 635–644.

Goldin T. 2016. India's drought below ground. *Nat Geosci,* **9:** 98.

Grigorova B, Vassileva V, Klimchuk D, Vaseva I, Demirevska K & Feller U. 2012.

Drought, high temperature, and their combination affect ultrastructure of chloroplasts and mitochondria in wheat (Triticum aestivum L.) leaves. *J Plant Interact,* **7:** 204–213.

Harhay MO. 2011. Water stress and water scarcity: A global problem. *Am J Public Health,* **101:** 1348–1349.

Hill Thomas M & Pimentel Martin. 2022. Available at: https://www.usip.org/publications/2022/06/morocco-reflects-global-south-dilemma-water-or-food.

Hu F, Zhang Y & Guo J. 2023. Effects of drought stress on photosynthetic physiological characteristics, leaf microstructure, and related gene expression of yellow horn. *Plant Signal Behav,* **18:** 2215025.

Hu W, Huang Y, Loka DA, Bai H, Liu Y, Wang S, et al. 2020. Drought-induced disturbance of carbohydrate metabolism in anthers and male abortion of two Gossypium hirsutum cultivars differing in drought tolerance. *Plant Cell Rep,* **39:** 195–206.

Ikegami K, Okamoto M, Seo M & Koshiba T. 2009. Activation of abscisic acid biosynthesis in the leaves of Arabidopsis thaliana in response to water deficit. *J Plant Res,* **122:** 235–243.

Imadi SR, Gul A, Dikilitas M, Karakas S, Sharma I & Ahmad P. 2016. Water stress: Types, causes, and impact on plant growth and development. pp. 343–355. *In:* Ahmad P. (ed.), Water Stress and Crop Plants. John Wiley & Sons, Ltd, Chichester, UK.

Jamro S, Channa FN, Dars GH, Ansari K & Krakauer NY. 2020. Exploring the evolution of drought characteristics in Balochistan, Pakistan. *Applied Sciences,* **10:** 913.

Kesawat MS, Satheesh N, Kherawat BS, Kumar A, Kim H-U, Chung S-M, et al. 2023. Regulation of reactive oxygen species during salt stress in plants and their crosstalk with other signaling molecules—Current perspectives and future directions. *Plants,* **12:** 864.

Khan A, Khan AL, Muneer S, Kim Y-H, Al-Rawahi A & Al-Harrasi A. 2019. Silicon and salinity: Crosstalk in crop-mediated stress tolerance mechanisms. *Front Plant Sci,* **10:** 1429.

Kumar D, Al Hassan M, Naranjo MA, Agrawal V, Boscaiu M & Vicente O. 2017. Effects of salinity and drought on growth, ionic relations, compatible solutes and activation of antioxidant systems in oleander (Nerium oleander L.). *PLoS One,* **12:** e0185017.

Kuzma Samantha, Saccoccia Liz & Chertock Merlena. 2023d. Available at: https://www.wri.org/insights/highest-water-stressed-countries.

Laskari M, Menexes G, Kalfas I, Gatzolis I & Dordas C. 2022. Water stress effects on the morphological, physiological characteristics of maize (Zea mays L.), and on environmental cost. *Agronomy,* **12:** 2386.

Lesk C, Rowhani P & Ramankutty N. 2016. Influence of extreme weather disasters on global crop production. *Nature,* **529:** 84–87.

Ma Y, Dias MC & Freitas H. 2020. Drought and salinity stress responses and microbe-induced tolerance in plants. *Front Plant Sci,* **11:** 591911.

Maddocks A, Young RS & Reig P. 2015. Ranking the World's Most Water-Stressed Countries in 2040. Available from: https://www.wri.org/insights/ranking-worlds-most-water-stressed-countries-2040.

Meddi H, Meddi M & Assani AA. 2014. Study of drought in seven Algerian Plains. *Arab J Sci Eng,* **39:** 339–359.

Mirbahar AA, Markhand GS, Mahar AR, Abro SA & Kanhar NA. 2009. Effect of water stress on yield and yield components of wheat (Triticum aestivum L.) varieties. *Pak J Bot,* **41:** 1303–1310.

Mishra V, Aadhar S, Asoka A, Pai S & Kumar R. 2016. On the frequency of the 2015 monsoon season drought in the Indo-Gangetic Plain. *Geophys Res Lett,* **43:** 102–112.

Miyan MA. 2015. Droughts in Asian least developed countries: Vulnerability and sustainability. *Weather Clim Extrem,* **7:** 8–23.

Moore JP, Nguema-Ona E, Chevalier L, Lindsey GG, Brandt WF, Lerouge P, et al. 2006. Response of the leaf cell wall to desiccation in the resurrection plant *Myrothamnus flabellifolius*. *Plant Physiol,* **141:** 651–662.

National Geographic Available at: https://education.nationalgeographic.org/resource/how-climate-change-impacts-water-access/.

NatureNews. 2022. Available at: https://naturenews.africa/worst-drought-since-2012-hits-mauritania-hardest-among-sahel-nations/.

Navarro CA, von Bernath D & Jerez CA. 2013. Heavy metal resistance strategies of acidophilic bacteria and their acquisition: Importance for biomining and bioremediation. *Biol Res,* **46:** 363–371.

NDTV WORLD. 2022. Available at: https://www.ndtv.com/world-news/pakistan-one-of-the-top-23-countries-facing-severe-drought-report-3076523.

NITI Ayog. 2019. Available at: https://social.niti.gov.in/uploads/sample/water_index_report2.pdf.

Osakabe Y, Osakabe K, Shinozaki K & Tran L-SP. 2014. Response of plants to water stress. *Front Plant Sci,* **5:** 1–8.

Rawat S, Ganapathy A & Agarwal A. 2022. Drought characterization over Indian sub-continent using GRACE-based indices. *Sci Rep,* **12:** 15432.

ReliefWeb. 2022a. Available at: https://reliefweb.int/report/syrian-arab-republic/syria-11-years-hunger-drought-and-collapsing-economy-threatens-even-more.

ReliefWeb. 2022b. Available at: https://reliefweb.int/report/mali/drought-stricken-mali-receives-first-ever-arc-limited-climate-risk-insurance-pay-out.

ReliefWeb. 2023. Available at: https://reliefweb.int/report/iraq/iraq-droughts-dref-operation-no-mdriq013-final-report.

Ribas-Carbo M, Taylor NL, Giles L, Busquets S, Finnegan PM, Day DA, et al. 2005. Effects of water stress on respiration in soybean leaves. *Plant Physiol,* **139:** 466–473.

Ritchie H & Roser M. 2017. Water use and stress. *Our World in Data*. Available from: https://ourworldindata.org/water-use-stress.

Schuppler U, He P-H, John PCL & Munns R. 1998. Effect of water stress on cell division and Cdc2-like cell cycle kinase activity in wheat leaves1. *Plant Physiol,* **117:** 667–678.

Shao H-B, Chu L-Y, Jaleel CA & Zhao C-X. 2008. Water-deficit stress-induced anatomical changes in higher plants. *C R Biol,* **331:** 215–225.

Shao RX, Xin LF, Zheng HF, Li LL, Ran WL, Mao J, et al. 2016. Changes in chloroplast ultrastructure in leaves of drought-stressed maize inbred lines. *Photosynthetica,* **54:** 74–80.

Song L, Jin J & He J. 2019. Effects of severe water stress on maize growth processes in the field. *Sustainability,* **11:** 5086.

Sun Y, Wang C, Chen HYH & Ruan H. 2020. Response of plants to water stress: A meta-analysis. *Front Plant Sci,* **11:** 978.

teleSUR. 2023. Available at: https://www.telesurenglish.net/news/Almost-80-Of-Mexicos-Territory-Affected-by-Drought-20230120-0019.html.

The World Bank. 2023. Available at: https://www.worldbank.org/en/country/india/brief/world-water-day-2022-how-india-is-addressing-its-water-needs.

Tribulato A, Toscano S, Di Lorenzo V & Romano D. 2019. Effects of water stress on gas exchange, water relations and leaf structure in two ornamental shrubs in the mediterranean area. *Agronomy,* **9:** 381.

Uddin MN, Hossain MA & Burritt DJ. 2016. Salinity and drought stress: Similarities and differences in oxidative responses and cellular redox regulation. pp. 86–101. *In:* Ahmad P. (ed.), Water Stress and Crop Plants, John Wiley & Sons, Ltd, Chichester, UK.

UNICEF. Available from: https://www.unicef.org/wash/water-scarcity

United Nations Available at: https://www.un.org/en/climatechange/science/climate-issues/water.

United Nations Available at: https://www.un.org/waterforlifedecade/index.shtml.

United Nations Convention to Combat Desertification. 2023. Available at: https://www.unccd.int/news-stories/special-feature/finding-hope-face-drought-bol-chad.

United Nations. 2022. Available at: https://news.un.org/en/story/2022/05/1119382.

USAID & IWMI Available at: https://menadrought.iwmi.org/wp-content/uploads/sites/44/2021/11/MENAdrought-project-Lebanon-country-profile.pdf.

Van Der Walt K, Burritt DJ & Nadarajan J. 2022. Impacts of rapid desiccation on oxidative status, ultrastructure and physiological functions of Syzygium maire (Myrtaceae) zygotic embryos in preparation for cryopreservation. *Plants,* **11:** 1056.

Wang L. 2014. Physiological and molecular responses to drought stress in rubber tree (Hevea brasiliensis Muell. Arg.). *Plant Physiology and Biochemistry,* **83:** 243–249.

Wang X, Chen Y, Fang G, Li Z & Liu Y. 2022. The growing water crisis in Central Asia and the driving forces behind it. *J Clean Prod,* **378:** 134574.

Wang Z, Li G, Sun H, Ma L, Guo Y, Zhao Z, et al. 2018. Effects of drought stress on photosynthesis and photosynthetic electron transport chain in young apple tree leaves. *Biol Open,* **7:** bio035279.

WMO UFC and U-D. 2014. Available at: https://www.droughtmanagement.info/literature/UNW-DPC_NDMP_Country_Report_Jordan_2014.pdf.

Yathisha NS, Barbara P, Gügi B, Yogendra K, Jogaiah S, Azeddine D, et al. 2020. Vegetative desiccation tolerance in Eragrostiella brachyphylla: Biochemical and physiological responses. *Heliyon,* **6:** e04948.

Zhang F-J, Zhang K-K, Du C-Z, Li J, Xing Y-X, Yang L-T, et al. 2015. Effect of drought stress on anatomical structure and chloroplast ultrastructure in leaves of sugarcane. *Sugar Tech,* **17:** 41–48.

Zhang J, Zhang S, Cheng M, Jiang H, Zhang X, Peng C, et al. 2018. Effect of drought on agronomic traits of rice and wheat: A meta-analysis. *Int J Environ Res Public Health,* **15**.

Zhao W, Liu L, Shen Q, Yang J, Han X, Tian F, et al. 2020. Effects of water stress on photosynthesis, yield, and water use efficiency in winter wheat. *Water (Basel),* **12:** 2127.

2 | Unraveling Cytological and Embryological Anomalies Amid Water Stress: Insights into Cytological and Embryological Aberrations Under Water Stress Conditions

Dwaipayan Sinha[1*], Satarupa Nath[2], Sukanya Chatterjee[3], Muhammad Majeed[4] and Robina Aziz[5]

[1] Department of Botany, Government General Degree College, Mohanpur, West Bengal, India
[2] Department of Botany, Rishi Bamkim Chandra College for Women, Naihati, West Bengal, India
[3] Department of Botany, Lady Brabourne College, Beniapukur, Kolkata, India
[4] Department of Botany, University of Gujrat, Pakistan
[5] Department of Botany, Govt. College Women University Sialkot, Pakistan

Water stress is a serious matter of concern throughout the globe. The plants suffer significantly from water stress due to their immobile nature. Water stress impacts plants in various ways influencing their morphology, physiology, and cytological parameters. The impact on the cytological parameter is fundamental, as it directly affects cell division and also serves as a major modulator of genes, thereby influencing the physiological stature of the plant. In addition, water stress at the cytological level directly impacts plant fertility, posing a threat to productivity that is closely linked to human food security. This chapter provides an overview of the cytological and embryological anomalies in plants under water-stressed conditions. Efforts are also made to illustrate enzyme action under water-stressed conditions and how it influences mitotic and meiotic processes.

*Corresponding author: dwaipayansinha@hotmail.com

1. Introduction

Water stress refers to the proportion of freshwater withdrawals in relation to freshwater renewable sources. Water stress does not necessarily indicate that a country has a water deficit. Instead, it reflects how close a region or country is to depleting its available renewable water resources. When water allocation exceeds the capacity of existing resources (i.e., surpassing 100 percent), it signifies that a nation is either procuring beyond the replenishment rate of aquifers or possesses a considerable volume of desalinated water production (Ritchie & Roser, 2024) Water scarcity is a form of water stress that can occur seasonally, annually, or multiple times per year. It occurs when water demand in river basins consistently exceeds the natural system's sustainable supply capacity. It can be computed as the ratio of freshwater resources that are renewable to the amount of water that is abstracted or used. Droughts are a temporary drop in average water availability caused by, for example, low rainfall and are classified as natural phenomena. Farmers encounter drought as a significant annual challenge. Drought is a natural phenomenon characterized by prolonged periods of insufficient freshwater supply for human needs and the health of ecosystems (Balint et al., 2013). Due to its frequent occurrence and absence of specific spatial or temporal constraints, drought may be challenging to detect (Salehi-Lisar & Bakhshayeshan-Agdam, 2016). It is a complex form of stress that mostly arises from a decrease in rainfall and the resulting prolonged period of dry weather (Gogoi & Tripathi, 2019). There exist four primary classifications of drought: Meteorological drought, which happens in areas where the weather is arid; Hydrological drought manifests from diminished and restricted water resources, notably on surface and groundwater levels, and becomes evident following prolonged periods of meteorological drought; Agricultural drought often correlates with decreased soil water levels, leading to crop disasters as a consequence, which has a major influence on the world's food output; and Socioeconomic drought, which refers to the breakdown of supply and demand for different commodities as a result of drought (Ahluwalia et al., 2021). In this chapter, drought, and water stress will both be discussed as situations that limit the amount of water available to plants and affect morphological, physiological, and cytological conditions.

To endure drought, plants deploy an array of adaptive mechanisms, including the production of reactive oxygen species (ROS), the exudation of stress-related hormones such as ethylene and abscisic acid (ABA), and modifications to root and shoot morphologies (Lee et al., 2012; Wang et al., 2012; Wei et al., 2015; Hussain et al., 2020; Kim et al., 2020; Li et al., 2020). These effects result from intricate physiological, biochemical, and cytological alterations occurring within plants during water scarcity. During drought stress, plants generate an excess of ROS, which surpasses the capacity of antioxidative mechanisms to neutralize. Consequently, the imbalance results in oxidative stress within plant cells. When cellular reactive oxygen species (ROS) production surges abruptly and intensively it can result in the oxidation of proteins and lipids, as well as DNA damage. These disturbances disrupt the delicate the balance within the cell (Filippou et al., 2011). Since drought stress-induced oxidative stress affects the nucleic acids, it is quite obvious that they have a profound effect on the chromosomal structures as well as the nucleus of the plant cell. Therefore, the objective of this chapter is to underscore the potential impacts of water scarcity and drought at the molecular level within plant cells. Moreover, attempts have been made to highlight the consequences of drought on the reproductive units of the plants or more appropriately pollens and carpels.

2. The Impact of Water Stress on Cell Division: A Comprehensive Analysis

Water stress is a substantial difficulty for crop productivity globally (Boutraa et al., 2010). As climate change worsens water scarcity in many areas, understanding how crop plants respond physiologically and at the molecular level to water and drought stress becomes increasingly vital (Seleiman et al., 2021). The mitotic index (MI) plays a crucial role in how plants respond to stress. It reflects the rate at which cells are dividing and is closely tied to plant growth (Qi & Zhang, 2020). Knowing how water stress affects the MI of crop plants is important for developing strategies to minimize its impact on crop yields. This section explores recent research on how water and drought stress influence cell division, including both mitosis and meiosis, in crop plants, helping to reveal the underlying mechanisms.

Elevated temperatures decrease plant water content by accelerating respiration, which can negatively impact the growth of plants and their yield. Plants have a biological mechanism to respond and adapt to water stress. This process entails the activation or inhibition of genes with various roles, including transcription factors, protein kinases, phosphatases, metabolic enzymes, and cis-acting elements (Kitsios & Doonan, 2011). Pea leaf tissues exhibit decreased rates of cellular division and expansion during drought stress, depending on the growth phase of the leaves (Lecoeur et al., 1995). The roots of maize plants experiencing drought stress exhibited elongated meristematic cells, indicating a delay in the initiation of mitosis. Additionally, the pace and extent of cellular division within the root meristem were diminished (Sacks et al., 1997). Root meristems have been observed to experience a reduction in mitotic activity when subjected to drought (Robertson et al., 1990; Chiatante et al., 1997). In sunflowers, exposure to drought along with ABA treatment led to a reduction in DNA synthesis and mitotic activity within the initial 6-hour timeframe. Decreased mitotic activity first appeared in the proximal areas of the meristem. Subsequently, a widespread decline in mitotic activity ensued throughout the meristem, succeeded by a marginal resurgence in the distal regions thereafter (Robertson et al., 1990). In maize, water stress elongated meristematic cells and decreased cell division rates, both across the majority of the meristem, whether measured by tissue length or on a per-cell basis (Sacks et al., 1997).

Over a 24-hour interval, the division of mesophyll cells in wheat seedlings under mild water stress plummeted to a mere 42% compared to their adequately watered counterparts, concurrently with a 50% reduction in the leaf expansion rate. In stark contrast to the 8 mm span observed in control leaves, the mesophyll cell division zone regressed to a mere 4 mm from the base. Notably, the presence of water stress precipitated a notable 50% decline in H1 histone kinase activity, notwithstanding unchanged enzyme levels. This decline in activity corresponded with a heightened share of the 34-kD Cdc2-like kinase being rendered inactive via tyrosine phosphorylation (Schuppler et al., 1998). In maize, a mild water scarcity caused a 45% decrease in cell division rate and p34cdc2 kinase activity across the leaf. However, it did not have an impact on the quantity of p34cdc2a (Granier et al., 2000). According to a study, in maize endosperm, a water deficit occurring between 5 and 13 days after pollination (DAP) results in a significant 90% reduction in the cell division rate of the endosperm. Additionally, it prevents the incorporation of [3H]-thymidine into DNA between 9 and 13 DAP. From 9 to 13 DAP in the control group, both the proportion of nuclei experiencing endoreduplication and the quantity of nuclear DNA increases consistently. Water scarcity has no effect on the expression levels of proteins associated with S-phase and cell division, such as PCNA, histone H3, and alpha-tubulin, until 13 DAP. However, the expression of ZmCdc2, a cyclin-dependent kinase (CDK) that regulates mitosis, was significantly suppressed from 9 to 13 DAP. Cellular replication and related activities are suppressed throughout the initial phases of the stress episode, while the duplication of genetic material and associated processes is not suppressed until the stress becomes more severe (Setter & Flannigan, 2001).

2.1 Investigating the Effect of Water Stress on Plant Mitosis

Mitosis, the quintessential biological phenomenon intrinsic to the augmentation, morphogenesis, and regenerative facets within multicellular entities, notably within the botanical realm, epitomizes a paramount evolutionary imperative. The MI, a pivotal metric in the cellular domain, serves as a quantifiable proxy elucidating the frequency of cellular division, the cornerstone of life's perpetuation. This index, emblematic of cellular dynamism, manifests as a numerical depiction, expressed in a percentage, delineating the contingent of cells currently undergoing the intricate choreography of mitotic partitioning, thus perpetuating the symphony of cellular propagation.

In the botanical domain, scrutiny of the MI finds its eminent application within meristematic enclaves, the bastions of cellular proliferation orchestrating the ontogeny and morphogenesis of botanical entities (Hayashi et al., 2013). Perturbations in this metric, emblematic of the ceaseless flux within cellular orchestration, echo the nuances of cellular cascades including the intricate ballet of cell cycle dynamics, the emergence of cellular identities through differentiation, and the responsiveness to myriad environmental cues (Ligasová et al., 2023). This index, thus, emerges as an invaluable lens through which the intricate tapestry of cellular kinetics is unraveled, affording insights into the ramifications of multifarious factors, encompassing the exigencies of environmental duress, upon the realm of cellular division and vegetal proliferation, transcending the botanical sphere to inform broader biological paradigms (Miettinen et al., 2019).

Drought stress has been demonstrated to impact the MI of crop plants in different ways (Seleiman et al., 2021). In general, long-term exposure to water stress or drought stress leads to a decrease in the MI, which indicates a halt in cell division activity (Özmen et al., 2022). This reduction in MI is attributed to several factors, including decreased cell elongation, disruption of the cell cycle, and induction of programmed cell death (PCD) (Cesare et al., 2017). Water stress-induced reductions in cell elongation result in smaller cell sizes and, consequently, a decrease in the overall MI (West G. et al., 2004). Moreover, water stress can disrupt the normal progression of the cell cycle, leading to delays or arrest at specific phases such as G1, S, or G2/M (Qi & Zhang, 2020). Additionally, persistent water stress or drought stress can activate PCD pathways, resulting in the death of meristematic cells and further lowering the MI in affected tissues (Petrov et al., 2015; Mira et al., 2017).

Several crop plants exhibit a reduced MI as a consequence of water and drought stress. Some examples include:

2.1.1 Zea Mays (Maize)

Maize exhibits sensitivity to water stress, whereby prolonged drought conditions precipitate a decline in the MI. Research indicates that water stress exerts inhibitory effects on the process of cell division within the root meristem of maize, consequently diminishing the MI and impeding root growth (Sheoran et al., 2022; Yang et al., 2023).

2.1.2 Triticum Aestivum (Wheat)

Wheat, akin to maize, undergoes a diminution in the MI when confronted with water stress. Drought stress, particularly during wheat's reproductive period, has an impact on the vigor of cell division within the burgeoning grains, thereby encumbering grain yield (Onyemaobi et al., 2021; Wan et al., 2022).

2.1.3 Oryza Sativa (Rice)

Rice displays pronounced susceptibility to water stress, particularly in its reproductive stage. Instances of water insufficiency precipitate a decline in the MI within rice roots and shoots, consequently impinging upon the overarching dimensions of plant growth as well as profitability (Hassan et al., 2023; Shafi et al., 2023).

2.1.4 Glycine Max (Soybean)

Soybean plants evince a decrement in the mitotic index (MI) under the duress of water stress, notably within the confines of the root apical meristem. Prolonged exposure to drought stress elicits a diminution in cell division activity within soybean roots, thereby impinging upon the pivotal processes of uptake of nutrients and water (Dong et al., 2019; Wang et al., 2022).

2.1.5 Solanum Lycopersicum (Tomato)

Tomato plants are vulnerable to the impacts of water stress, precipitating a reduction in the MI within both the root and shoot apical meristems. This attenuation of cell division in tomato roots under water stress conditions poses a threat to root development and the acquisition of vital nutrients (González-García et al., 2023; Turan et al., 2023).

2.1.6 Hordeum Vulgare (Barley)

Water stress during the vegetative phase of barley cultivation precipitates a decline in the MI within both root and shoot meristems. The concomitant decrease in cell division activity within the root system exerts deleterious repercussions on nutrient assimilation and the overarching dimensions of the growth of plants (Anjum et al., 2003; Sallam et al., 2019; Elakhdar et al., 2022).

2.1.7 Gossypium Spp (Cotton)

Cotton plants are susceptible to the adverse effects of drought stress, particularly during the pivotal phases of flowering and boll development, which engender a diminution in the MI within developing bolls. The resultant attenuation in cell division activity within ovule primordia exerts consequential impacts on boll dimensions and the lint yield (Meshram et al., 2022; Zafar et al., 2023).

2.1.8 Solanum Tuberosum (Potato)

Water stress during the critical phase of tuber formation in potato plants precipitates a decline in the MI within developing tubers. The ensuing reduction in cell division activity within stolon tips exerts consequential effects on tuber initiation, ultimately impacting yield (Kawakami et al., 2006).

2.1.9 Sorghum Bicolor (Sorghum)

Drought stress throughout the critical period of reproduction of sorghum cultivation precipitates a reduction in the MI within developing panicles. The resultant decrease in cell division activity within panicle meristems exerts profound repercussions on grain development and harvest (Prasad et al., 2021; Chadalavada et al., 2022).

2.1.10 Beta Vulgaris (Sugar Beet)

Water stress during the key stage of root growth in sugar beet plants causes a decrease in the MI within growing taproots. The ensuing reduction in cell division activity within root meristems exerts discernible effects on root elongation and ultimately impacts sugar yield (Fitters et al., 2022; Alavilli et al., 2023).

These examples underscore the pervasive influence of water and drought stress on the MI of crop plants, accentuating the imperative of comprehending and alleviating these ramifications to safeguard sustainable crop production within water-constrained environments.

2.2 The Molecular Processes Underpinning the Impact of Water Stress and Drought Stress on Mitotic Index: Assessing How Water Stress Affects Plant Mitosis

At the molecular level, water stress induces complex signaling networks involving various stress-responsive genes and pathways (Mahmood et al., 2019). Transcription factors, protein kinases, and hormone signaling pathways play important roles in regulating these responses (Hoang et al., 2017). Water stress and drought usually result in the activation of stress-responsive genes that encode proteins involved in osmotic control, ROS detoxification, and cell cycle progression (You & Chan, 2015; Seleiman et al., 2021). For instance, genes encoding late embryogenesis abundant (LEA) proteins, dehydrins, and antioxidant enzymes typically undergo upregulation in response to water stress and drought conditions (Chen Y. et al., 2019). The interaction between ABA signalling and cell cycle control plays a role in how water stress affects the MI of crop plants (Sah et al., 2016).

The molecular intricacies dictating the influence of water stress and drought stress on the MI of crop plants are multifaceted, encompassing an array of signaling pathways and regulatory networks. Some pivotal molecules that affect mitotic mechanisms during drought and water stress are discussed in the next section.

2.2.1 Non-enzymatic Molecules

2.2.1.1 Abscisic Acid (ABA) Signaling

Abscisic acid acts as a pivotal phytohormone in orchestrating plant reactions to water and drought stress (Zhang et al., 2020; Mukherjee et al., 2023). Elevated ABA levels, precipitated by water stress, initiate downstream signaling cascades that modulate gene expression and elicit physiological adaptations. These ABA-mediated signaling pathways intricately intersect with the regulatory mechanisms governing the cell cycle, exerting a discernible influence on the MI of crop plants under circumstances of water stress (Zhao et al., 2014; Beier et al., 2023).

2.2.1.2 Osmotic Stress Signaling Pathways

Osmotic stress, arising from water deficit conditions, initiates signaling pathways pivotal for regulating osmotic adjustment and cellular hydration. These pathways entail the production and accretion of osmoprotectants, including proline and sugars, which serve to uphold cellular turgor and osmotic equilibrium. Moreover, the signaling pathways activated by osmotic stress can intricately intersect with the regulatory machinery governing the cell cycle, thereby modulating the MI of crop plants (Qi & Zhang, 2020; Choudhary et al., 2023).

2.2.1.3 Reactive Oxygen Species (ROS) Signaling

Water stress and drought stress can cause the accumulation of ROS in plant cells, leading to oxidative stress (Cruz de Carvalho, 2008; Qamer et al., 2021). ROS signaling pathways are crucial for regulating the expression of genes that encode antioxidant enzymes and stress-responsive proteins responsible for ROS detoxification. The interplay between ROS signaling pathways and cell cycle regulatory mechanisms can intricately influence the MI of crop plants during water stress (Zhang et al., 2016; Villalpando-Rodriguez & Gibson, 2021).

2.2.1.4 Transcriptional Regulation

Water stress and drought stress elicit alterations in gene expression profiles within crop plants, orchestrated by transcription factors and regulatory elements sensitive to stress cues (Joshi et al., 2016). Transcriptional regulation intricately governs the gene expressions implicated in osmotic adjustment, ROS detoxification, and cell cycle modulation, thereby exerting a discernible impact on the MI (Bashir et al., 2021).

2.2.1.5 Hormonal Crosstalk

Interactions among divergent hormone signaling pathways, encompassing the intricate interplay of ABA, ethylene, cytokinins, and auxins, stand as pivotal orchestrators of plant responses to the rigors of water stress and drought stress (Verma et al., 2016; Kim et al., 2022). These intricate hormonal signaling cascades intricately intertwine with the regulatory machinery governing the cell cycle, thereby wielding a discernible influence over the MI of crop plants amid stress-laden environments (Shimotohno et al., 2021).

These molecular mechanisms epitomize a complex network of interactions intricately regulating the MI of crop plants as a consequence of the exigencies of water and drought stress. A profound comprehension of these mechanisms is indispensable for devising strategies aimed at fortifying crop resilience and bolstering productivity within water-limited environments.

Enzymes implicated in orchestrating the MI of plant cells, especially those pertinent to cell cycle advancement and mitotic initiation, are susceptible to modulation by water stress. Though the precise impact of water stress on individual enzymes may fluctuate contingent upon the plant species, as well as the gravity and duration of the stress endured, below are several enzymes that could be influenced.

2.2.2 Enzymes Involved in Modulation of Mitotic Index Under Stressed Condition

2.2.2.1 Protein Kinase Signaling Cascades

Protein kinases serve as pivotal mediators in transducing stress signals and regulating cellular responses to water stress and drought stress (Chakradhar et al., 2019). Through intricate signaling cascades, protein kinases phosphorylate target proteins concerned in the perception of stress, signal transduction, and subsequent cellular responses. These signaling pathways can intricately intersect with cell cycle regulatory mechanisms, thereby exerting a discernible influence on the MI of crop plants under a stressed environment.

2.2.2.2 Cyclin-dependent Kinases (CDKs)

Water stress has demonstrated an ability to perturb both the activity and expression levels of Cyclin-Dependent Kinases (CDKs), which serve as paramount regulators governing the progression of the cell cycle (Kitsios & Doonan, 2011). These disruptions in CDK activity have the potential to intricately alter the temporal dynamics of mitotic entry and subsequent progression, thereby wielding a discernible influence over the MI of plant cells amidst conditions of water stress (Basu et al., 2022).

2.2.2.3 Cyclins

Water stress exerts a discernible influence on the expression and stability of cyclin proteins, potentially altering their availability to engage in active complexes with CDKs. Such modifications in the cyclin-CDK complex assembly have the propensity to perturb the customary flow of the cell cycle, thereby impinging upon the MI of plant cells (Tarn & Lai, 2011).

2.2.2.4 Cyclin-dependent Kinase Inhibitors (CKIs)

Water stress possesses the capacity to intricately modulate both the expression and activity of Cyclin-dependent Kinase Inhibitors (CKIs), pivotal regulators overseeing the activity of CDK and the advancement of the cell cycle. Alterations in CKI levels have the potential to disrupt the delicate equilibrium between CDK activity and inhibition, thereby influencing the temporal dynamics of mitotic entry and progression in response to water stress (Osakabe et al., 2014).

2.2.2.5 Mitotic Kinases

Mitotic kinases, including Polo-like kinases (PLKs), Aurora kinases, and NIMA-related kinases (NEKs), play multifaceted roles in orchestrating diverse facets of mitotic progression and chromosome segregation. Water stress has the potential to exert an influence on the activity or expression of these kinases, thereby impacting various mitotic events and ultimately shaping the MI of plant cells (Archambault & Glover, 2009; Zitouni et al., 2014).

2.2.2.6 Protein Phosphatases

Protein phosphatases, notably exemplified by protein phosphatase 1 (PP1) and protein phosphatase 2A (PP2A), function as counteracting agents against protein kinases by catalyzing the dephosphorylation of target proteins. Water stress may intricately impact the activity or expression of these protein phosphatases, thus precipitating a disruption in the intricate equilibrium of phosphorylation events crucial for the orchestration of mitotic progression and precise control of the cell cycle (Ramos et al., 2019).

2.2.2.7 Anaphase-promoting Complex/Cyclosome (APC/C)

The activity of the anaphase-promoting complex/cyclosome (APC/C), pivotal for orchestrating the breakdown of essential cell cycle regulatory proteins during mitosis, may undergo modulation in response to water stress. Alterations in APC/C activity have the potential to perturb the timing of mitotic exit and consequently influence the MI of plant cells when subjected to water stress conditions (Willems & De Veylder, 2022).

In essence, water stress has the efficiency to disrupt the customary regulation of cell cycle advancement and mitotic occurrences by impinging upon the activity, expression, or stability of enzymes crucially involved in these pathways. This disruption, in turn, could precipitate alterations in the MI of plant cells when subjected to water stress conditions.

2.3 Investigating the Effect of Water Stress on Plant Meiosis

In the life cycle of plants, meiosis is essential. Furthermore, it signifies the shift from a diploid sporophyte to a haploid gametophyte, while also allowing for genetic reassortment (Bhatt et al., 2001). An initial investigation revealed that the combination of high warmth and dryness worsened the decrease in bivalent production in asynaptic rye (Prakken, 2010). Water stress can cause meiotic chromosomal abnormalities in rice, including laggards, micronuclei, univalents, and a partial interruption of meiosis in extreme stress situations (Namuco & O'Toole, 1986).

2.3.1 Pollen Sterility: Androecium Anomaly

Pollen sterility brought on by drought stress is a negative factor that decreases crop yields globally. Water deficit affects both the meiotic as well as mitotic stages of the reproductive process in anthers. Drought at these phases causes pollen sterility by changing the anthers' types and structures, leading to the breakdown of certain meiocytes, misaligned microspores, an expanded middle layer, and abnormal vacuolated tapeta (Yu et al., 2019). During pollen development, the tapetum goes through a PCD phase to generate enzymes for releasing microspores from tetrads, mediating microspore development, and depositing pollen coating components (Sun et al., 2007; Wilson and Zhang, 2009; Zhu et al., 2011). In peach, it was discovered that disrupting ROS homeostasis causes aberrant tapetum and microspore formation, resulting in male sterility (Cai et al., 2021). In wheat, water scarcity during meiosis hinders the development of both male and female reproductive cells, perhaps leading to the formation of infertile pollen and/or egg cells, decreased viability of gametes, or other abnormalities in reproductive structures that lower seed output (Smith

& Zhao, 2016; Fábián et al., 2019). In wheat water stress is reported to cause abnormal binucleate pollen grains in addition to trinucleate, tetranucleate, and mononucleate types. It was further observed that the abnormality of pollen increased with an increase in water stress (Alam et al., 2011). Another study observed that drought stress generated by PEG in wheat caused the tapetum to detach from the anther wall, leading to programmed degeneration of the tapetum (Dong et al., 2017). In a distinct investigation, it was noted that drought-stressed wheat exhibited abnormal trinucleate pollen grain which was deficient in starch content and arrested at the early stage of gametogenesis. In addition,entirely sterile pollen grains were also found in drought-stressed wheat plants. Further, collective heat and drought stress during the meiosis of microspore mother cells resulted in morphological abnormalities in relation to equal microspore divisions, non-separation of tetrads, pollens fused, development of microspore within the same lumen and multinucleate pollen grains (Jäger et al., 2008). In a previous investigation on wheat, the initial impacts of stress were observed between the meiosis and young microspore stages. These effects included the deterioration of meiocytes, disorientation of reproductive cells, and anomalous formation of vacuoles in tapetal cells. Moreover, in later developmental stages, other effects were observed, including the buildup of starch in the connective tissue in unusual locations, enlargement of the middle layer of endothelial cells, and the deposition of substances similar to sporopollenin in the anther loculus. Consequently, the ensuing pollen grains were devoid of both starch and intine (Lalonde et al., 1997). In maize, the application of drought and heat stress during anthesis had a negative impact on pollen germination (Bheemanahalli et al., 2022). A different research found that heat and water stress influenced the dehiscence of anthers in maize plants. Furthermore, there was a reduction in the number of pollen grains on the stigma in the stressed plants. Moreover, spikelet fertility was also dropped in the stressed plants (Rang et al., 2011). Water stress in rice and wheat is closely associated with male sterility due to a substantial decrease in starch accumulation. Nevertheless, in these instances, there is an increased accumulation of soluble carbohydrates (Sheoran and Saini, 1996).

3. Water Stress and Accumulation of Metabolites in Reproductive Units

The developing microspores serve as a highly efficient site for photosynthesis and store photoassimilates. At the early stage of the microspore's development, the amount of stored carbohydrates is limited. However, during the process of pollen maturation initiation (PMI), the developing spores go through a fast phase of starch production and rapidly amass large quantities of starch (Datta et al., 2002). In plants, the accumulation of sugars in the anthers serves multiple purposes. Firstly, these sugars provide energy for the processes of microsporogenesis and pollen maturation. Additionally, they participate in the formation of pollen tubes by delivering energy. Furthermore, these sugars act as osmolytes, helping to shield pollen against desiccation and other abiotic stresses. For example, sucrose helps to maintain the integrity of the pollen membrane. Carbohydrates are crucial for both the growth of pollen and play a pivotal role in determining its viability and capability to germinate (De Storme & Geelen, 2014). In wheat, water stress-induced male sterility occurs when there is a lack of water for 4 days throughout the meiosis stage. This leads to a considerable decrease in the amount of soluble sugars in the spores and changes in how starch is distributed in the various layers of the anther (Dorion et al., 1996). In tomato plants, the reduced ability of pollen to fertilize under continuous high-temperature stress is intricately related with a reduction in the amount of sugars stored in the protective layers of the anthers and the forming pollen. However, there is an upsurge in the sugar content found in the fluid within the locules (Pressman, 2002). Heat stress may also induce water stress, making the two conditions intricately related. Water stress-induced male sterility in rice and wheat is closely associated with a substantial reduction in starch buildup. Nevertheless, in these instances, there is an increased accumulation of soluble carbohydrates (Sheoran & Saini, 1996; Nguyen et al., 2010).

4. Effect of Water Stress on Female Gametophyte Development

Water deficiencies in numerous plant species reduce seed production or crop yields. Under optimal growth conditions, an ample amount of ovules are begin to maximize reproduction. Due to suboptimal growth conditions, some of the ovules in most plants may inevitably abort. During environmental stress, the demand for nutrients for reproduction can exceed the available carbon and nitrogen resources provided to the ovule. As a result, the gynoecium or the plant itself may not be able to provide sufficient nutrition for all the ovules that are initiated, leading to an excess in their quantity (Sun et al., 2004). Following water stress, the photosynthetic rate in maize reduces considerably, resulting in a reduction of sugar transportation into the embryo sac. This limitation of resources hampers seed development (Westgate & Boyer, 1985; Schussler & Westgate, 1995). An early study revealed that a lack of water during the development of the maize embryo sac resulted in significant abnormalities in the structure of the female gametophyte. The occurrence of aberrant ovule development ranged from 15% to 45% in response to the severity and length of drought stress, whereas only 2.5% of the ovules in irrigated plants showed abnormalities. Due to a lack of water, the stigma and style experienced gradual elongation (Moss & Downey, 1971). In maize, drought stress can meaningfully impede the development of female reproductive organs. Maize experiences embryo abortion soon after pollination because of the stress caused by drought (Andersen et al., 2002). An investigation of the expression of gene patterns in fertilized maize ovaries subjected to water shortage conditions demonstrated the activation of many processes, including ABA signaling and senescence, in response to drought (Kakumanu et al., 2012).

5. Effect of Water Stress on Endosperm Development

The endosperm is an essential part of the seed that sustains the embryo's growth and acts as a reserve storage. Drought stress at the beginning of seed development diminishes the kernel/seed's capacity to accumulate starch by restricting the formation of endosperm cells and amyloplasts. This leads to a decrease in grain weight and a drop in the endosperm's capacity to efficiently store starch, both in terms of speed and duration (Nicolas et al., 1985; Saini & Westgate, 1999). Drought stress in wheat caused a delay in the endosperm developmental stage that transits from the syncytial to the cellularized stage. Alongside a reduction in seed size and delayed endosperm development, a specific group of genes associated with cytoskeleton organization exhibited improper regulation in growing seeds subjected to drought stress. Furthermore, it is worth mentioning that drought stress significantly inhibited the production of wheat storage protein genes, including gliadins, glutenins, and avenins, within a mere 3rd day of pollination (Begcy & Walia, 2015). Transcriptome analysis has shown of that drought stress induces differential gene expressions in barley caryopses, particularly related to the induction of storage proteins and protein synthesis pathways. This led to an augmentation in the endosperm's buildup of protein (Li et al., 2020). Drought stress in wheat reduced endosperm cell survival, accelerated nuclear deformation, and increased mitochondria disintegration. The drought treatments exhibited higher nucleic acid hydrolase activity and lower nucleic acid concentrations. Consequently, the drought treatment caused an earlier occurrence of the peak in DNA fragmentation. The expression of four ABA-related genes was dramatically enhanced during drought stress. Conversely, the occurrence of drought had a substantial impact on the expression of four genes associated with ETH receptors and one gene associated with PCD (dad1) (Li et al., 2018). A recent study analyzed the physiological specifications, transcriptome, and proteome of two winter wheat varieties, Xindong No. 18 (XD18) and Xindong No. 22 (XD22), which have different degrees of drought tolerance, under progressive and severe drought stress (PADS). Under PADS conditions, the survival of endosperm cells in XD18 and XD22 was reduced. Additionally, there was a 21% and 9% increase in the number of nuclei with DNA breakage, respectively. Furthermore, the chromatin became condensed, and there was evidence of nuclear envelope rupture and cytoplasm leakage, compared to the control group. The nuclei count in XD18 and XD22 decreased by 6.07% and 7.5% respectively. Additionally, the DNA hydrolase activity rose by 24.7% and 28.9%, while H_2O_2 content rose by 54.3% and 38.7%. MDA content also climbed by 44.1% and 33.8%. A total of 21,677 differentially expressed genes (DEGs) and 3388 differentially expressed proteins (DEPs) were identified. Among these, 48 proteins associated with PCD were discovered, with the ubiquitin protein showing significant expression in

XD22. The combined study of transcriptomics and proteomics revealed an up-regulation in the expression of genes linked with DNA repair and membrane activities, thereby corroborating the findings of the physiological analysis (Li et al., 2024).

6. Conclusion

Water stress is always harmful to the overall growth of plants. Apart from the deleterious morphological and physiological anomaly that set in during the stressed condition, the impact on the genetic and developmental machinery of the plant is equally pronounced. The development of the plant has a close relationship with the cell division pattern. In addition, reproductive development is also vital for the perpetuation of the species. As evident, drought and water scarcity remarkably hinder the cell division pattern both in the mitotic and the meiotic level. Even proper endosperm development is also not spared from the clutches of drought. All these effects directly are a warning sign not only to the future existence of the plant under current environmental conditions, but also to humans as the food resource is largely dependent on the productivity of the former. Thus it's high time for us to devise the mitigation strategies that might give a favorable resolution by reducing the deleterious effects of the stressful environment. In addition, sincere efforts are also required at the molecular and physiological levels to identify the candidate genes that might confer tolerance to the crops from water stress. Optimizing the fertility of the plant under stressful conditions should be another action plan to ensure the production volume for the booming global population and overall benefit of mankind.

References

Ahluwalia O, Singh PC & Bhatia R. 2021. A review on drought stress in plants: Implications, mitigation and the role of plant growth promoting rhizobacteria. *Resources, Environment and Sustainability,* **5:** 100032.

Alam ASM, Kabir G & Islam M. 2011. Effect of water stress on pollen grain abnormality and sterility in 21 near isogenic lines of wheat (Triticum aestivum L.). *Journal of the Bangladesh Society for Agricultural Science and Technology,* **8:** 99–104.

Alavilli H, Yolcu S, Skorupa M, Aciksoz SB & Asif M. 2023. Salt and drought stress-mitigating approaches in sugar beet (Beta vulgaris L.) to improve its performance and yield. *Planta,* **258:** 30.

Andersen MN, Asch F, Wu Y, Jensen CR, Næsted H, Mogensen VO, et al. 2002. Soluble invertase expression is an early target of drought stress during the critical, abortion-sensitive phase of young ovary development in maize. *Plant Physiol,* **130:** 591–604.

Anjum F, Yaseen M, Rasool E, Wahid A & Anjum S. 2003. Water stress in barley (Hordeum vulgare L.) I. Effect on morphological characters. *Pak J Agric Sci,* **40:** 43–44.

Archambault V & Glover DM. 2009. Polo-like kinases: Conservation and divergence in their functions and regulation. *Nat Rev Mol Cell Biol,* **10:** 265–275.

Balint Z, Mutua F, Muchiri P & Omuto CT. 2013. Monitoring Drought with the Combined Drought Index in Kenya. pp. 341–356. Elsevier.

Bashir SS, Hussain A, Hussain SJ, Wani OA, Zahid Nabi S, Dar NA, et al. 2021. Plant drought stress tolerance: Understanding its physiological, biochemical and molecular mechanisms. *Biotechnology & Biotechnological Equipment,* **35:** 1912–1925.

Basu S, Greenwood J, Jones AW & Nurse P. 2022. Core control principles of the eukaryotic cell cycle. *Nature,* **607:** 381–386.

Begcy K & Walia H. 2015. Drought stress delays endosperm development and misregulates genes associated with cytoskeleton organization and grain quality proteins in developing wheat seeds. *Plant Science,* **240:** 109–119.

Beier MP, Jinno C, Noda N, Nakamura K, Sugano S, Suzuki Y, et al. 2023. ABA signaling converts stem cell fate by substantiating a tradeoff between cell polarity, growth and cell cycle progression and abiotic stress responses in the moss Physcomitrium patens. *Front Plant Sci,* **14.**

Bhatt AM, Canales C & Dickinson HG. 2001. Plant meiosis: The means to 1N. *Trends Plant Sci,* **6:** 114–121.

Bheemanahalli R, Ramamoorthy P, Poudel S, Samiappan S, Wijewardane N & Reddy KR. 2022. Effects of drought and heat stresses during reproductive stage on pollen germination, yield, and leaf reflectance properties in maize (<scp> *Zea mays* </scp> L.). *Plant Direct,* **6.**

Boutraa T, Akhkha A, Al-Shoaibi AA & Alhejeli AM. 2010. Effect of water stress on growth and water use efficiency (WUE) of some wheat cultivars (Triticum durum) grown in Saudi Arabia. *Journal of Taibah University for Science,* **3:** 39–48.

Cesare E Di, Verrico A, Miele A, Giubettini M, Rovella P, Coluccia A, et al. 2017. Mitotic cell death induction by targeting the mitotic spindle with tubulin-inhibitory indole derivative molecules. *Oncotarget,* **8:** 19738–19759.

Chadalavada K, Guna K, Ranjitha Kumari BD & Senthil Kumar T. 2022. Drought stress in sorghum: Impact on grain quality. pp. 113–134. *In:* Climate Change and Crop Stress, Elsevier.

Chakradhar T, Reddy RA & Chandrasekhar T. 2019. Protein kinases and phosphatases in stress transduction. pp. 533–547. *In:* Plant Signaling Molecules, Elsevier.

Chen Y, Li C, Zhang B, Yi J, Yang Y, Kong C, Lei C & Gong M. 2019. The role of the Late Embryogenesis-Abundant (LEA) protein family in development and the abiotic stress response: A comprehensive expression analysis of potato (Solanum Tuberosum). *Genes*, **10(2):** 148. https://doi.org/10.3390/genes10020148.

Chiatante D, Rocco M, Maiuro L, Scippa GS, Di Martino C & Bryant JA, et al. 1997. Cell division and DNA topisomerase I activity in root meristems of pea seedlings during water stress. *Plant Biosystems – An International Journal Dealing with all Aspects of Plant Biology,* **131:** 163–173.

Choudhary S, Wani KI, Naeem M, Khan MMA & Aftab T. 2023. Cellular responses, osmotic adjustments, and role of osmolytes in providing salt stress resilience in higher plants: Polyamines and nitric oxide crosstalk. *J Plant Growth Regul,* **42:** 539–553.

Cruz de Carvalho MH. 2008. Drought stress and reactive oxygen species: Production, scavenging and signaling. *Plant Signal Behav,* **3:** 156–165.

Datta R, Chamusco KC & Chourey PS. 2002. Starch biosynthesis during pollen maturation is associated with altered patterns of gene expression in maize. *Plant Physiol,* **130:** 1645–1656.

De Storme N & Geelen D. 2014. The impact of environmental stress on male reproductive development in plants: Biological processes and molecular mechanisms. *Plant, Cell & Environment,* **37(1):** 1–18. https://doi.org/10.1111/pce.12142.

Dong B, Zheng X, Liu H, Able JA, Yang H, Zhao H, et al. 2017. Effects of drought stress on pollen sterility, grain yield, abscisic acid and protective enzymes in two winter wheat cultivars. *Front Plant Sci,* **8**

Dong S, Jiang Y, Dong Y, Wang L, Wang W, Ma Z, et al. 2019. A study on soybean responses to drought stress and rehydration. *Saudi J Biol Sci,* **26:** 2006–2017.

Dorion S, Lalonde S & Saini HS. 1996. Induction of male sterility in wheat by meiotic-stage water deficit is preceded by a decline in invertase activity and changes in carbohydrate metabolism in anthers. *Plant Physiol,* **111:** 137–145.

Elakhdar A, Solanki S, Kubo T, Abed A, Elakhdar I, Khedr R, et al. 2022. Barley with improved drought tolerance: Challenges and perspectives. *Environ Exp Bot,* **201:** 104965.

Fábián A, Sáfrán E, Szabó-Eitel G, Barnabás B & Jäger K. 2019. Stigma functionality and fertility are reduced by heat and drought co-stress in wheat. *Front Plant Sci,* **10**.

Filippou P, Antoniou C & Fotopoulos V. 2011. Effect of drought and rewatering on the cellular status and antioxidant response of *Medicago truncatula* plants. *Plant Signal Behav,* **6:** 270–277.

Fitters TFJ, Mooney SJ & Sparkes DL. 2022. Impact of water availability on root growth of sugar beet varieties. *Soil Use Manag,* **38:** 1033–1043.

Gogoi A & Tripathi B. 2019. 42% India's land area under drought, worsening farm distress in election year. *India Spend.* Available from: https://www.indiaspend.com/42-indias-land-area-under-drought-worsening-farm-distress-in-election-year/

González-García MP, Conesa CM, Lozano-Enguita A, Baca-González V, Simancas B, Navarro-Neila S, et al. 2023. Temperature changes in the root ecosystem affect plant functionality. *Plant Commun,* **4:** 100514.

Granier C, Inzé D & Tardieu F. 2000. Spatial distribution of cell division rate can be deduced from that of p34cdc2 kinase activity in maize leaves grown at contrasting temperatures and soil water conditions. *Plant Physiol,* **124:** 1393–1402.

Hassan MA, Dahu N, Hongning T, Qian Z, Yueming Y, Yiru L, et al. 2023. Drought stress in rice: Morpho-physiological and molecular responses and marker-assisted breeding. *Front Plant Sci,* **14**.

Hayashi K, Hasegawa J & Matsunaga S. 2013. The boundary of the meristematic and elongation zones in roots: Endoreduplication precedes rapid cell expansion. *Sci Rep,* **3:** 2723.

Hoang XLT, Nhi DNH, Thu NBA, Thao NP & Tran L-SP. 2017. Transcription factors and their roles in signal transduction in plants under abiotic stresses. *Curr Genomics,* **18.**

Hussain HA, Men S, Hussain S, Zhang Q, Ashraf U, Anjum SA et al. 2020. Maize tolerance against drought and chilling stresses varied with root morphology and antioxidative defense system. *Plants,* **9:** 720.

Jäger K, Fábián A & Barnabás B. 2008. Effect of water deficit and elevated temperature on pollen development of drought sensitive and tolerant winter wheat (Triticum aestivum L.) genotypes. *Acta Biologica Szegediensis,* **52:** 67–71.

Joshi R, Wani SH, Singh B, Bohra A, Dar ZA, Lone AA, et al. 2016. Transcription factors and plants response to drought stress: Current understanding and future directions. *Front Plant Sci,* **7**.

Kakumanu A, Ambavaram MMR, Klumas C, Krishnan A, Batlang U, Myers E, et al. 2012. Effects of drought on gene expression in maize reproductive and leaf meristem tissue revealed by RNA-Seq. *Plant Physiol,* **160:** 846–867.

Kawakami J, Iwama K & Jitsuyama Y. 2006. Soil water stress and the growth and yield of potato plants grown from microtubers and conventional seed tubers. *Field Crops Res,* **95:** 89–96.

Kim G, Ryu H & Sung J. 2022. Hormonal crosstalk and root suberization for drought stress tolerance in plants. *Biomolecules,* **12**.

Kim Y, Chung YS, Lee E, Tripathi P, Heo S & Kim K-H. 2020. Root response to drought stress in rice (Oryza sativa L.). *Int J Mol Sci,* **21:** 1513.

Kitsios G & Doonan JH. 2011. Cyclin dependent protein kinases and stress responses in plants. *Plant Signal Behav,* **6:** 204–209.

Lalonde S, Beebe DU & Saini HS. 1997. Early signs of disruption of wheat anther development associated with the induction of male sterility by meiotic-stage water deficit. *Sex Plant Reprod,* **10:** 40–48.

Lecoeur J, Wery J, Turc O & Tardieu F. 1995. Expansion of pea leaves subjected to short water deficit: Cell number and cell size are sensitive to stress at different periods of leaf development. *J Exp Bot,* **46:** 1093–1101.

Lee S, Seo PJ, Lee H & Park C. 2012. A NAC transcription factor NTL4 promotes reactive oxygen species production during drought-induced leaf senescence in Arabidopsis. *The Plant Journal,* **70:** 831–844.

Li C, Bao Y, Guo W, Li C & Li C. 2024. Multi-omics analysis of response mechanism of programmed cell death in wheat endosperm to post-anthesis drought stress. *Environ Exp Bot,* **218:** 105614.

Li C, Li C, Wang B, Zhang R, Fu K, Gale WJ, et al. 2018. Programmed cell death in wheat (Triticum aestivum L.) endosperm cells is affected by drought stress. *Protoplasma,* **255:** 1039–1052.

Li F, Chen X, Yu X, Chen M, Lu W, Wu Y, et al. 2020. Novel insights into the effect of drought stress on the development of root and caryopsis in barley. *Peer J,* **8:** e8469.

Ligasová A, Frydrych I & Koberna K. 2023. Basic methods of cell cycle analysis. *Int J Mol Sci,* **24:** 3674.

Mahmood T, Khalid S, Abdullah M, Ahmed Z, Shah MKN, Ghafoor A, et al. 2019. Insights into drought stress signaling in plants and the molecular genetic basis of cotton drought tolerance. *Cells,* **9:** 105.

Meshram JH, Singh SB, Raghavendra KP & Waghmare VN. 2022. Drought stress tolerance in cotton: Progress and perspectives. pp. 135–169. *In:* Climate Change and Crop Stress, Elsevier.

Miettinen TP, Kang JH, Yang LF & Manalis SR. 2019. Mammalian cell growth dynamics in mitosis. *Elife,* **8**.

Mira MM, Huang S, Kapoor K, Hammond C, Hill RD & Stasolla C, et al. 2017. Expression of Arabidopsis class 1 phytoglobin (AtPgb1) delays death and degradation of the root apical meristem during severe PEG-induced water deficit. *J Exp Bot,* **68:** 5653–5668.

Moss GI & Downey LA. 1971. Influence of drought stress on female gametophyte development in corn (Zea mays L.) and subsequent grain yield. *Crop Sci,* **11:** 368–372.

Mukherjee A, Dwivedi S, Bhagavatula L & Datta S. 2023. Integration of light and ABA signaling pathways to combat drought stress in plants. *Plant Cell Rep,* **42:** 829–841.

Namuco OS & O'Toole JC. 1986. Reproductive stage water stress and sterility. I. Effect of stress during meiosis. *Crop Sci,* **26:** 317–321.

Nguyen GN, Hailstones DL, Wilkes M & Sutton BG. 2010. Drought stress: Role of carbohydrate metabolism in drought-induced male sterility in rice anthers. *J Agron Crop Sci,* **196:** 346–357.

Nicolas ME, Gleadow RM & Dalling MJ. 1985. Effect of post-anthesis drought on cell division and starch accumulation in developing wheat grains. *Ann Bot,* **55:** 433–444.

Onyemaobi O, Sangma H, Garg G, Wallace X, Kleven S, Suwanchaikasem P, et al. 2021. Reproductive stage drought tolerance in wheat: Importance of stomatal conductance and plant growth regulators. *Genes (Basel),* **12:** 1742.

Osakabe Y, Osakabe K, Shinozaki K & Tran L-SP. 2014. Response of plants to water stress. *Front Plant Sci,* **5.**

Özmen S, Tabur S, Öney-Birol S & Özmen S. 2022. Molecular responses of exogenous polyamines under drought stress in the barley plants. *Cytologia (Tokyo),* **87:** D-21-00069.

Petrov V, Hille J, Mueller-Roeber B & Gechev TS. 2015. ROS-mediated abiotic stress-induced programmed cell death in plants. *Front Plant Sci,* **6.**

Prakken R. 2010. Studies of Asynapsis in rye. *Hereditas,* **29:** 475–493.

Prasad VBR, Govindaraj M, Djanaguiraman M, Djalovic I, Shailani A, Rawat N, et al. 2021. Drought and high temperature stress in sorghum: Physiological, genetic, and molecular insights and breeding approaches. *Int J Mol Sci,* **22:** 9826.

Pressman E. 2002. The effect of heat stress on tomato pollen characteristics is associated with changes in carbohydrate concentration in the developing anthers. *Ann Bot,* **90:** 631–636.

Qamer Z, Chaudhary MT, Du X, Hinze L & Azhar MT. 2021. Review of oxidative stress and antioxidative defense mechanisms in Gossypium hirsutum L. in response to extreme abiotic conditions. *Journal of Cotton Research,* **4:** 9.

Qi F & Zhang F. 2020. Cell cycle regulation in the plant response to stress. *Front Plant Sci,* **10.**

Ramos F, Villoria MT, Alonso-Rodríguez E & Clemente-Blanco A. 2019. Role of protein phosphatases PP1, PP2A, PP4 and Cdc14 in the DNA damage response. *Cell Stress,* **3:** 70–85.

Rang ZW, Jagadish SVK, Zhou QM, Craufurd PQ & Heuer S. 2011. Effect of high temperature and water stress on pollen germination and spikelet fertility in rice. *Environ Exp Bot,* **70:** 58–65.

Ritchie H & Roser M. 2024. Available at: https://ourworldindata.org/water-use-stress.

Robertson JM, Yeung EC, Reid DM & Hubick KT. 1990. Developmental responses to drought and abscisic acid in sunflower roots. *J Exp Bot,* **41:** 339–350.

Sacks MM, Silk WK & Burman P. 1997. Effect of water stress on cortical cell division rates within the apical meristem of primary roots of maize. *Plant Physiol,* **114:** 519–527.

Sah SK, Reddy KR & Li J. 2016. Abscisic acid and abiotic stress tolerance in crop plants. *Front Plant Sci,* **7.**

Saini HS & Westgate ME. 1999. Reproductive development in grain crops during drought. *Advances in Agronomy,* **68:** 59–96.

Salehi-Lisar SY & Bakhshayeshan-Agdam H. 2016. Drought stress in plants: Causes, consequences, and tolerance. pp. 1–16. *In:* Hossain M., Wani S., Bhattacharjee S., Burritt D. & Tran L. (eds), Drought Stress Tolerance in Plants, Vol 1, Springer International Publishing, Cham.

Sallam A, Alqudah AM, Dawood MFA, Baenziger PS & Börner A. 2019. Drought stress tolerance in wheat and barley: Advances in physiology, breeding and genetics research. *Int J Mol Sci,* **20:** 3137.

Schuppler U, He P-H, John PCL & Munns R. 1998. Effect of water stress on cell division and Cdc2-like cell cycle kinase activity in wheat leaves1. *Plant Physiol,* **117:** 667–678.

Schussler JR & Westgate ME. 1995. Assimilate flux determines kernel set at low water potential in maize. *Crop Sci,* **35:** 1074–1080.

Seleiman MF, Al-Suhaibani N, Ali N, Akmal M, Alotaibi M, Refay Y, et al. 2021. Drought stress impacts on plants and different approaches to alleviate its adverse effects. *Plants,* **10:** 259.

Setter TL & Flannigan BA. 2001. Water deficit inhibits cell division and expression of transcripts involved in cell proliferation and endoreduplication in maize endosperm. *J Exp Bot,* **52:** 1401–1408.

Shafi S, Shafi I, Zaffar A, Zargar SM, Shikari AB, Ranjan A, et al. 2023. The resilience of rice under water stress will be driven by better roots: Evidence from root phenotyping, physiological, and yield experiments. *Plant Stress,* **10:** 100211.

Sheoran IS & Saini HS. 1996. Drought-induced male sterility in rice: Changes in carbohydrate levels and enzyme activities associated with the inhibition of starch accumulation in pollen. *Sex Plant Reprod,* **9:** 161–169.

Sheoran S, Kaur Y, Kumar S, Shukla S, Rakshit S & Kumar R, et al. 2022. Recent advances for drought stress tolerance in maize (Zea mays L.): Present status and future prospects. *Front Plant Sci,* **13**.

Shimotohno A, Aki SS, Takahashi N & Umeda M. 2021. Regulation of the plant cell cycle in response to hormones and the environment. *Annu Rev Plant Biol,* **72:** 273–296.

Smith AR & Zhao D. 2016. Sterility caused by floral organ degeneration and abiotic stresses in Arabidopsis and cereal grains. *Front Plant Sci,* **7**.

De Storme N & Geelen D. 2014. The impact of environmental stress on male reproductive development in plants: Biological processes and molecular mechanisms. *Plant Cell Environ,* **37:** 1–18.

Sun K, Hunt K & Hauser BA. 2004. Ovule abortion in Arabidopsis triggered by stress. *Plant Physiol,* **135:** 2358–2367.

Sun YJ, Hord CLH, Chen C Bin & Ma H. 2007. Regulation of Arabidopsis early anther development by putative cell-cell signaling molecules and transcriptional regulators. *J Integr Plant Biol,* **49:** 60–68.

Tarn W-Y & Lai M-C. 2011. Translational control of cyclins. *Cell Div,* **6:** 5.

Turan M, Ekinci M, Argin S, Brinza M & Yildirim E. 2023. Drought stress amelioration in tomato (Solanum lycopersicum L.) seedlings by biostimulant as regenerative agent. *Front Plant Sci,* **14**.

Verma V, Ravindran P & Kumar PP. 2016. Plant hormone-mediated regulation of stress responses. *BMC Plant Biol,* **16:** 86.

Villalpando-Rodriguez GE & Gibson SB. 2021. Reactive Oxygen Species (ROS) regulates different types of cell death by acting as a rheostat. *Oxid Med Cell Longev,* **2021:** 1–17.

Wan C, Dang P, Gao L, Wang J, Tao J, Qin X, et al. 2022. How does the environment affect wheat yield and protein content response to drought? A meta-analysis. *Front Plant Sci,* **13**.

Wang X, Wu Z, Zhou Q, Wang X, Song S & Dong S, et al. 2022. Physiological response of soybean plants to water deficit. *Front Plant Sci,* **12**.

Wang Y, Wan L, Zhang L, Zhang Z, Zhang H, Quan R, et al. 2012. An ethylene response factor OsWR1 responsive to drought stress transcriptionally activates wax synthesis related genes and increases wax production in rice. *Plant Mol Biol,* **78:** 275–288.

Wei L, Wang L, Yang Y, Wang P, Guo T & Kang G, et al. 2015. Abscisic acid enhances tolerance of wheat seedlings to drought and regulates transcript levels of genes encoding ascorbate-glutathione biosynthesis. *Front Plant Sci,* **6**.

West G, Inzé D & Beemster GT. 2004. Cell cycle modulation in the response of the primary root of Arabidopsis to salt stress. *Plant Physiol,* **135(2):** 1050–1058.

Westgate ME & Boyer JS. 1985. Carbohydrate reserves and reproductive development at low leaf water potentials in maize. *Crop Sci,* **25:** 762–769.

Willems A & De Veylder L. 2022. The plant anaphase-promoting complex/cyclosome. *Annu Rev Cell Dev Biol,* **38:** 25–48.

Wilson ZA & Zhang D-B. 2009. From Arabidopsis to rice: Pathways in pollen development. *J Exp Bot,* **60:** 1479–1492.

Yang Z, Cao Y, Shi Y, Qin F, Jiang C & Yang S, et al. 2023. Genetic and molecular exploration of maize environmental stress resilience: Toward sustainable agriculture. *Mol Plant,* **16:** 1496–1517.

You J & Chan Z. 2015. ROS regulation during abiotic stress responses in crop plants. *Front Plant Sci,* **6**.

Yu J, Jiang M & Guo C. 2019. Crop pollen development under drought: From the phenotype to the mechanism. *Int J Mol Sci,* **20:** 1550.

Zafar S, Afzal H, Ijaz A, Mahmood A, Ayub A, Nayab A, et al. 2023. Cotton and drought stress: An updated overview for improving stress tolerance. *South African Journal of Botany,* **161:** 258–268.

Zhang H, Zhao Y & Zhu J-K. 2020. Thriving under stress: How plants balance growth and the stress response. *Dev Cell,* **55:** 529–543.

Zhang J, Wang X, Vikash V, Ye Q, Wu D, Liu Y, et al. 2016. ROS and ROS-mediated cellular signaling. *Oxid Med Cell Longev,* **2016:** 1–18.

Zhao FY, Wang K, Zhang SY, Ren J, Liu T & Wang X, et al. 2014. Crosstalk between ABA, auxin, MAPK signaling, and the cell cycle in cadmium-stressed rice seedlings. *Acta Physiol Plant,* **36:** 1879–1892.

Zhu J, Lou Y, Xu X & Yang Z-N. 2011. A genetic pathway for tapetum development and function in Arabidopsis. *J Integr Plant Biol,* **53:** 892–900.

Zitouni S, Nabais C, Jana SC, Guerrero A & Bettencourt-Dias M. 2014. Polo-like kinases: Structural variations lead to multiple functions. *Nat Rev Mol Cell Biol,* **15:** 433–452.

3 | Anatomical Anomaly of Plants During Water Stress

Dwaipayan Sinha[1*], Supti Das[2], Srijonee Choudhury[3] Sukanya Chatterjee[4], Somnath Kar[5] and Asha Kadam[6]

[1] Department of Botany, Government General Degree College, Mohanpur, West Bengal, India
[2] Department of Biochemistry, University of Calcutta, Kolkata, India
[3] Department of Botany, Kalyani University, Kalyani, West Bengal, India
[4] Department of Botany, Lady Brabourne College, P-1/2, Beniapukur, West Bengal, India
[5] Department of Botany, Holy Cross College, Jubatara, Agartala, India
[6] P.G. Department of Botany, Dada Patil Mahavidyalaya, Karjat, India

Water stress is one of the burning global concerns. The plants being immobile suffer the worst. Having to stay rooted, during stress, the plants first try to combat stress and then start to cope giving rise to numerous physiological, morphological, and hence anatomical changes. As an impact of water stress, the anatomical adaptations identified, including alterations in leaf morphology, stomatal behavior, vascular tissue structure, and cellular organization are so basic that they directly emphasize the complexity of plant water relations and the probabilistic nature of plant physiology. These changes allow the plant to efficiently make the best use of the water that is available, maintain cellular turgor pressure, and alleviate the detrimental effects of water stress on growth and development. The study of variable anatomical anomalies in different plants due to water stress also has significance in agriculture, ecology, and environmental conservation. By recognizing the structural alterations that facilitate plants to thrive under water-limited conditions, traits with potential agronomic value can be identified for breeding drought-tolerant crop varieties and for selecting resilient species for ecosystem restoration and land rehabilitation efforts. This chapter is an attempt to overview the anatomical alteration of plants in response to water stress.

*Corresponding author: dwaipayansinha@hotmail.com

1. Introduction

The ability—or lack thereof—to meet the needs of the environment and humanity for clean water is referred to as water stress. Compared to scarcity, water stress is a more complete notion that takes into account a larger range of issues. Numerous physical aspects of water resources are included in the analysis, including the quantity, quality, and accessibility of the water, all of which are typically influenced by the cost of water and the sufficiency of the infrastructure. Both water use and water withdrawals yield valuable data that provide insights into the degree of water scarcity (Liu et al., 2017). Water stress arises when the water demand surpasses the available supply within a specific timeframe, or when poor water quality restricts its usability. Water stress diminishes the amount and quality of freshwater resources. Human water consumption worldwide has increased by a factor of six in the last century and continues to rise at approximately 1% per year, driven by population growth and the advancement of economies. Water scarcity poses substantial risks to individuals, especially food security, energy security, and conflict (Vanham et al., 2021). Urbanization and climate change are exacerbating water scarcity in cities globally, as demand surpasses availability (He et al., 2021). Plant development, maturity, and productivity all depend on water. Plants that are stressed by water lose several of their physiological characteristics, most notably their ability to photosynthesize. Extended stress significantly hinders plant growth and yield (Osakabe et al., 2014). Drought may be viewed as a combination of several challenging situations that lead to imbalances in osmotic and ionic levels in plants. Given the essentiality of water for plant existence, the continued viability of plants is uncertain in times of drought (Mukarram et al., 2021). A water deficit stress, whether permanent or temporary, significantly hinders the growth and spread of both artificial and natural vegetation, as well as the yield of agricultural crops, exceeding any other environmental element (Shao et al., 2009). Plants possess several defensive mechanisms to mitigate water stress and minimize the negative impacts of drought. The mechanisms may include adjustments to the behavior of the stomata, changes in the osmotic balance and antioxidant metabolism, and regulation of ion levels (Mukarram et al., 2021). Water deprivation disrupts the plant's anatomical structure, affecting the control of water flow via the vascular bundles. Drought stress mainly affects the cellular structure (Shafqat et al., 2021). When plant cells experience water scarcity, they typically exhibit increased vascular tissue, thicker cell walls, reduced cell size, and, in extreme cases, cell wall and membrane rupture (Fan et al., 2006; Beltrano & Ronco, 2008; Litvin et al., 2016; Yang et al., 2021). Anatomical modifications may arise as a means of safeguarding plants during periods of stress. This chapter provides an overview of the anatomical changes that plants undergo in response to water stress. Efforts have been made to provide an in-depth discussion of the mechanistic aspects of the anatomical changes at the cellular and molecular levels.

2. Effect of Water Stress on Stem

2.1 Effect on Cuticle and Epidermis of Stem

Cuticles are found on the aerial parts of plants, primarily on the leaves and stems, serving to shield these areas from drying out. Though several reports have been obtained regarding the anatomical changes in the cuticular cells of leaves along with their biochemical composition with special reference to wax and alkanes, not many reports have been obtained regarding the changes in cuticles present in the stem. Experimental evidence on the impact of water restriction on epidermal cells in the stem is limited, despite reports of its effects below the cuticle. According to a study conducted in the stem epidermis area of *Astragalus gombiformis*, it was reported that epidermal cells do not increase even in severe drought stress (Boughalleb et al., 2014). Four different citrus species were grown under varying levels of water stress, including two very tolerant genotypes (Brazilian sour orange and Gadha dahi) and two very sensitive genotypes (Rangpur Poona nucellar and Sunki × bentake). When subjected to moderate to severe water stress all genotypes exhibited a reduction in the size and length of the stem's epidermal cells (Shafqat et al., 2021). In another study, a substantial reduction in cell size was reported in the epidermal cells present in the secondary branches of mango under drought stress (Zaharah & Razi, 2009).

To assess the drought tolerance of five ecotypes of *Panicum antidotale* that were acclimated to a variety of environments, drought stress was applied to their adaptive anatomical components. Every ecotype exhibited an increase in epidermis thickness at 50% field capacity, except for two ecotypes from habitats not under drought stress. The ecotype from a salinity-stressed habitat showed the maximum increase in epidermis thickness (Nawaz et al., 2013).

Morpho-anatomical features were studied in two marigold (*T. erecta*) cultivars, Inca (drought resilient) and Bonanza (drought sensitive) (Zulfiqar et al., 2020). The study went on to explain that, despite the Inca cultivar not exhibiting any decline in growth, the development and growth of the drought-sensitive variety Bonanza were significantly reduced. There are reports that both *T. erecta* cultivars have larger epidermal areas.

Another study found that during mild to severe drought stress, the thickness of epidermal cells in wheat plants increased to limit water loss and survive extremely dry conditions (El-Nady et al., 2012). Impact of water conditions on stem and root structures of Durum wheat (*Triticum durum* Desf.) was investigated in a study. Water shortage had varying effects on the epidermal cells of 10 selected genotypes, although severe water stress caused a reduction in the diameter of epidermal cells in several genotypes (Labdelli et al., 2014).

2.2 Effect of Water Stress in the Cortical Region of the Stem

Drought stress affects the cortical region of the stem by reducing the cell area of the cortical cells which leads to reducing cortical thickness. Investigations into the secondary branch structure of the drought-stressed mango (*Mangifera indica*) cv. Chokanan revealed that water stress markedly reduced the cortical thickness (Zaharah & Razi, 2009).

The anatomical changes in the cortex of the stem of *Glycine max* were studied under conditions of drought (Makbul et al., 2011). According to the study, unstressed plants have smaller parenchymatous cells in their stems and leaves than stressed plants do. Drought conditions are hypothesized to have an impact on the thin-walled parenchymatous cells found in the cortical portion of the stem. Under drought stress, the fibers extend out as a continuous layer between the cortex and phloem in stressed plants, whereas, in unstressed plants, they appear as clustered layers, explaining why they proliferate. It was suggested that sclerenchymatous fibers offer a protective edge against water loss. This study further indicated that the fibers which are sclerenchymatous, provide an advantage against the loss of water. At 50% field capacity, it was found that the cortical cell area in the stem exhibited variable behavior, increasing in certain ecotypes and decreasing in others. *Astragalus gombiformis* exhibited reduced thickness of cortical cells under moderate to prolonged exposure to drought (Boughalleb et al., 2014).

Water scarcity conditions reduced the cortical area in the cultivars of marigold (*Tagetes erecta*), including Inca (drought-resilient), and Bonanza (drought-sensitive). Notably, cv. Bonanza exhibited a decrease in shoot cortex area while cv. Inca outperformed cv. Bonanza in terms of growth and biomass during water stress (Zulfiqar et al., 2020). Given that the cortex parenchyma is where plants store the majority of their water, issues caused by environmental dangers may harm both water conservation and the disintegration of the cortical region (Pena-Valdivia & Sanchez Urdaneta, 2009). Nonetheless, during drought stress, cv. Bonanza demonstrated comparatively greater damage to its cortical parenchyma than the cv. Inca.

A study found a decrease in the thickness of cortical cells in four citrus cultivars during moderate and severe drought stress. Two of the cultivars were very drought-tolerant genotypes, whereas the other two were highly susceptible genotypes. The study found that the decrease in cortical area thickness was caused by the contraction of cortical cells during water or drought stress. The tolerant genotypes showed a 43.4% decrease in cortical cell thickness (Shafqat et al., 2021).

2.3 Effect of Water Stress on Vascular Bundle of Stem

Decreased conductivity in the water pathway through soil, plants, and the atmosphere is a consequence of the water stress response, which can reduce water loss in the atmosphere. Drought stress causes embolism in plants which causes both immediate and long-term effects including stomatal closure, modifications to the rate of photosynthesis, decreased growth, decreased productivity, and even plant death. Reducing xylem transport capability, also interferes with the plant's water continuity (Sperry et al., 1998; Brodribb & Cochard, 2009).

Other vascular traits of trees, like the size of the conduit, the structure of the pit membrane, and the density of the wood, can also be impacted by drought (Jansen et al., 2009). In a study, the inter-vessel pit membrane of 26 hardwood tree species reported that smaller pits with thicker, less porous pit membranes are found in drier regions; this phenomenon may be an adaptive reaction (Jansen et al., 2009).

The anatomical characteristics of Spanish populations of *Pinus sylvestris* were examined in a study (Martín et al., 2010). The findings suggested that drier conditions cause gymnosperms to generate larger tracheid lumens, thicker cell walls, and more ray tracheids, all of which increase the efficiency of water transfer. Drier conditions stimulate the development of narrow vessel components in angiosperms, which may lower the risk of embolism. Nevertheless, this benefit may come at the expense of a reduction in xylem transport capacity (Martín et al., 2010).

Drought stress decreases vascular diameter in herbaceous plants similar to woody ones. Common zinnia has been found to increase the polymerization of lignin in response to dry conditions, leading to thickening of the cell wall (Twumasi et al., 2005). Similar results were also reported in case of clover (Li et al., 2013) and sugarcane(Santos et al., 2015).

It is further reported in a study that drought stress decreases xylem vessel diameter. Given that water transport rises with xylem radius, this could be adaptive since the consumption of water would be lower for early-season growth and conserved for other purposes, such as grain filling (Lynch et al., 2014).

It was investigated that the tropical evergreen tree *Moringa oleifera* was affected by water stress. Drought has been shown to decrease or postpone the vascular cambium's cell division by lowering the cambial cells' turgor pressure, which prevents plant growth. Woody plants are particularly sensitive to water deficits (Patel et al., 2014).

The structural responses of vascular bundles in four varieties of maize stem were studied under water deficit conditions. The first internode of the basal stem exhibited a significant reduction in the space spanned by vascular bundles, xylem, and phloem, along with a drop in the number and diameter of vessels in comparison with the control group experiencing water scarcity. The number of vascular bundles increased as the size of each vascular bundle decreased inside the unit's range of vision. The vascular bundle structure of the ear internode changed in response to water scarcity, similar to the first internode of the basal stem. The impact on vascular bundles decreased and was mostly concentrated in the phloem region. During the seedling stage, a water deficit led to a rise in various characteristics of the first internode of the ear stem's vascular bundle, including the xylem area, phloem area, number of vessels, and vessel diameter (Wang et al., 2014). *Astragalus gombiformis* subjected to prolonged water scarcity showed a significant decrease in both xylem tissue width and xylem vessel diameter. Increased drought levels led to a significant decrease in phloem and phloem sclerenchyma width in *A. gombiformis* stems (Boughalleb et al., 2014).

3. Effect of Water Stress on Leaf

3.1 Effect of Water Stress on Stomata

Stomata are tiny pores located on the surface of leaves that regulate transpiration and the intake of carbon dioxide. The capacity to recognize a variety of stressors from the internal and external environment allows stomatal guard cells to quickly respond by shutting down under adverse circumstances (Agurla et al., 2018). Most plants respond to drought stress by closing their stomata. During osmotic stress, stomatal closure mainly serves to reduce water loss, therefore improving drought resistance (Lim et al., 2015). Stomatal closure is mostly affected by soil moisture levels rather than the leaf hydration state. It is mainly controlled by chemical signals, like the production of abscisic acid (ABA) in dehydrating roots (Pirasteh-Anosheh et al., 2016). The relationship between drought and Abscisic acid is intimate and triggers many plant processes that aid in drought adaptation, influenced by ABA (Zhu et al., 2014). Plants react to water stress by closing their stomata through a sophisticated signaling mechanism. ABA is the primary signaling chemical responsible for the process of drought-mediated stomatal closure (Lim et al., 2015). ABA triggers the efflux of anions and potassium via complex signaling pathways that involve ion channels in the plasma membrane of guard cells. As a result, the decrease in turgor pressure within the guard cells triggers the closure of the stomata (Munemasa et al., 2015). Plants synthesize ABA in reaction to water scarcity, and it is rapidly produced in both the roots and shoots (Hu et al., 2016). ABA is transported to guard cells via ABA transporters (Liu et al., 2022). Concurrently, roots generate a signal peptide known as CLE25 and convey it to the outer surface of guard cells via vascular tissues. The CLE25 peptide is recognized in the leaves by the BAM1 and BAM3 receptors, distinguished by leucine-rich repeat domains. These triggers initiate further signaling pathways that result in an upregulation of the ABA biosynthesis gene NCED3. Increased NCED3 expression in the leaves and shoots leads to the accumulation of ABA, resulting in stomatal closure (McLachlan et al., 2018). The PYR/PYL/RCAR receptors sense the build-up of ABA in guard cells and block the PP2Cs, resulting in the activation of SnRK2s via autophosphorylation (Ng et al., 2014). SnRK2s directly target several proteins involved in stomatal closure, including SLAC1, KAT1, RBOHF, and PIP2;1 (Hasan et al., 2022). SnRK2s boost the production of apoplastic O2· in guard cells by triggering RBOHF. The $O_2^{\cdot -}$ is converted into H_2O_2, which is recognized by the receptor kinase HPCA1. HPCA1 recognizes and transduces the H_2O_2 signal into the guard cell via calcium ions. Hydrogen peroxide (H_2O_2) induces an elevation in the concentration of calcium ions (Ca^{2+}) in the cytoplasm of the guard cells (Liu et al., 2022). Furthermore, under the direction of ABA-mediated SnRK2s, an aquaporin PIP2; 1 can carry apoplastic H_2O_2 into the cytoplasm (Rodrigues et al., 2017). The presence of hydrogen peroxide (H_2O_2) in the cytoplasm controls the process of stomatal closure by influencing the activity of certain proteins, such as ATGPX3, ABI2, and ABI1 (Liu et al., 2022).

In addition to stomatal closure, drought stress can also cause variations in stomatal number and size in plants. A decrease in the length of the stomata was seen in *Ocimum basillicum*, accompanied by an increase in density (Driesen et al., 2023). In rice cultivar KDML105 drought stress resulted in the decrease of stomatal density and was negatively correlated with the size of stomata. In addition, there was a marginal difference in the length of the guard cells in plants subjected to drought (Lertngim et al., 2022). In a different research, drought stress caused a notable decrease in the length of stomata on the lower surface of the leaf. Although drought stress did not impact stomatal width, it was broader on the lower surface of the leaf compared to the upper surface. The stomata holes on the upper portion of the leaf decreased by 14% under mild stress conditions compared to well-watered plants. No substantial effect was detected with MS. Both mild stress and moderate stress treatments resulted in a decrease in stomata pore perimeter on the abaxial leaf surface by 28.2% and 33.4%, respectively, compared to well-watered plants. The stomatal pore area per unit leaf area was 132-235% more on the lower surface of the leaf compared to the upper surface (Lei et al., 2018). In *Leymus chinensis*, it was shown that stomatal size reduced as the water deficit rose. This advocates that severe water stress leads to a reduction in stomatal size, as indicated by the almost linear connections identified between stomatal size and water potential (Xu & Zhou, 2008). When rice was stressed by drought, the density of stomata decreased modestly on both sides of the leaf: 1.99% on the adaxial and 1.07% on the abaxial leaf surface. Guard cells showed slight changes in both length and breadth when exposed to drought-induced stress (Lertngim et al., 2022).

3.2 Effect of Water Stress on Leaf Epidermis and Cuticle

The scarcity of water also impacts the epidermal cells of a leaf. In a study, it was observed that *Shepherdia × utahensis* had a silvery appearance as it was cultivated in drier soils. Plants cultivated under water scarcity conditions exhibited an increased density of trichomes and a decrease in the size of leaf epidermal cells. The reduction of water content in the substrate resulted in a proportional increase in leaf trichome density, trichome coverage fraction, and epidermal cell density. However, the presence of stomata, the radius of trichomes, and the size of epidermal cells all exhibited a linear reduction as the volumetric water content of the substrate decreased (Chen et al., 2022). In an alternative investigation, it was shown that when *Saccharum* 'UT12' and *Saccharum* 'UT13' were grown in drought, their lamina thickness reduced, while the thickness of their cell walls and cuticles on the adaxial and abaxial sides of their lamina epidermal cells increased (Taratima et al., 2020). Drought stress significantly increased the thickness of the leaf sclerenchyma, the area of the leaf bulliform cells, and the thickness of the leaf adaxial epidermis in *Avena sativa* (Ghafoor et al., 2019).

Plant cuticular waxes serve a crucial function as a protective shield against environmental pressures. Their production, transportation, deposition, and composition are all influenced by a range of environmental conditions (Shepherd & Wynne Griffiths, 2006). Several investigations have shown that drought circumstances increase the amount and change the nature of cuticular wax in Arabidopsis, tobacco (Cameron et al., 2006), wheat (Guo et al., 2016), and maize (Li et al., 2019). During an experiment, it was shown that Arabidopsis in the control group exhibited an 80% increase in the amount of wax per unit leaf area when they were deprived of sufficient water. Furthermore, the plants treated with water deficiency showed a 49% rise in cuticle thickness. Overall, the rise in the number of longer-chain elements led to a 93% increase in the total amount of wax in plants experiencing water deficiency. This increase can be mostly attributed to alkanes (Kosma et al., 2009). In Arabidopsis, it was shown that the MYB96 protein functions as a transcriptional activator of genes that encode enzymes responsible for the synthesis of cuticular wax, namely those involved in the condensation of very long-chain fatty acids. The activation occurs through direct binding of MYB96 to specific sequence motifs found in the promoters of these genes (Seo et al., 2011). Water stress in *Pinus pinaster* led to increased deposition of cuticular wax, with upregulation of genes involved in cuticular wax production (Le Provost et al., 2013). It has been found that rice contains the Drought-Induced Wax Accumulation 1 (DWA1) gene. 2,391 amino acids make up the lengthy protein that this gene encodes. The protein is composed of multiple enzymatic components, including an allene oxide synthase-like domain, an oxidoreductase-like domain, and a prokaryotic non-ribosomal peptide synthetase-like module with an AMP-binding domain. DWA1 showed a tendency to be expressed in vascular tissues and epidermal layers, with a notable increase in expression under drought-stress conditions. Under drought stress, the dwa1 mutant showed decreased cuticular wax buildup, resulting in significant alterations in the plant's cuticular wax composition. As a result, the mutant showed increased vulnerability to drought. The mutant had lower amounts of very long-chain fatty acids, but plants overexpressing DWA1 showed higher quantities of very long-chain fatty acids relative to the wild type (Zhu & Xiong, 2013). *Nicotiana glauca* employs lipid transfer proteins (LTPs) to move lipids across the extracellular matrix to aid in cuticular wax synthesis. Through the application of northern analysis, it was revealed that the tree tobacco LTP gene transcripts experienced a significant 6-fold increase following three instances of drying. This finding serves as more proof of the involvement of LTP in the process of cuticle deposition. The amount of leaf cuticular wax grew progressively following each of the three episodes of dehydration stress (Cameron et al., 2006). Tomato foliage and fruit include VLC alkanes as their main wax components. In a study, five ECERIFERUM 1 (CER1) homologs and two ECERIFERUM 3 (CER3) homologs in tomato, termed SlCER1-1 to SlCER1-5 and SlCER3-1 and SlCER3-2, respectively, were discovered. Abiotic stresses induce tissue and organ-specific gene expression patterns. SlCER1-1 was located in the endoplasmic reticulum (ER), the main site of wax synthesis. Silencing SlCER1-1 in tomatoes resulted in a considerable reduction in n-alkanes and branched alkanes. Under drought stress, wild-type tomato plants produced considerably more n-alkanes and branched alkanes than SlCER1-1 RNAi tomato plants. Furthermore, silencing SlCER1-1 enhanced the cuticular permeability of leaves and fruits. The study indicates that SlCER1-1 plays a role in producing wax alkanes in tomatoes, which greatly enhances their ability to withstand drought and prolongs the storage life of the fruit (Wu et al., 2022a).

3.3 Effect of Water Stress on Mesophyll Cells

The spongy mesophyll and palisade are two distinct layers of photosynthetic tissue found on the majority of plant leaves. While the spongy mesophyll's structure is poorly defined and usually thought to be a random assortment of irregularly shaped cells, the palisade mesophyll is made up of densely packed columnar cells (Borsuk et al., 2022). In a study on *Ulmus szechuanica*, it was observed that during drought conditions, there were noticeable changes in the ultrastructure of mesophyll and chloroplast morphology. A rise was seen in both the quantity of chloroplasts and the percentage of cell area. Ultimately, the cell wall surrounding the ring shrinks and the chloroplasts take on a spherical shape. Additionally, there is a reduction in both the size of the chloroplasts and the number of grains of starch. Moreover, there was an expansion and a propensity towards disarray in the arrangement of thylakoids in chloroplasts, which resulted in a reduced photosynthetic capability in the leaves (Jin et al., 2022). In *Phedimus aizoon*, as reported by a study, the control plants' mesophyll cells showed signs of relaxation and good moisture, with individual cells being of significant size. Nevertheless, with the escalation of drought stress, the configuration of mesophyll cells underwent a progressive transition from a loose to a compact arrangement, accompanied by a steady reduction in size and water loss. Particularly during intense drought conditions, the intercellular gaps within the mesophyll cells exhibited a reduction, leading to a drop in cell volume. Concurrently, in optimal moisture circumstances, the chloroplasts within the mesophyll cells were uniformly distributed along the cell wall. As drought stress increased, the chloroplasts moved toward the center of the cells. During intense stress, the chloroplast envelope disintegrates, resulting in partial fusion of the chloroplasts and the absence of fully intact chloroplasts within the cells (Liu et al., 2021). In a separate research, it was shown that Juglans regia and *Juglans microcarpa* experienced a decrease in the volume of their mesophyll cells due to drought. However, the resulting increase in porosity and gas-phase diffusion was not observed due to a drop in biochemical activity, which led to a reduction in stomatal and mesophyll conductance (Momayyezi et al., 2022). A study comparing *Quercus acutissima* and *Quercus serrata* found significant differences in the thickness of the palisade and spongy mesophyll between treatments of well-watered conditions and water stress. Seedlings exposed to water stress exhibited a thicker palisade mesophyll compared to appropriately watered samples in both genera. In control plants, the mesophyll cells in the leaves showed larger and more relaxed parenchyma cells compared to water-stressed plants, especially in the case of *Q. serrata*. The spongy mesophyll cells in water-deficit-challenged *Q. serrata* seedlings saw a significant shift in thickness. More precisely, the discrete cells exhibited a reduction in size of roughly 40–50% (Hernandez & Park, 2022). In *Croton blanchetianus*, plants developing in the summer regime had larger palisade and spongy cell thickness than those growing in the winter regime (Mendes et al., 2022). In *Arachis hypogaea*, drought stress results in a decrease in palisade cell numbers per unit area as well as its length (Makbul et al., 2011). In *Nicotiana tabacum*, drought hardening resulted in a pervasive lack of organization among its cells. Significant disarray, distortion, and increased gaps between cells were particularly noted in the palisade and spongy tissues (Khan et al., 2020).

3.4 Effect of Water Stress on Vascular Bundles of The Leaf

Water deprivation also affects the leaf's vascular bundles. During the process of leaf desiccation in *Podocarpus grayi*, a significant number of tracheids undergo a transformation from a circular shape to a flat shape within the water potential range of –1.5 to –2.8 MPa. Under water stress conditions, the proportion of collapsed tracheids increased, whereas leaf hydraulic conductivity decreased, indicating a direct impact on water transport efficiency (Brodribb & Cochard, 2009). During periods of water scarcity in Avena sativa, there was a considerable decrease in the leaf vascular bundle region, midrib width, and phloem area (Ghafoor et al., 2019). In *Ctenanthe setosa*, drought stress induced a reduction in the diameter of the vessel as well as the leaf vein (Kutlu et al., 2009). Similar results were also obtained in wheat where drought stress-induced reduction in the dimensions of phloem and vascular bundle (Mabrouk El-Afry et al., 2012). Similar outcomes were observed in cowpeas, showing that water stress reduced the thickness of the midrib vascular bundle (Farouk & Amany, 2012). Within the genus *Pinus*, drought-related reductions in xylem pressure caused tracheids to gradually collapse below a threshold pressure, which is linked to the development of cavitation in the stems. There is a trade-off between the xylem's efficiency, its vulnerability to collapse, and the cost of wall reinforcement. Species with thicker walls and lower tracheid diameters showed a larger negative threshold pressure (Cochard et al., 2004).

4. Effect of Water Stress on Roots

4.1 Effect of Water Stress on Root Hairs and Root Epidermis

Root hairs are specialized appendages that serve a critical role in the absorption of nutrients in plants, emphasizing the importance of root hairs (Grierson et al., 2014). These cylindrical projections formed by the outer layer of cells on a plant's surface enhance the area available for absorption, making it easier for the plant to take water (Cui et al., 2018). Water stress significantly impacts root hairs and epidermis, affecting plant water and nutrient uptake (Cai & Ahmed, 2022).

Water stress induces substantial modifications in root hair growth, enhancing the plant's capacity to adapt to restricted water supplies. In Arabidopsis roots achieve a 10-fold increase in total absorptive surface area (Grierson et al., 2014). Nevertheless, when subjected to stressful circumstances such as inadequate nutrients or limited water availability, the growth of root hairs and epidermal cells may undergo modifications. A meta-analysis found that water deficit promotes deeper rooting across many species while suppressing lateral root branching (Kou et al., 2022). The water stress increases the growth of root hairs, expanding the root's surface area, which improves its ability to absorb water and nutrients (Filipović, 2021). Plants rely on this adaptive response to counteract the negative impacts of water scarcity and maintain their physiological functions, enhancing their ability to withstand water-stressed situations and ensure survival. Under water stress conditions, there was a notable increase in both the density and length of root hairs, as well as a rise in the root-to-shoot ratio. Fibrous root systems like wheat respond similarly with increased root hair length and density and root-to-shoot ratio (Kou et al., 2022). The density of root hairs increased by 49.4%, while the length of root hairs and root-to-shoot ratio increased by 35.8% and 19.8%, respectively (Kou et al., 2022).

The deepening of water shortage leads to considerable changes in the root piliferous zone, affecting water absorption efficiency (Stocker et al., 2023). It has been discovered that when there is a substantial scarcity of water, there are substantial modifications in the architecture of the roots' piliferous zone (Labdelli et al., 2014). In another study, it was observed that the diameter of the roots, as shown in durum wheat, is exposed to water scarcity, which might affect the ability to absorb water (Kou et al., 2022). *Fagopyrum esculentum* i.e. Buckwheat exhibits how root hairs and the root epidermis react to water stress. Buckwheat, being a stress-tolerant crop, shows adaptation mechanisms in root structure and physiology to deal with water constraints. Buckwheat may develop higher root hair compactness and structural changes in the root epidermis to improve water absorption and sustain development when facing water stress (Rangappa et al., 2024). These results underscore the significance of root morphology in how plants react to water stress, underscoring the need to comprehend and control root structural changes to enhance water absorption efficiency.

4.2 Effect of Water Stress on Root Cortex

Water stress initiates a series of physiological reactions in plants, with the root cortex being a key location for these responses. The alteration in root cortex morphology caused by water stress is essential for the survival of plants. Under water stress conditions, plants may augment the thickness of cortical cell walls to bolster structural integrity and reduce water loss (Kou et al., 2022). Furthermore, alterations in the distribution of vascular tissue in the cortex enable adequate water transportation under water scarcity conditions (Labdelli et al., 2014). Research has shown changes in the osmotic potential, ion movements, and hormone levels in the root cortex when plants are exposed to water stress

(Kar, 2011). In maize, water stress induces a drop in cytokinin levels in the root cortex, influencing root growth and development (S. Lisar et al., 2012).

The root cortex experiences notable structural changes when faced with water stress to improve water absorption and reduce water loss (Yang et al., 2020). It is suggested that the root cortex undergoes a size reduction in drought, decreasing its capacity to absorb water and nutrients effectively (Kang et al., 2022). Studies on *Arabidopsis thaliana* have shown that aerenchyma forms in the root cortex under water-stressed circumstances, aiding in oxygen diffusion and nutrient transport (Yamauchi et al., 2019). In rice, water stress leads to the formation of a thicker root cortex layer, enhancing water retention and increasing drought resistance (Henry et al., 2012). Furthermore, water stress impacts the dimensions and arrangement of root cortical cells, disrupting tissue integrity and hydraulic conductivity (Yang et al., 2011).

The root cortex reacts to water stress using physiological processes to maintain cellular homeostasis and ensure the plant's life. These reactions include changes in osmotic potential, ion absorption, and hormone signaling pathways. Research has shown that plants can gather Osmoprotectants, such as proline and sugars, in the root cortex during drought. This helps to reduce water loss and alleviate osmotic stress (Wu et al., 2022b). Moreover, water stress triggers alterations in ion transportation mechanisms, impacting nutrient absorption and maintaining ion balance in the root cortex (Wu et al., 2022b).

Water stress at the molecular level initiates intricate signaling cascades and gene regulatory networks in the root cortex, coordinating adaptive responses. Transcriptomic studies have revealed differences in the expression of genes linked to stress signaling, water transportation, and antioxidant defense pathways in the brain underwater during stressful situations (Kwasniewski et al., 2016). Transcriptomic studies have also identified key genes in the root cortex of plants such as soybean and wheat that detect water stress, regulate osmotic levels, and scavenge reactive oxygen species (Mickelbart et al., 2015). In addition, the dynamic control of gene expression in response to water stress is influenced by epigenetic regulatory mechanisms (Chiatante et al., 2000). DNA methylation and histone acetylation have a function in regulating the expression of genes in the root cortex of wheat plants during water stress (Li et al., 2021).

4.3 Effect of Water Stress on Vascular Bundles of Root

With the aid of root hairs, roots draw water from the soil. This water then travels radially via the root's vascular bundles and is transferred to the leaves employing the stem. Conditions of water deficiency affect the anatomical characteristics of the root system, which interferes with water conduction. A positive association between vessel size and water volume was observed in *Paspalum dilatatum*. Larger vessels are more susceptible to vessel embolism when tension builds up during a water deficit, leading to cavitation (Atkinson & Taylor, 1996). In this ecotype, intense sclerification surrounding the vascular tissue may lower the danger of vessel embolism. Additionally, intake of nutrients and water, particularly in settings with limited moisture, may be advantageous for the ecotype's ecological success (Vasellati, 2001).

Water conduction in monocotyledons differs from dicotyledons due to the absence of vascular cambium. This results in water stress reducing the diameter of xylem vessels in monocots and increasing it in dicots (Gorsuch et al., 2001). In *Oryza sativa*, drought causes a decrease in late metaxylem vessel diameter, which has been proposed as a way to lower the danger of cavitation (Henry et al., 2012).

A study examined the impact of drought on five different ecotypes of *Panicum antidotale* Retz. that have been acclimated to various environments to evaluate their drought tolerance based on adaptive morphological features. The metaxylem vessel area and the total vascular region area of the ecotype exposed to drought stress showed a considerable increase (Nawaz et al., 2013). This adaptation is considered significant in drought-stressed situations due to its ability to enhance water conduction with low resistance (Nawaz et al., 2013).

In *Astragalus gombiformis*, the diameter of the xylem vessels decreased as the drought intensified. Xylem vessel diameter decreased by 19.8%, 39.6%, and 36% in plants subjected to water shortage conditions for 10, 20, and 30 days, respectively, compared to the control group (Boughalleb et al., 2014). Oat plants cultivated under 60% field capacity (water-deficit stress) showed an increase in the diameter of xylem vessels and metaxylem area, indicating alterations in root anatomical features (Ghafoor et al., 2019).

Studies were conducted on the impact of drought stress on the anatomical structure and root morphology of lucerne (*Medicago sativa*) types that had varying levels of drought tolerance. *M. sativa* cv. Longzhong, which is highly resistant to drought, Longdong, which is somewhat resistant, and Gannong No. 3, which is susceptible to drought, were subjected to 15 days of drought stress. Root tip number, root dry weight, total root length, total root surface area, xylem area, vascular bundle diameter, and the regions of vascular bundle and phloem in roots all grew steadily with increasing stress exposure duration. In the meantime, only the vascular bundles' areas and diameters continued to grow in Longdong. Gannong No. 3 exhibited a continual increase in total root length, total root surface area, and the ratio of cortical thickness to root diameter. However, the dry weight of the root, primary and secondary xylem vessel diameters, vascular bundle diameter, and vascular bundle area progressively decreased. Gannong No. 3 exhibited considerably shorter total root length, total root surface area, and total root volume compared to Longzhong and Longdong after six days of stress. When the stress exposure time was further exceeded, Gannong No. 3 (drought-sensitive) plants' internal conducting tissue changed, limiting their capacity to carry water (Zhang et al., 2018).

5. Conclusion

In conclusion, the anatomical anomalies observed in plants during water stress highlight the intricate adaptations and responses that plants have evolved to cope with environmental challenges. Through a thorough investigation into the structural alterations occurring in various plant tissues under water stress conditions, this paper has shed light on the multifaceted nature of plant responses to water scarcity. The anatomical changes identified, including alterations in leaf morphology, stomatal behavior, vascular tissue structure, and cellular organization, underscore the complexity of plant-water relations and the dynamic nature of plant physiology. These adaptations serve as mechanisms for plants to optimize water use efficiency, maintain cellular turgor pressure, and mitigate the detrimental effects of water stress on growth and development. Moreover, the diversity of anatomical responses observed across different plant species, genotypes, and environmental conditions highlights the plasticity and versatility of plant anatomical traits in response to water stress. Understanding these variations is essential for elucidating the underlying physiological and molecular mechanisms governing plant-water relations and for developing strategies to enhance plant resilience to drought and water scarcity in a changing climate. Furthermore, the insights gained from studying anatomical anomalies during water stress have significant implications for agriculture, ecology, and environmental conservation. By deciphering the structural adaptations that enable certain plant species to thrive under water-limited conditions, researchers can identify traits with potential agronomic value for breeding drought-tolerant crop varieties and for selecting resilient species for ecosystem restoration and land rehabilitation efforts. Overall, this paper underscores the importance of considering plant anatomy as a key component of plant responses to water stress and highlights the need for interdisciplinary approaches that integrate anatomical, physiological, molecular, and ecological perspectives to advance our understanding of plant-water relations and to inform sustainable water management practices in agriculture and natural resource conservation.

References

Agurla S, Gahir S, Munemasa S, Murata Y & Raghavendra AS. 2018. Mechanism of stomatal closure in plants exposed to drought and cold stress. *Adv Exp Med Biol,* **1081:** 215–232. doi: 10.1007/978-981-13-1244-1_12. PMID: 30288712.

Atkinson CJ & Taylor JM. 1996. Effects of elevated CO_2 on stem growth, vessel area and hydraulic conductivity of oak and cherry seedlings. *New Phytologist,* **133:** 617–626.

Beltrano J & Ronco MG. 2008. Improved tolerance of wheat plants (Triticum aestivum L.) to drought stress and rewatering by the arbuscular mycorrhizal fungus Glomus claroideum: Effect on growth and cell membrane stability. *Brazilian Journal of Plant Physiology,* **20:** 29–37.

Borsuk AM, Roddy AB, Théroux-Rancourt G & Brodersen CR. 2022. Structural organization of the spongy mesophyll. *New Phytologist,* **234:** 946–960.

Boughalleb F, Abdellaoui R, Ben-Brahim N & Neffati M. 2014. Anatomical adaptations of Astragalus gombiformis Pomel. under drought stress. *Open Life Sci,* **9:** 1215–1225.

Brodribb TJ & Cochard H. 2009. Hydraulic failure defines the recovery and point of death in water-stressed conifers. *Plant Physiol,* **149:** 575–584.

Cai G & Ahmed MA. 2022. The role of root hairs in water uptake: Recent advances and future perspectives. *J Exp Bot,* **73:** 3330–3338.

Cameron KD, Teece MA & Smart LB. 2006. Increased accumulation of cuticular wax and expression of lipid transfer protein in response to periodic drying events in leaves of tree tobacco. *Plant Physiol,* **140:** 176–183.

Chen X, Shi X, Ai Q, Han J, Wang H & Fu Q. 2022. Transcriptomic and metabolomic analyses reveal that exogenous strigolactones alleviate the response of melon root to cadmium stress. *Hortic Plant J,* **8:** 637–649.

Chiatante D, Di Iorio A, Maiuro L & Scippa SG. 2000. Effect of water stress on root meristems in woody and herbaceous plants during the first stage of development. pp. 245–258. *In:* The Supporting Roots of Trees and Woody Plants: Form, Function and Physiology, Springer Netherlands, Dordrecht.

Cochard H, Froux F, Mayr S & Coutand C. 2004. Xylem wall collapse in water-stressed pine needles. *Plant Physiol,* **134:** 401–408.

Cui S, Suzaki T, Tominaga-Wada R & Yoshida S. 2018. Regulation and functional diversification of root hairs. *Semin Cell Dev Biol,* **83:** 115–122.

Driesen E, De Proft M & Saeys W. 2023. Drought stress triggers alterations of adaxial and abaxial stomatal development in basil leaves increasing water-use efficiency. *Hortic Res,* **10.**

El-Afry MM. (2012). Anatomical studies on drought-stressed wheat plants (Triticum aestivum L.) treated with some bacterial strains. *Acta Biologica Szegediensis,* **56(2):** 165–174.

El-Nady M, El-Afry M, Belal E & Metwaly M. 2012. Improvment of wheat plants growth and productivity under drought stress conditions by using some Endophytic barcterial strains. *Journal of Plant Production,* **3:** 2091–2106.

Fan L, Linker R, Gepstein S, Tanimoto E, Yamamoto R & Neumann PM. 2006. Progressive inhibition by water deficit of cell wall extensibility and growth along the elongation zone of maize roots is related to increased lignin metabolism and progressive stelar accumulation of wall phenolics. *Plant Physiol,* **140:** 603–612.

Farouk S & Amany A. 2012. Improving growth and yield of cowpea by foliar application of chitosan under water stress. *Egyptian Journal of Biology,* **14.**

Filipović A. 2021. Water Plant and Soil Relation under Stress Situations. Soil Moisture Importance. IntechOpen.

Ghafoor R, Akram NA, Rashid M, Ashraf M, Iqbal M & Lixin Z. 2019. Exogenously applied proline induced changes in key anatomical features and physio-biochemical attributes in water stressed oat (Avena sativa L.) plants. *Physiology and Molecular Biology of Plants,* **25:** 1121–1135.

Gorsuch DM, Oberbauer SF & Fisher JB. 2001. Comparative vessel anatomy of arctic deciduous and evergreen dicots. *Am J Bot,* **88:** 1643–1649.

Grierson C, Nielsen E, Ketelaarc T & Schiefelbein J. 2014. Root Hairs. *Arabidopsis Book,* **12:** e0172.

Guo J, Xu W, Yu X, Shen H, Li H, Cheng D, et al. 2016. Cuticular wax accumulation is associated with drought tolerance in wheat near-isogenic lines. *Front Plant Sci,* **7.**

Hasan Md M, Liu X-D, Waseem M, Guang-Qian Y, Alabdallah NM, Jahan MS, et al. 2022. ABA activated SnRK2 kinases: An emerging role in plant growth and physiology. *Plant Signal Behav,* **17.**

He, Chunyang, Zhifeng Liu, Jianguo Wu, Xinhao Pan, Zihang Fang, Jingwei Li & Brett A Bryan. 2021. Future global urban water scarcity and potential solutions. *Nature Communications,* **12(1):** 4667. doi:10.1038/s41467-021-25026-3

Henry A, Cal AJ, Batoto TC, Torres RO & Serraj R. 2012. Root attributes affecting water uptake of rice (Oryza sativa) under drought. *J Exp Bot,* **63:** 4751–4763.

Hernandez JO & Park BB. 2022. The leaf trichome, venation, and mesophyll structural traits play important roles in the physiological responses of oak seedlings to water-deficit stress. *Int J Mol Sci,* **23:** 8640.

Hu B, Cao J, Ge K & Li L. 2016. The site of water stress governs the pattern of ABA synthesis and transport in peanut. *Sci Rep,* **6:** 32143.

Jansen S, Choat B & Pletsers A. 2009. Morphological variation of intervessel pit membranes and implications to xylem function in angiosperms. *Am J Bot,* **96:** 409–419.

Jin Y, Chang J, Shi Y, Zhang Q & Xia C. 2022. Effects of drought stress on growth, photosynthesis and ultrastructure of mesophyll cell of Ulmus Szechuanica Fang seedlings. *Bangladesh J Bot,* **51:** 461–468.

Kambiranda DM, Hemanth KN, Vasanthaiah, Ramesh K. & Mehboob Sheikh. 2011. Impact of drought stress on peanut (Arachis hypogaea L.) productivity and food safety. *Plants and Environment*, InTech.

Kang J, Peng Y & Xu W. 2022. Crop root responses to drought stress: Molecular mechanisms, nutrient regulations, and interactions with microorganisms in the rhizosphere. *Int J Mol Sci,* **23:** 9310.

Kar RK. 2011. Plant responses to water stress: Role of reactive oxygen species. *Plant Signal Behav,* **6:** 1741–1745.

Khan R, Ma X, Shah S, Wu X, Shaheen A, Xiao L, et al. 2020. Drought-hardening improves drought tolerance in Nicotiana tabacum at physiological, biochemical, and molecular levels. *BMC Plant Biol,* **20:** 486.

Kosma DK, Bourdenx B, Bernard A, Parsons EP, Lü S, Joubès J, et al. 2009. The impact of water deficiency on leaf cuticle lipids of Arabidopsis. *Plant Physiol,* **151:** 1918–1929.

Kou X, Han W & Kang J. 2022. Responses of root system architecture to water stress at multiple levels: A meta-analysis of trials under controlled conditions. *Front Plant Sci,* **13**.

Kutlu N, Terzi R, Tekeli Ç, Şenel G, Battal P & Kadioğlu A. 2009. Changes in anatomical structure and levels of endogenous phytohormones during leaf rolling in Ctenanthe setosa under drought stress. *Turkish Journal of Biology*, **33(2):** 115-122.

Kwasniewski M, Daszkowska-Golec A, Janiak A, Chwialkowska K, Nowakowska U, Sablok G, et al. 2016. Transcriptome analysis reveals the role of the root hairs as environmental sensors to maintain plant functions under water-deficiency conditions. *J Exp Bot,* **67:** 1079–1094.

Labdelli A, Adda A, Halis Y & Soualem S. 2014. Effects of water regime on the structure of roots and stems of durum wheat (*Triticum durum* Desf.). *J Bot,* **2014:** 1–7.

Le Provost G, Domergue F, Lalanne C, Ramos Campos P, Grosbois A, Bert D, et al. 2013. Soil water stress affects both cuticular wax content and cuticle-related gene expression in young saplings of maritime pine (Pinus pinaster Ait). *BMC Plant Biol,* **13:** 95.

Lei ZY, Han JM, Yi XP, Zhang WF & Zhang YL. 2018. Coordinated variation between veins and stomata in cotton and its relationship with water-use efficiency under drought stress. *Photosynthetica,* **56:** 1326–1335.

Lertngim N, Ruangsiri M, Klinsawang S, Raksatikan P, Thunnom B, Siangliw M, et al. 2022. Photosynthetic plasticity and stomata adjustment in chromosome segment substitution lines of rice cultivar KDML105 under drought stress. *Plants,* **12:** 94.

Li C, Li L, Reynolds MP, Wang J, Chang X, Mao X, et al. 2021. Recognizing the hidden half in wheat: Root system attributes associated with drought tolerance. *J Exp Bot,* **72:** 5117–5133.

Li L, Du Y, He C, Dietrich CR, Li J, Ma X, et al. 2019. Maize glossy6 is involved in cuticular wax deposition and drought tolerance. *J Exp Bot,* **70:** 3089–3099.

Li Z, Peng Y & Ma X. 2013. Different response on drought tolerance and post-drought recovery between the small-leafed and the large-leafed white clover (Trifolium repens L.) associated with antioxidative enzyme protection and lignin metabolism. *Acta Physiol Plant,* **35:** 213–222.

Lim C, Baek W, Jung J, Kim J-H & Lee S. 2015. Function of ABA in stomatal defense against biotic and drought stresses. *Int J Mol Sci,* **16:** 15251–15270.

Lisar SYS, Motafakkerazad R, Hossain MM & Rahman IMM. 2012. Water Stress in Plants: Causes, Effects and Responses. Water Stress, InTech.

Litvin AG, van Iersel MW & Malladi A. 2016. Drought stress reduces stem elongation and alters gibberellin-related gene expression during vegetative growth of tomato. *Journal of the American Society for Horticultural Science,* **141:** 591–597.

Liu H, Song S, Zhang H, Li Y, Niu L, Zhang J, et al. 2022. Signaling transduction of ABA, ROS, and Ca^{2+} in plant stomatal closure in response to drought. *Int J Mol Sci.,* **23:** 14824.

Liu J, Yang H, Gosling SN, Kummu M, Flörke M, Pfister S, et al. 2017 Jun. Water scarcity assessments in the past, present, and future. *Earth's Future,* **5(6):** 545–559. https://doi.org/10.1002/2016EF000518

Liu Y, He Z, Xie Y, Su L, Zhang R, Wang H, et al. 2021. Drought resistance mechanisms of Phedimus aizoon L. *Sci Rep,* **11:** 13600.

Lynch JP, Chimungu JG & Brown KM. 2014. Root anatomical phenes associated with water acquisition from drying soil: Targets for crop improvement. *J Exp Bot,* **65:** 6155–6166.

Makbul S, Güler NS, Durmuş N & Güven S. 2011. Changes in anatomical and physiological parameters of soybean under drought stress. *Turkish Journal of Botany,* **35(4):** 369–377. https://doi.org/10.3906/bot-1002-7.

Martín JA, Esteban LG, de Palacios P & Fernández FG. 2010. Variation in wood anatomical traits of Pinus sylvestris L. between Spanish regions of provenance. *Trees,* **24:** 1017–1028.

McLachlan DH, Pridgeon AJ & Hetherington AM. 2018. How Arabidopsis talks to itself about its water supply. *Mol Cell,* **70:** 991–992.

Mendes KR, Batista-Silva W, Dias-Pereira J, Pereira MPS, Souza EV, Serrão JE, et al. 2022 Jan. Leaf plasticity across wet and dry seasons in Croton blanchetianus (Euphorbiaceae) at a tropical dry forest. *Sci Rep,* **12(1):** 954. doi: 10.1038/s41598-022-04958-w. PMID: 35046463; PMCID: PMC8770520.

Mickelbart MV, Hasegawa PM & Bailey-Serres J. 2015. Genetic mechanisms of abiotic stress tolerance that translate to crop yield stability. *Nat Rev Genet,* **16:** 237–251.

Momayyezi M, Borsuk AM, Brodersen CR, Gilbert ME, Théroux-Rancourt G, Kluepfel DA, et al. 2022. Desiccation of the leaf mesophyll and its implications for CO_2 diffusion and light processing. *Plant Cell Environ,* **45:** 1362–1381.

Mukarram M, Choudhary S, Kurjak D, Petek A & Khan MMA. 2021. Drought: Sensing, signalling, effects and tolerance in higher plants. *Physiol Plant,* **172:** 1291–1300.

Munemasa S, Hauser F, Park J, Waadt R, Brandt B & Schroeder JI. 2015. Mechanisms of abscisic acid-mediated control of stomatal aperture. *Curr Opin Plant Biol,* **28:** 154–162.

Nawaz T, Hameed M, Ashraf M, Batool S & Naz N. 2013. Modifications in root and stem anatomy for water conservation in some diverse blue panic (*Panicum antidotale* Retz.) ecotypes under drought stress. *Arid Land Research and Management,* **27:** 286–297.

Ng LM, Melcher K, Teh BT & Xu HE. 2014. Abscisic acid perception and signaling: Structural mechanisms and applications. *Acta Pharmacol Sin,* **35:** 567–584.

Osakabe Y, Osakabe K, Shinozaki K & Tran L-SP. 2014. Response of plants to water stress. *Front Plant Sci,* **5**.

Patel VR, Pramod S & Rao KS. 2014. Cambial activity, annual rhythm of xylem production in relation to phenology and climatic factors and lignification pattern during xylogenesis in drum-stick tree (Moringa oleifera). *Flora – Morphology, Distribution, Functional Ecology of Plants,* **209:** 556–566.

Peña-Valdivia CB & Sánchez-Urdaneta AB. 2009. Effects of substrate water potential in root growth of Agave salmiana Otto ex Salm-Dyck seedlings. *Biol Res,* **42**.

Pirasteh-Anosheh H, Saed-Moucheshi A, Pakniyat H & Pessarakli M. 2016. Stomatal responses to drought stress. pp. 24–40. *In:* Water Stress and Crop Plants, Wiley.

Rangappa K, Kumar AU, Choudhury B, Moirangthem P, Layek J, Rajkhowa D, et al. 2024. Buckwheat: Potential Stress-Tolerant Crop for Mid-Hills of Eastern Himalaya under Changing Climate. IntechOpen. doi: 10.5772/intechopen.112096.

Rodrigues O, Reshetnyak G, Grondin A, Saijo Y, Leonhardt N, Maurel C, et al. 2017. Aquaporins facilitate hydrogen peroxide entry into guard cells to mediate ABA- and pathogen-triggered stomatal closure. *Proceedings of the National Academy of Sciences,* **114:** 9200–9205.

Santos AB dos, Bottcher A, Kiyota E, Mayer JLS, Vicentini R, Brito M dos S, et al. 2015. Water stress alters lignin content and related gene expression in two sugarcane genotypes. *J Agric Food Chem,* **63:** 4708–4720.

Seo PJ, Lee SB, Suh MC, Park M-J, Go YS & Park C-M. 2011. The MYB96 transcription factor regulates cuticular wax biosynthesis under drought conditions in Arabidopsis. *Plant Cell,* **23:** 1138–1152.

Shafqat W, Mazrou YSA, Sami-ur-Rehman, Nehela Y, Ikram S, Bibi S, et al. 2021. Effect of three water regimes on the physiological and anatomical structure of stem and leaves of different citrus rootstocks with distinct degrees of tolerance to drought stress. *Horticulturae,* **7:** 554.

Shao H-B, Chu L-Y, Jaleel CA, Manivannan P, Panneerselvam R & Shao M-A. 2009. Understanding water deficit stress-induced changes in the basic metabolism of higher plants – Biotechnologically and sustainably improving agriculture and the ecoenvironment in arid regions of the globe. *Crit Rev Biotechnol* **29:** 131–151.

Shepherd T & Wynne Griffiths D. 2006. The effects of stress on plant cuticular waxes. *New Phytologist,* **171:** 469–499.

Sperry JS, Adler FR, Campbell GS & Comstock JP. 1998. Limitation of plant water use by rhizosphere and xylem conductance: Results from a model. *Plant Cell Environ,* **21:** 347–359.

Stocker BD, Tumber-Dávila SJ, Konings AG, Anderson MC, Hain C & Jackson RB. 2023. Global patterns of water storage in the rooting zones of vegetation. *Nat Geosci,* **16:** 250–256.

Taratima W, Ritmaha T, Jongrungklang N, Maneerattanarungroj P & Kunpratum N. 2020. Effect of stress on the leaf anatomy of sugarcane cultivars with different drought tolerance (Saccharum officinarum, Poaceae). *Rev Biol Trop,* **68**.

Twumasi P, van Ieperen W, Woltering EJ, Emons AMC, Schel JHN, Snel JFH, et al. 2005. Effects of water stress during growth on xylem anatomy, xylem functioning and vase life in three zinnia elegans cultivars. *Acta Hortic,* 303–312.

Vanham D, Alfieri L, Flörke M, Grimaldi S, Lorini V, de Roo A, et al. 2021. The number of people exposed to water stress in relation to how much water is reserved for the environment: A global modelling study. *Lancet Planet Health,* **5:** e766–e774.

Vasellati V. 2001. Effects of flooding and drought on the anatomy of Paspalum dilatatum. *Ann Bot,* **88:** 355–360.

Wang N, Liu J & Zhou Y. 2014. Structural responses of vascular bundles in different maize stems to water deficit during seedling stage: A compensating development. *Research on Crops,* **15:** 532.

Wu H, Liu L, Chen Y, Liu T, Jiang Q, Wei Z, et al. 2022a. Tomato SlCER1–1 catalyzes the synthesis of wax alkanes, increasing drought tolerance and fruit storability. *Hortic Res,* **9**.

Wu J, Wang J, Hui W, Zhao F, Wang P, Su C, et al. 2022b. Physiology of plant responses to water stress and related genes: A review. *Forests,* **13:** 324.

Xu Z & Zhou G. 2008. Responses of leaf stomatal density to water status and its relationship with photosynthesis in a grass. *J Exp Bot,* **59:** 3317–3325.

Yamauchi T, Abe F, Tsutsumi N & Nakazono M. 2019. Root cortex provides a venue for gas-space formation and is essential for plant adaptation to waterlogging. *Front Plant Sci,* **10**.

Yang J, Isabel Ordiz M, Jaworski JG & Beachy RN. 2011. Induced accumulation of cuticular waxes enhances drought tolerance in Arabidopsis by changes in development of stomata. *Plant Physiology and Biochemistry,* **49:** 1448–1455.

Yang X, Lu M, Wang Y, Wang Y, Liu Z & Chen S. 2021. Response mechanism of plants to drought stress. *Horticulturae,* **7:** 50.

Yang Y, Guo Y, Zhong J, Zhang T, Li D, Ba T, et al. 2020. Root physiological traits and transcriptome analyses reveal that root zone water retention confers drought tolerance to Opisthopappus taihangensis. *Sci Rep,* **10:** 2627.

Zaharah SS & Razi IM. 2009. Growth, stomata aperture, biochemical changes and branch anatomy in mango (Mangifera indica) cv. Chokanan in response to root restriction and water stress. *Sci Hortic,* **123:** 58–67.

Zhang C, Shi S, Wang B & Zhao J. 2018. Physiological and biochemical changes in different drought-tolerant alfalfa (Medicago sativa L.) varieties under PEG-induced drought stress. *Acta Physiol Plant,* **40:** 25.

Zhu L, Guo J, Zhu J & Zhou C. 2014. Enhanced expression of EsWAX1 improves drought tolerance with increased accumulation of cuticular wax and ascorbic acid in transgenic Arabidopsis. *Plant Physiology and Biochemistry,* **75:** 24–35.

Zhu X & Xiong L. 2013. Putative megaenzyme DWA1 plays essential roles in drought resistance by regulating stress-induced wax deposition in rice. *Proceedings of the National Academy of Sciences,* **110:** 17790–17795.

Zulfiqar F, Younis A, Riaz A, Mansoor F, Hameed M, Akram NA, et al. 2020. Morpho-anatomical adaptations of two Tagetes erecta L. cultivars with contrasting response to drought stress. *Pak J Bot,* **52:** 801–810.

4 | Physiological Anomalies in Plants during Water Stress

Praveen Kumar Shukla, Jyoti Vishwakarma, Pradeep Kumar, Rajesh Saini and Kavindra Nath Tiwari*

Department of Botany, MMV, Banaras Hindu University, Varanasi, India

Water stress is becoming a serious ecological concern affecting all of mankind due to the disturbance of ecological equilibrium and anomalies in the global climate. Water stress is the main environmental stress caused by temperature fluctuations, changes in light intensity, and inadequate rainfall. Plants create more reactive oxygen species (ROS) in response to various water stresses. For essential processes, plants need a specific amount of ROS, and variations in their concentration affect the physiological functioning of the plant altogether. Plant physiology is negatively affected by water stress in numerous ways, most notably photosynthetic capability. Prolonged stress drastically reduces plant growth and production. Perhaps several key underlying elements of the environment determining the growth of vegetation, development, and production processes involve stress generated by water. Assessing how climate change and unpredictability may affect agricultural output requires a deep understanding of the consequences of drought. Drought stress affects several systems, including the physiological, growth, developmental, productive, and quality of agricultural systems. In this chapter, some of the most important traits associated with water stress tolerance are discussed and emphasized.

1. Introduction

Stress is a physiologically changed state brought on by factors that are likely to upset the balance. According to Gaspar et al. (2002), stress is a chemical or physical alteration brought on by stressful conditions. The phrase stress has several meanings, including the physiological description as well as the suitable term as an outcome in various situations. Plants frequently experience a range of challenges when growing in the wild, such as heat, oxidative stresses,

*Corresponding author: kntiwaribhu@gmail.com

drought, cold temperatures, salt, floods, and toxic heavy metals. The continued existence of crop cultivation and the sustainability of food production depend on addressing the global concerns including salt stress and water shortages (Nikolaou et al., 2020).

Water scarcity exerts a negative influence on several areas of plant physiology, including the most important photosynthetic process. Plant developments as well as yield are both substantially hampered when the period of stress is extended. In order to respond to various climatic challenges, plants also establish extensive morphological as well as molecular modifications. The biological and genetic mechanisms leading to water-stress resistance and also water-use capacity have gained significant research focus (Osakabe et al., 2014).

Water-related stress serves major critical climatic challenges among vegetation, causing significant biological alterations. The process of photosynthesis is hampered under water stress, which drastically inhibits the carbon metabolic processes widely in agricultural crops (Bjorkman and Powles, 1984; Tezara et al., 1999; Griffiths and Parry, 2002; Lawlor, 2002).

Water shortage has a negative influence on agricultural crops and productivity. As a consequence, growing crops that exhibit more effective survivorship and development despite water-related stresses is an essential goal in plant reproduction. Water use efficiency (WUE) is a significant selection factor for the measurement of plant growth under water scarcity conditions. Plants, in fact, adapt a variety of genetic processes that minimize product utilization as well as modify their development in response to unfavorable ecological influences (Ahuja et al., 2010; Ha et al., 2014; Nishiyama et al., 2011; Osakabe et al., 2011; Skirycz and Inze, 2010).

Plants experience a broad spectrum of morphological and enzymatic aberrations when they are under stress, including adjustments to their absolute water percentage, the photosynthesis procedure, glucose utilization, sequences of amino acids, protein synthesis, and enzyme activity. Numerous researches have been analyzed regarding the effects of stress linked to water on plant photosynthesis. However, research concerning the process by which drought affects nutrient intake and how that affects plant physiology has several unresolved aspects (Farooq et al., 2009). Due to its direct impact on the physicochemical characteristics of soils and consequent reduction in their mobility and absorption of mineral nutrients, drought stress can result in nutritional deficiencies in plants, irrespective of whether they are cultivated in highly fertile soils (Ahanger et al., 2016) (Figure 1).

A major stress on plant metabolism is water deficiency, which affects developmental metrics, plasma membrane composition, and both the quantity and quality of total protein content. A significant amount of data indicates that water stress reduces the amount of a few proteins synthesized by plants and also maintains and promotes the production of other proteins (Salekdeh et al., 2002). Ultimately, selecting drought-tolerant plants will benefit from a deep comprehension of the processes that plants perform to adjust to water constraints and continue growing and producing during dry spells (Chaitanya et al., 2003).

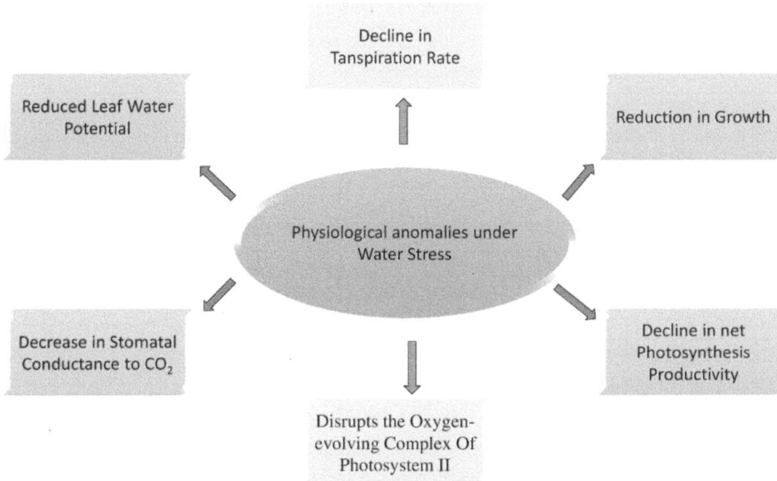

Figure 1: Various physiological dysfunctions that plants encounter under water stress.

Due to decreased precipitation and shifting rainfall patterns, there has been a significant increase in drought globally. Drought stress exerts the greatest impact on plant productivity during the breeding or growing periods. Drought damages crop development, breeding, and biological functions, reducing the yield (Fahad et al., 2017).

Plants that sense a minor drop in soil water content and accelerate their development to finish their lifespan prior to prolonged drought hits are said to be engaging in drought escape. The most frequent adaptations that plants gain while using this technique are early blooming and the remobilization of photoassimilates around seeds (Kooyers, 2015; Shavrukov et al., 2017).

2. Reduction in Growth

Growth in plants is the result of a complicated interaction among both the sources and the sinking constraints of a plant's two primary organs, the shoot and root system, that maintains functional homeostasis in plant tissues. Water deficiency, either long-term or short-term, possesses essential characteristics in the development of plants, and their proliferation in all climates. One of the biological processes most dependent on water retention is cell growth, which is thought to be caused by a drop in turgor pressure. Growth is caused by meristematic cell divisions producing daughter cells as well as resulting in immense expansion among the new cells. When faced with extreme water shortage, higher plants' ability to elongate their cells can be prevented by blocking water flow through their xylary elements into neighboring cells (Nonami, 1998).

The disruption of mitosis caused by drought, also reduces the growth (Hussain et al., 2008). A deficiency of water diminishes the soil's water retention ability, hence reducing the number, size and the foliage of the plant. The expansion in leaf area is dependent on temperature, leaf turgor, and the way it absorbs resources for growth. The decrease in the leaf area caused by drought affects the photosynthesis process (Rucker et al., 1995). It was observed that plant height, leaf surface area, and stem breadth are also drastically decreased with increasing water stress. Plants under water stress accelerate the leaf senescence (Manivannan et al., 2007a).

Thornley and Johnson (1990) reported that growth of plants under stress is stunted by increasing glucose allocation within the root systems and reducing the leaf expansion (Sarlikioti et al., 2010).

Water scarcity factors are referred to as a condition through which the water potential of a plant decreases, so basic cell functions are affected. Decline in the water content of the cells induces wilting reduction in stomatal closure, turgor pressure, and impaired cell growth (Ahanger et al., 2017).

Water stress in soybean reduces the height of the plant (Specht et al., 2001). Under water stress, citrus seedlings height was reduced by 25% (Wu et al., 2008) in comparison to the control. This effect was also observed in other crops like *Vigna unguiculata*, *Solanum tuberosum*, *Abelmoschus esculentus*, *Petroselinum crispum* (Jaleel et al., 2009).

Water scarcity produces architectural changes in roots that optimize water and nutrient absorption. Root structure in crop plants under drought has received a lot of attention (Lynch 2018). According to a study, in changing environmental conditions plants have roots of different phenotypes. Plants establish more deeper roots to increase efficiency of water utilization (Klein et al., 2020). The most essential feature for obtaining nutrients and water from poor soil is the root growth angle (Lynch, 2018).

Abiotic stress components in the soil also affect root gravitational pull and the hydrotropic movement of the roots (Takahashi et al., 2003). Water stress is primarily responsible for the breakdown of amyloplast, which causes roots to lose their ability to detect gravity and utilize hydrotropism for effective water acquisition. This could be a physiological reaction of roots to dryness (Takahashi et al., 2003).

In *Phoenix dactylifera*, when a root system grows, more water is absorbed and higher proline concentrations maintain the required osmotic pressure (Djibril et al., 2005). Previously, it was demonstrated that water related stress stimulates the progressive growth of roots among *Catharanthus roseus* and *Helianthus annuus* (Jaleel et al., 2008a,c; Tahir et al., 2002).

In *Populus* species, during either extreme of water stress conditions, root dry weight was reduced (Wullschleger et al., 2005). Greater ABA concentrations in the roots and shoots during a drought were associated with a greater root-to-shoot ratio (Sharp and LeNoble, 2002; Manivannan et al., 2007b).

3. Effect on Photosynthesis

Water stress influences photosynthesis, an essential physiological process that leads to a decline in the carbon fixing efficiency. It leads to decreased electron transport from photosynthesis and carbon absorption, which results in a loss in the net productivity (Chaves et al., 2009; Lawlor and Tezara, 2009). The biochemical processes underlying photosynthesis mechanisms during drought stress are quite intricate. They are influenced by various factors, including water availability, environmental cues, stress duration and developmental stage of the plants. Under stress conditions, carbon absorption in green plants is primarily repressed by decreased carbon dioxide release due to the closure of stomatal pores and reduced mesophyll conductance (Evans et al., 2009). This also results from some other limitations not related to stomata (Lawlor and Tezara, 2009). Miyazawa et al. (2008) discovered that under drought, chlorophyll permeability decreased due to a disruption in the normal activity of aquaorins, which participate in carbon dioxide dispersion to the site of carboxylation in the Calvin cycle by RuBisCO. Long-term exposure to stress due to water in fully developed adult plants contributes to a drop in RuBisCO's carboxylation process efficacy and the actions of several other enzymes related to the Calvin cycle (Maroco et al., 2002; Lawlor, et al., 2009). An overabundance of Calvin cycle enzymes (Triose Phosphate Isomerase (TPI; EC 5.3.1.1), Sedoheptulose-1,7-bisphophatase (EC 3.1.3.37), Fructose 1,6-bisphosphate aldolase (FBA; EC 4.1.2.13), Phosphoglycerate Kinase (PGK; EC 2.7.2.3), RuBisCO) decreased due to water stress, according to physiological, biochemical, and proteomic approaches (Kottapalli et al., 2009; Gazanchian et al., 2007; Ford et al., 2011; Mirzaei et al., 2014; Faghani et al., 2015).

Deficiency of water severely affects the chlorophyll content which is influenced by chloroplast membranes. Under stress swelling and deformation of the thylakoid lamellae took place resulting in the production of lipid drops (Kaiser, 1982). Under a water deficit situation in *Gossypium herbaceum* L., a decline in the chlorophyll content may be responsible for a decrease in various photosynthetic properties (Deeba et al., 2012). Water stress alters vital physiological processes like photosynthesis, resulting in reduced stomatal flexibility as well as changed enzyme functions involved in the photosynthetic mechanism (Karimi et al., 2015).

The decreased water availability directly affects the photosynthetic rate by altering photosynthetic metabolism and reducing carbon dioxide (CO_2) production-mediated stomatal closure (Chaves et al., 2009; Flexas et al., 2006). Frequently it has been observed that photosynthetic activities of plants rapidly reduced during water shortages (Pic et al., 2002).

Stomatal closing and opening is directly influenced by the water limit (Lawlor and Tezara, 2009; Cornic, 2000). Limited water caused reduction in both the photosynthetic rate and photosystem II (PSII) quantum yield, because the stomata had not expanded properly (Chaves et al., 2002). So, the process of carbon dioxide assimilation significantly declined, and the light energy taken by the leaf chlorophyll could not ordinarily be used for photosynthetic electron transport. The resultant effect is that when a proportion of this energy is wasted for the generation of heat energy, non photochemical quenching (NPQ) proceeds more quickly (Baker and Rosenqvist, 2004; Krause and Weis, 1991).

The process of photosynthesis is especially susceptible to water shortages. As the soil hydric levels reduce, stomatal apertures contract to preserve water (Lawlor and Tezara, 2009). Thus, stomata regulate the water level in the plants (Guo et al., 2010). Plants exhibit a reduction in transpiration and photochemical synthesis during periods of dryness (Endres, 2007; Hessini et al., 2009; Saibo et al., 2009), as shown in Figure 2. This decrease is even seen in environmental conditions with extremely large transpiration demands (Feng and Cao, 2005). For example, water stress triggers the closure of stomatal pores as well as a reduction in intracellular carbon dioxide (CO_2) levels inside leaves, which enhances the activity of oxygenase over carboxylase and raises the photorespiration to the photosynthesis ratio (Flexas and Medrano, 2002).

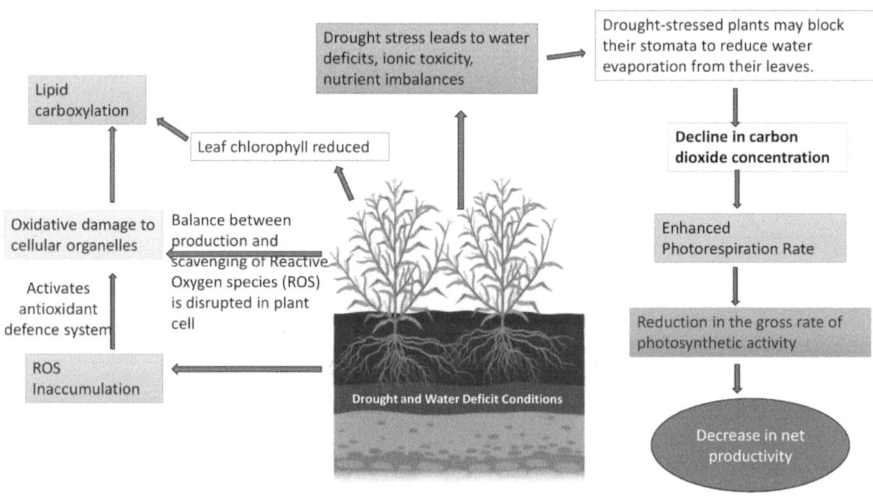

Figure 2: Influence of water-related stress upon photosynthetic mechanism and ROS scavenging activity.

In addition to the impact of stomatal closure, additional factors that hinder photosynthetic activity include chlorophyll degradation, a reduction in the specific activity of enzyme (RuBisCO), and a reduction in Photosystem II's (PS II) photochemical efficacy under conditions of water-associated stress (Parkash and Singh, 2020).

4. Water Stress Impact on Nutritional Content

Reduced transpiration rates under water stress cause plants to absorb fewer nutrients (da Silva et al., 2011). Additionally, low cell membrane permeability and transport efficiency result in a decrease in plant root intake capacity (Hsiao and Xu, 2000). Thus, nutrient absorption through the soil solution is directly corresponding to the level of water present in the soil as well as the condition of soil. The effectiveness of the transportation of vital nutrients from the ground's soil matrices to the recipient root zone decreases when the soil water level falls (Hsiao and Xu, 2000).

The reduction of nutrient absorption in plants caused by water stress is referred to as decreased surface area, length, and hydraulic conductivity of the root. Restricted root hair growth also leads to increased production of ABA by dehydrated roots. ABA, in turn, triggers the closure of stomatal apertures, resulting in a slower transpiration rate.

Plants require inorganic ions, which are absorbed from mineral-rich nutritious soil, for their development and proliferation. Insufficient soil water content during a drought stress slows the transportation of mineral nutrients like Ca and Mg from soil particles to the rhizosphere. This affects the nutrient flow by slowing absorption and the flow rate from the roots underneath, which are in close proximity to the soil, to the aerial parts of plants such as leaves. Drought stress triggers rapid stomatal closure, inhibiting transpiration rates. Consequently, the transfer of mineral nutrients from the root system to the growing aerial shoots decreases. As a result, drought-related stress minimizes nutrient availability as well as mobility in the soil surface and plants (Singh and Sale, 1998; da Silva et al., 2011).

5. Effect on Photochemical Efficiency

One of the important components of chloroplasts is chlorophyll, and the proportional amount of chlorophyll positively correlates directly with the efficiency of photosynthesis (Anjum et al., 2011). Water stress constituted an influence on the concentrations of carotenoids along with chlorophyll '*a*' and chlorophyll '*b*' (Farooq et al., 2009; Anjum et al., 2003). *Catharanthus roseus* and *Gossypium* species have lower chlorophyll contents under drought stress (Jaleel et al., 2008a-d; Massacci et al., 2008). When exposed to prolonged water concentration both, *Vaccinium myrtillus* and *Helianthus* species exhibited a decrease in relative total chlorophyll concentration in their leaves (Tahkokorpi et al., 2007; Kiani et al., 2008). This reduction in chlorophyll content can impact photosynthesis and overall plant health.

A slight decrease in the proportionate concentration of moisture as well as water potential amongst the leaf tissue of a plant has been demonstrated to trigger an impairment in the foliage photosynthetic process in shrubs and trees (Lawlor and Cornic, 2002). There is ongoing debate on whether stomatal pore closure or metabolic degradation prevents photosynthesis when plants are dry (Anjum et al., 2003; Lawson et al., 2003). The generation of free radicals like ROS within the thylakoids is among the main reasons for water stress which lowers the cell membrane proportions of carotenoids and chlorophylls (Kiani et al., 2008; Reddy et al., 2004).

Chlorophyll fluorescence can be used to measure the activity of Photosystem II. The quantity of chlorophyll fluorescence indicates how much light energy is emitted by PS II. Chlorophyll may either squander the light energy it absorbs as heat energy, re-emit it to the environment (chlorophyll fluorescence), or use it to power photosynthesis (photochemistry). Because of their interdependence, the efficiency of each of these three processes decreases as its efficiency rises. Photochemical quenching involves using light energy for photochemistry, non-photochemical quenching involves dispersing light energy as heat energy. According to Murchie et al. (2013), photochemistry decreases CO_2 assimilation while producing energy. Thus, photosynthesis is greatly influenced by photochemistry (Figure 3).

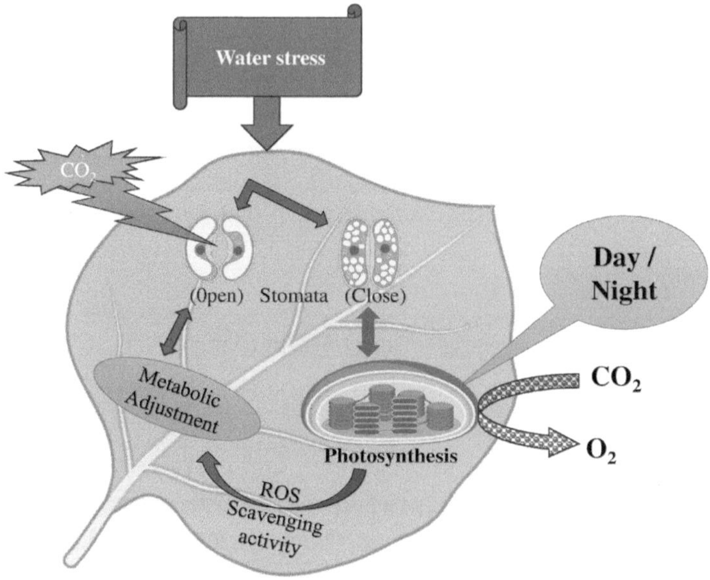

Figure 3: Photochemical efficiency imbalance under water stress.

A photoprotective process called heat energy dissipation transfers excess excitation energy out of complexes containing chlorophyll to prevent harmful free radicals from accumulating (Demmig and Adams, 2006). As a result, chlorophyll fluorescence investigations provide quantitative data on photochemical efficiency and heat energy dissipation.

Mild and moderate water deficits do not significantly damage the optimal photochemical activity of PS II, unlike other environmental stresses. Depending on the kind of plant, the photosynthetic system's effective operating range varies depending on water concentrations. This concept was given by Kawamitsu et al. (2000), who described in their relative analysis that sunflower leaves abruptly stopped photosynthesizing when the water concentration reduced below 75%. In marine alga *Fucus vesiculosus*, photosynthesizing declined slightly after water losses of up to 20%.

There are many ecological types of plants. Some are desiccation-tolerant and can rapidly recover PSII activity after being rewatered following a dry period. Others are sensitive to desiccation and cannot endure prolonged dehydration. In this context, two points need to be emphasized: first, photochemical activity of PSII is totally reduced in both desiccation-sensitive as well as extremely dry poikilohydric plants; second, PS II of plants that are desiccation-prone may be able to regain their photochemical capacity even in extremely severe conditions of desiccation (Bukhov and Carpentier, 2004).

Stomatal aperture regulation is essential for the exchange of gases at the vegetation-atmosphere interface. Plants lost 90% of their water content due to transpiration through stomatal pores. Stomatal restriction is known to be a significant contributing cause behind the decrease in photosynthetic efficiency when water becomes scarce. Non-stomatal limitations, such as reductions in PSII photochemistry efficiency, CO_2 availability in the chloroplast, and Rubisco activity, advance progressively with the duration and intensity of water stress (da Silva and Arrabaca, 2004; Xu Z et al., 2009). When the drought stress level becomes severe, RuBisCO activity as well as its photochemical efficiency get restricted, potentially reducing the net effect to zero; simultaneously, other negative biochemical-physiological metabolisms such as peroxidation could be promoted, diminishing plant productivity (Xu et al., 2010)

6. Effect on Light Reaction

Photosystem II was proven to be highly reactive to water stress. Recent experimental investigations have shown that water stress significantly disrupts both the photosystem II reaction center (He et al., 1995) as well as the oxygen-evolving complex of photosystem II (Subrahmanyam et al., 2006). Further research, in contrast, demonstrated that blocked CO_2-dependent O_2 evolution as well as net carbon dioxide assimilation triggered by water stress would be acquired through high external atmospheric CO_2 levels. This indicates that biochemical modifications are not responsible for inhibiting carbon dioxide assimilation even though stomatal pores actually serve as a primary function in reducing CO_2 assimilation during water stress. The finding suggests that water stress has a negligible impact on photosystem II photo-chemistry that lends credence to this view (Cornic and Briantais, 1991; Liang et al., 1997).

Water stress effectively reduces photosynthetic rates because it decreases CO_2 production as a response to the closing of the stomatal aperture and/or alters the metabolic photosynthesis process (Lawlor, 2002). When water stress affects photosynthetic rates, excess light has an adverse impact on photosynthesis (Osakabe et al., 2014; Li et al., 2009). Research has indicated that there is a significant connection among the various responses concerned with drought stress and excess light, with 70% of the genes activated by drought also being triggered by excessive light (Chan et al., 2010; Estavillo et al., 2011; Kimura et al., 2002). Furthermore, excessive light stimulates a few photochemical and metabolic processes that produce ROS, that include free radicals for example hydrogen peroxide (H_2O_2), singlet oxygen ($^1O_2$), superoxide (O_2^-). Plant photosynthetic processes are negatively impacted by these activities (Li et al., 2009).

Plants are capable of tracking chloroplast function using plastid-to-nucleus signals and retrograde communication. This signaling pathway is able to regulate gene expression within the chloroplast. According to Chan et al. (2010), retrograde signalling is crucial for controlling chloroplastic functions as well as adaptive reactions to atmospheric perturbations. Chlorophyll precursors within plants stimulated by high light, such Mg-protoporphyrin IX (Mg-Proto), function as an analogous indication that stimulates the expression of nuclear genes.

ALX8 mutants showed drought-tolerant traits and constitutive upward regulation of a quarter (25%) of the excess light-regulating transcriptome. These findings imply that the SAL1/ALX8/FRY1 signaling networks can be involved in both drought and excess light signaling, and that nuclear gene expression can be changed in both scenarios by the SAL1-PAP retrograde route (Estavillo et al., 2011; Rossel et al., 2006; Wilson et al., 2009).

7. Effect on C3 and C4 Cycles

In all three pathways C3, C4 and CAM, CO_2 assimilation processes during photosynthesis are different. The C4 carbon fixation was more efficient than the C3 carbon fixation methods under drought conditions. The Kranz anatomy is a distinctive trait of C4 plants' leaves. Compared to C3 plants, its water usage efficiency (WUE) also seems to be significantly higher. Under a water-stress condition, the C4 plant may absorb CO_2 and generate additional organic matter, helping the plant to avoid premature drying out. The CO_2 absorption processes of various species are distinct; thus carbon metabolism pathways of plants are substantially modified by varying external factors. In contrast, alterations during plant maturation as well as production, developmental conditions, nutrient intake, and metabolic regulators might cause a parallel modification in the pathways involved in plant carbon dioxide (CO_2) fixation. Under drought situations, ABA concentrations are high in plants, which enhances the C4 pathway's function. Likewise, the C3 pathway could be modified directly into the CAM pathway (Huber and Sankhla, 1976). Thus, as the surrounding atmosphere changes, certain species show distinct photosynthetic morphologies which modify their absorption methods, as demonstrated by Winter's research on *Mesembryanthemum crystallinum* L. (Aizoaceae) and several orchid varieties. Without stress, they function in C3 mode while under stress, either salt or drought, they function in CAM mode (Winter et al., 2005; Silvera et al., 2005; Winter et al., 2007). C3-CAM shift occurs throughout the seedling process in the cactus plant. Once the ground incorporates sufficient moisture in it, the mechanism of C3 photosynthesis can stimulate plant development soon after germination. Facultative CAM ingredients may increase the growth of constitutive CAM and help plants flourish in water-related stress scenarios (Winter et al., 2011).

8. Effect on Crassulacean Acid Metabolism (CAM)

CAM is a drought stress response. It is a consequence of CO_2 uptake mediated by phosphoenolpyruvate carboxylase (PEPC) during dark conditions. A significant range of flexibility in shifting atmospheric conditions becomes possible through the CAM pathway. Drought effectively ceases gas exchange in CAM plants throughout the day, which is beneficial to the environment due to the extremely minimal water loss through transpiration in leaves (Kluge and Ting, 2012).

In CAM plants stomata that open during the night and CO_2 is utilized to produce malic acid. This acid is known to be stored in vacuoles by the action of the PEPC enzyme. During the daytime, stomatal closure occurs, and MAL decarboxylation releases CO_2. In addition to avoiding transpiration loss during stomatal opening during the day, CAM fixes CO_2 via stomatal opening at night, which resolves the imbalance between CO_2 absorption and transpiration through the stomatal pore during drought conditions (Yang et al., 2021).

9. Effect on RuBisCO

Ribulose-1,5-bisphosphate carboxylase/oxygenase commonly abbreviated as RuBisCO aids in the absorption of CO_2 by photoautotrophs. It is crucial for a plant's photosynthesis. Apart from its carboxylase function, RuBisCO functions additionally as an oxygenase, competing with ribulose-1,5-bisphosphate (RuBP) for the CO_2 and O_2. Accordingly, the initial reaction in the photorespiration cycle is regulated by the oxygenase enzyme, whereas the carboxylase enzyme stimulates the photosynthesis process. The equilibrium between these two competing processes is controlled by the kinetic properties of the RuBisCO enzyme as well as the amount of both O_2 and CO_2 at the enzyme's site (Galmes et al., 2005).

RuBisCO, the most prevalent protein on Earth, is a crucial enzyme in the fixation and oxygenation of CO_2. Its molecular mass is between 55 and 56 kDa for the major subunit and between 14 and 16 kDa for the small subunit. It is made up of 8 small (RSS) and 8 large (RLS) subunits (Ma et al., 2009). It is still highly complicated to understand and deal with RuBisCO activity and quantity control (Feller et al., 2008; Houtz and Portis, 2003).

RuBisCO acts as a strong regulator of metabolites (such as branched chain sugars, tightly bound inhibitors, and others), RuBisCO activase, and phosphate ions (Keys et al., 1995; Portis, 1995; Servaites & Geiger, 1995). Under Mediterranean field circumstances, total RuBisCO activity of tobacco increased in the early morning but decreased considerably at by noon when drought stress was severe (Parry et al., 1993).

Three variables affect the functioning of the RuBisCO enzyme: (i) the concentration of inhibitors which bind effectively; (ii) the concentration of RuBisCO; and (iii) the process of activation of RuBisCO across its association with CO_2, which causes a lysine remain in its active region to be carbamylated

(Roy and Andrew, 2000). There is proof suggesting the concentration of RuBisCO and its carbamylation states have inverse correlations (Ruuska et al., 1998; Eichelmann et al., 2009; Quick et al., 1991). RuBisCO concentrations are known to alter under water stress (Holaday et al., 1992). Consequently, modifications in carbamylation as well as activation states are primarily responsible for variation in RuBisCO activity during water-related stress.

Photosynthesis decreases in very harsh drought circumstances due to an abrupt downturn in the biological function of the RuBisCO enzyme (Bota et al., 2004). The rate of biosynthesis and breakdown regulates the quantity of RuBisCO in leaves. With an approximate half-life of 3 hours 10 minutes, the RuBisCO holoenzyme is extremely stable even in dry environments (Hoekstra et al., 2001). Conversely, in plants experiencing severe drought the quantity of RuBisCO small subunit mRNA rapidly decreased in this condition, suggesting a reduction in synthesis (Vu et al., 1999).

It was discovered that tight-binding inhibitors as well as variations in CO_2 and Mg^{2+} activation, were the important aspects for overall decline in the RuBisCO enzyme activity in tobacco plants during drought stress conditions (Parry et al., 2002). Reddy et al. (2004) reported that a significant drop in photosynthesis amid a period of drought is linked to a drop in photosystem II quantum efficiency among higher plants, the degree of RuBP recovery as well as the carboxylation of 5-bisphosphate through RuBisCO.

10. Conclusion

Stress impacts the body and mind, affecting plants' growth and productivity. Water scarcity negatively impacts photosynthetic processes and agricultural productivity. Improving crop development and survival in water-related stressors is crucial for plant reproduction. Stress causes a variety of phenotypic and enzymatic alterations in plants, such as adjustments to the amount of water in the plant, photosynthesis, glucose use, amino acid sequences, protein synthesis, and enzyme activity. Lack of water has a major impact on plant growth and multiplication by decreasing soil water retention, leaf area expansion, and root gravitational pull. Stress from water also impacts photosynthesis, which lowers the efficiency of fixing carbon.

Water scarcity impacts carbon absorption and photosynthesis, lowering net productivity. It affects chlorophyll content, photosynthetic rate, and stomatal closure. Soil hydric levels decrease stomatal apertures, making photosynthesis vulnerable. Water stress reduces nutrient absorption in plants due to reduced transpiration rates and root surface area. Drying roots' increased ABA production leads to stomatal aperture closure, reduced transpiration, and restricted root hair growth. Photosystem II activity is measured using chlorophyll fluorescence. Water stress lowers photosynthetic rates by inhibiting CO_2 production and modifying metabolic photosynthesis. The C4 carbon fixation method is more effective under drought conditions. Under water stress, C4 plants absorb CO_2 and produce more organic matter, avoiding drying. Drought conditions improve C4 activity and CAM pathway adaptability. Drought stops gas exchange in CAM plants, reducing water loss. Extreme drought decreases RuBisCO mRNA, indicating decreased synthesis.

References

Ahanger MA, Morad-Talab N, Abd-Allah EF, Ahmad P & Hajiboland R. 2016. Plant growth under drought stress: Significance of mineral nutrients. *Water Stress and Crop Plants: A Sustainable Approach,* **2:** 649–668.

Ahanger MA, Tomar NS, Tittal M, Argal S & Agarwal R. 2017. Plant growth under water/salt stress: ROS production; antioxidants and significance of added potassium under such conditions. *Physiology and Molecular Biology of Plants,* **23:** 731–744.

Ahuja I, de Vos RC, Bones AM & Hall RD. 2010. Plant molecular stress responses face climate change. *Trends in Plant Science,* **15(12):** 664–674.

Anjum F, Yaseen M, Rasul E, Wahid A & Anjum S. 2003. Water stress in barley (Hordeum vulgare L.) II. Effect on chemical composition and chlorophyll contents. *Pakistan Journal of Agricultural,* **40(1–2):** 45–49.

Baker NR & Rosenqvist E. 2004. Applications of chlorophyll fluorescence can improve crop production strategies: An examination of future possibilities. *Journal of Experimental Botany,* **55(403):** 1607–1621.

Bota J, Medrano H & Flexas J. 2004. Is photosynthesis limited by decreased Rubisco activity and RuBP content under progressive water stress? *New Phytologist,* **162(3):** 671–681.

Bjorkman O & Powles SB. 1984. Inhibition of photosynthetic reactions under water stress: Interaction with light level. *Planta,* **161:** 490–504.

Bukhov NG & Carpentier R. 2004. Effects of water stress on the photosynthetic efficiency of plants. pp. 623–635. *In:* Chlorophyll a Fluorescence: A Signature of Photosynthesis, Dordrecht: Springer Netherlands.

Chaitanya KV, Jutur PP, Sundar D & Ramachandra Reddy A. 2003. Water stress effects on photosynthesis in different mulberry cultivars. *Plant Growth Regulation,* **40:** 75–80.

Chan KX, Crisp PA, Estavillo GM & Pogson BJ. 2010. Chloroplast-to-nucleus communication: Current knowledge experimental strategies and relationship to drought stress signaling. *Plant Signaling & Behavior,* **5(12):** 1575–1582.

Chaves MM, Pereira JS, Maroco J, Rodrigues ML, Ricardo CPP, Ośorio ML, et al. 2002. How plants cope with water stress in the field? Photosynthesis and growth. *Annals of Botany,* **89:** 907–916.

Chaves MM, Flexas J & Pinheiro C. 2009. Photosynthesis under drought and salt stress: Regulation mechanisms from whole plant to cell. *Annals of Botany,* **103(4):** 551–560.

Cornic G & Briantais JM. 1991. Partitioning of photosynthetic electron flow between CO_2 and O_2 reduction in a C3 leaf (*Phaseolus vulgaris* L.) at different CO_2 concentrations and during drought stress. *Planta,* **183:** 178–184.

Cornic G. 2000. Drought stress inhibits photosynthesis by decreasing stomatal aperture – not by affecting ATP synthesis. *Trends in Plant Science,* **5(5):** 187–188.

da Silva EC, Nogueira RJMC, da Silva MA & de Albuquerque MB. 2011. Drought stress and plant nutrition. *Plant Stress,* **5(1):** 32–41.

Da Silva JM & Arrabaca MC. 2004. Photosynthesis in the water-stressed C4 grass *Setaria sphacelata* is mainly limited by stomata with both rapidly and slowly imposed water deficits. *Physiologia Plantarum,* **121(3):** 409–420.

Deeba F, Pandey AK, Ranjan S, Mishra A, Singh R, Sharma YK, et al. 2012. Physiological and proteomic responses of cotton (*Gossypium herbaceum* L.) to drought stress. *Plant Physiology and Biochemistry,* **53:** 6–18.

Demmig-Adams B & Adams III WW. 2006. Photoprotection in an ecological context: The remarkable complexity of thermal energy dissipation. *New Phytologist,* **172(1):** 11–21.

Djibril S, Mohamed OK, Diaga D, Diégane D, Abaye BF, Maurice S, et al. 2005. Growth and development of date palm (*Phoenix dactylifera* L.) seedlings under drought and salinity stresses. *African Journal of Biotechnology,* **4(9)**.

Eichelmann H, Talts E, Oja V, Padu E & Laisk A. 2009. RuBisCO in planta k cat is regulated in balance with photosynthetic electron transport. *Journal of Experimental Botany,* **60(14):** 4077–4088.

Endres L. 2007. Daily and seasonal variation of water relationship in sugar ale (*Annona squamosa* L.) under different irrigation regimes at semi-arid Brazil. *Scientia Horticulturae,* **113(2):** 149–154.

Estavillo GM, Crisp PA, Pornsiriwong W, Wirtz M, Collinge D, Carrie C, et al. 2011. Evidence for a SAL1-PAP chloroplast retrograde pathway that functions in drought and highlight signaling in *Arabidopsis*. *The Plant Cell,* **23(11):** 3992–4012.

Evans JR, Kaldenhoff R, Genty B & Terashima I. 2009. Resistances along the CO_2 diffusion pathway inside leaves. *Journal of Experimental Botany,* **60(8):** 2235–2248.

Faghani E, Gharechahi J, Komatsu S, Mirzaei M, Khavarinejad RA, Najafi F, et al. 2015. Comparative physiology and proteomic analysis of two wheat genotypes contrasting in drought tolerance. *Journal of Proteomics,* **114:** 1–15.

Fahad S, Bajwa AA, Nazir U, Anjum SA, Farooq A, Zohaib A, et al. 2017. Crop production under drought and heat stress: Plant responses and management options. *Frontiers in Plant Science,* 1147.

Farooq M, Wahid A, Kobayashi NSMA, Fujita DBSMA & Basra SMA 2009. Plant drought stress: Effects mechanisms and management. *Sustainable Agriculture,* 153–188.

Feller U, Anders I & Mae T. 2008. RuBisCOlytics: Fate of RuBisCO after its enzymatic function in a cell is terminated. *Journal of Experimental Botany,* **59(7):** 1615–1624.

Feng YL & Cao KF. 2005. Photosynthesis and photoinhibition after night chilling in seedlings of two tropical tree species grown under three irradiances. *Photosynthetica,* **43:** 567–574.

Flexas J & Medrano H. 2002. Drought inhibition of photosynthesis in C3 plants: Stomatal and nonstomatal limitations revisited. *Annals of Botany,* **89(2):** 183–189.

Flexas J, Bota J, Galmés J, Medrano H & Ribas-Carbó M. 2006. Keeping a positive carbon balance under adverse conditions: Responses of photosynthesis and respiration to water stress. *Physiologia Plantarum,* **127(3):** 343–352.

Ford KL, Cassin A & Bacic A. 2011. Quantitative proteomic analysis of wheat cultivars with differing drought stress tolerance. *Frontiers in Plant Science,* **2:** 44.

Frederick JR, Alm DM, Hesketh JD & Below FE. 1990. Overcoming drought-induced decreases in soybean leaf photosynthesis by measuring with CO_2-enriched air. *Photosynthesis Research,* **25:** 49–57.

Galmes J, Flexas J, Keys AJ, Cifre J, Mitchell RA, Madgwick PJ, et al. 2005. RuBisCO specificity factor tends to be larger in plant species from drier habitats and in species with persistent leaves. *Plant Cell & Environment,* **28(5):** 571–579.

Gaspar T, Franck T, Bisbis B, Kevers C, Jouve L, Hausman JF, et al. 2002. Concepts in plant stress physiology. Application to plant tissue cultures. *Plant Growth Regulation,* **37:** 263–285.

Gazanchian A, Hajheidari M, Sima NK & Salekdeh GH. 2007. Proteome response of *Elymus elongatum* to severe water stress and recovery. *Journal of Experimental Botany,* **58(2):** 291–300.

Gimenez C, Mitchell VJ & Lawlor DW. 1992. Regulation of photosynthetic rate of two sunflower hybrids under water stress. *Plant Physiology,* **98(2):** 516–524.

Griffiths H & Parry MAJ. 2002. Plant responses to water stress. *Annals of Botany,* **89(7):** 801–802.

Gunasekera D & Berkowitz GA 1992. Heterogenous stomatal closure in response to leaf water deficits is not a universal phenomenon. *Plant Physiology,* **98(2):** 660–665.

Guo XY, Zhang XS & Huang ZY. 2010. Drought tolerance in three hybrid poplar clones submitted to different watering regimes. *Journal of Plant Ecology,* **3(2):** 79–87.

Ha CV, Leyva-González MA, Osakabe Y, Tran UT, Nishiyama R, Watanabe Y, et al. 2014. Positive regulatory role of strigolactone in plant responses to drought and salt stress. *Proceedings of the National Academy of Sciences,* **111(2):** 851–856.

He JX, Wang J & Liang HG. 1995. Effects of water stress on photochemical function and protein metabolism of photosystem II in wheat leaves. *Physiologia Plantarum,* **93(4):** 771–777.

Hessini K, Martínez JP, Gandour M, Albouchi A, Soltani A & Abdelly C. 2009. Effect of water stress on growth osmotic adjustment cell wall elasticity and water-use efficiency in *Spartina alterniflora. Environmental and Experimental Botany,* **67(2):** 312–319.

Hoekstra FA, Golovina EA & Buitink J. 2001. Mechanisms of plant desiccation tolerance. *Trends in Plant Science,* **6(9):** 431–438.

Holaday AS, Ritchie SW & Nguyen HT. 1992. Effects of water deficit on gas-exchange parameters and ribulose 1,5-bisphosphate carboxylase activation in wheat. *Environmental and Experimental Botany,* **32(4):** 403–410.

Houtz RL & Portis Jr AR. 2003. The life of ribulose 1,5-bisphosphate carboxylase/oxygenase—Posttranslational facts and mysteries. *Archives of Biochemistry and Biophysics,* **414(2):** 150–158.

Hsiao TC & Xu LK. 2000. Sensitivity of growth of roots versus leaves to water stress: Biophysical analysis and relation to water transport. *Journal of Experimental Botany,* **51(350):** 1595–1616.

Huber W & Sankhla N. 1976. C4 pathway and regulation of the balance between C4 and C3 metabolism. pp. 335–363. *In:* Water and Plant Life: Problems and Modern Approaches, Berlin Heidelberg: Springer Berlin Heidelberg.

Hussain AI, Anwar F, Sherazi STH & Przybylski R. 2008. Chemical composition antioxidant and antimicrobial activities of basil (*Ocimum basilicum*) essential oils depends on seasonal variations. *Food Chemistry,* **108(3):** 986–995.

Jaleel C, Abdul P, Manivannan B, Sankar A, Kishorekumar R, Gopi R, et al. 2007. Induction of drought stress tolerance by ketoconazole in *Catharanthus roseus* is mediated by enhanced antioxidant potentials and secondary metabolite accumulation. *Colloids and Surfaces B: Biointerfaces,* **60(2):** 201–206.

Jaleel CA, Gopi R & Panneerselvam R. 2008a. Growth and photosynthetic pigments responses of two varieties of *Catharanthus roseus* to triadimefon treatment. *Comptes Rendus Biologies,* **331(4):** 272–277.

Jaleel CA, Gopi R, Manivannan P, Gomathinayagam M, Sridharan R & Panneerselvam R. 2008b. Antioxidant potential and indole alkaloid profile variations with water deficits along different parts of two varieties of *Catharanthus roseus*. *Colloids and Surfaces B: Biointerfaces,* **62(2):** 312–318.

Jaleel CA, Gopi R, Sankar B, Gomathinayagam M & Panneerselvam R. 2008c. Differential responses in water use efficiency in two varieties of *Catharanthus roseus* under drought stress. *Comptes Rendus Biologies,* **331(1):** 42–47.

Jaleel CA, Manivannan P, Lakshmanan GMA, Gomathinayagam M & Panneerselvam R. 2008d. Alterations in morphological parameters and photosynthetic pigment responses of *Catharanthus roseus* under soil water deficits. *Colloids and Surfaces B: Biointerfaces,* **61(2):** 298–303.

Jaleel CA, Manivannan P, Wahid A, Farooq M, Al-Juburi HJ, Somasundaram R, et al. 2009. Drought stress in plants: A review on morphological characteristics and pigments composition. *International Journal of Agriculture and Biology,* **11(1):** 100–105.

Kaiser WM. 1982. Correlation between changes in photosynthetic activity and changes in total protoplast volume in leaf tissue from hygro- meso- and xerophytes under osmotic stress. *Planta,* **154(6):** 538–545.

Karimi S, Yadollahi A, Arzani K, Imani A & Aghaalikhani M. 2015. Gas-exchange response of almond genotypes to water stress. *Photosynthetica,* **53:** 29–34.

Kawamitsu Y, Driscoll T & Boyer JS. 2000. Photosynthesis during desiccation in an intertidal alga and a land plant. *Plant and Cell Physiology,* **41(3):** 344–353.

Keys AJ, Major L & Parry MA. 1995. Is there another player in the game of RuBisCO regulation? *Journal of Experimental Botany,* **46:** 1245–1251.

Kiani SP, Maury P, Sarrafi A & Grieu P. 2008. QTL analysis of chlorophyll fluorescence parameters in sunflower (*Helianthus annuus* L.) under well-watered and water-stressed conditions. *Plant Science,* **175(4):** 565–573.

Kimura M, Yamamoto YY, Seki M, Sakurai T, Sato M, Abe T, et al. 2003. Identification of *Arabidopsis* genes regulated by high light–stress using cDNA microarray. *Photochemistry and Photobiology,* **77(2):** 226–233.

Klein SP, Schneider HM, Perkins AC, Brown KM & Lynch JP 2020. Multiple integrated root phenotypes are associated with improved drought tolerance. *Plant Physiology,* **183(3):** 1011–1025.

Kluge M & Ting IP. 2012. Crassulacean Acid Metabolism: Analysis of an Ecological Adaptation (Vol. 30). Springer Science & Business Media.

Kooyers NJ. 2015. The evolution of drought escape and avoidance in natural herbaceous populations. *Plant Science,* **234:** 155–162.

Kottapalli KR, Rakwal R, Shibato J, Burow G, Tissue D, Burke J, et al. 2009. Physiology and proteomics of the water-deficit stress response in three contrasting peanut genotypes. *Plant Cell & Environment,* **32(4):** 380–407.

Krause AG & Weis E. 1991. Chlorophyll fluorescence and photosynthesis: The basics. *Annual Review of Plant Biology,* **42(1):** 313–349.

Lawlor DW. 2002. Limitation to photosynthesis in water-stressed leaves: Stomata vs. metabolism and the role of ATP. *Annals of Botany,* **89(7):** 871–885.

Lawlor DW & Cornic G. 2002. Photosynthetic carbon assimilation and associated metabolism in relation to water deficits in higher plants. *Plant Cell & Environment,* **25(2):** 275–294.

Lawlor DW & Tezara W. 2009. Causes of decreased photosynthetic rate and metabolic capacity in water-deficient leaf cells: A critical evaluation of mechanisms and integration of processes. *Annals of Botany,* **103(4):** 561–579.

Lawson T, Oxborough K, Morison JI & Baker NR. 2003. The responses of guard and mesophyll cell photosynthesis to CO_2 O_2 light and water stress in a range of species are similar. *Journal of Experimental Botany,* **54(388):** 1743–1752.

Li Z, Wakao S, Fischer BB & Niyogi KK. 2009. Sensing and responding to excess light. *Annual Review of Plant Biology,* **60:** 239–260.

Liang J, Zhang J & Wong MH. 1997. Can stomatal closure caused by xylem ABA explain the inhibition of leaf photosynthesis under soil drying? *Photosynthesis Research,* **51:** 149–159.

Lynch JP. 2018. Rightsizing root phenotypes for drought resistance. *Journal of Experimental Botany,* **69(13):** 3279–3292.

Ma Z, Cooper C, Kim HJ & Janick-Buckner D. 2009. A study of rubisco through western blotting and tissue printing techniques. *CBE—Life Sciences Education,* **8(2):** 140–146.

Majumdar S, Sibdas G, Bernard RG, Erwin B & Dumbroff. 1991. Activities of chlorophyllase phosphoenolpyruvate carboxylase and ribulose-1,5-bisphosphate carboxylase in the primary leaves of soybean during senescence and drought. *Physiologia Plantarum,* **81(4):** 473–480.

Manivannan P, Jaleel CA, Kishore Kumar A, Sankar B, Somasundaram R, Sridharan R, et al. 2007a. Changes in antioxidant metabolism of *Vigna unguiculata* (L.) Walp. by propiconazole under water deficit stress. *Colloids and Surfaces B: Biointerfaces,* **57(1):** 69–74.

Manivannan P, Jaleel CA, Sankar B, Kishore Kumar A, Somasundaram R, Lakshmanan GA, et al. 2007b. Growth biochemical modifications and proline metabolism in *Helianthus annuus* L. as induced by drought stress. *Colloids and Surfaces B: Biointerfaces,* **59(2):** 141–149.

Maroco JP, Rodrigues ML, Lopes C & Chaves MM. 2002. Limitations to leaf photosynthesis in field-grown grapevine under drought—Metabolic and modelling approaches. *Functional Plant Biology,* **29(4):** 451–459.

Massacci A, Nabiev SM, Pietrosanti L, Nematov SK, Chernikova TN, Thor K, et al. 2008. Response of the photosynthetic apparatus of cotton (Gossypium hirsutum) to the onset of drought stress under field conditions studied by gas-exchange analysis and chlorophyll fluorescence imaging. *Plant Physiology and Biochemistry,* **46(2):** 189–195.

Mirzaei M, Soltani N, Sarhadi E, George IS, Neilson KA, Pascovici D, et al. 2014. Manipulating root water surely elicits major shifts in the shoot proteome. *Journal of Proteome Research,* **13(2):** 517–526.

Miyazawa SI, Yoshimura S, Shinzaki Y, Maeshima M & Miyake C 2008. Deactivation of aquaporins decreases internal conductance to CO_2 diffusion in tobacco leaves grown under long-term drought. *Functional Plant Biology,* 553–564.

Murchie EH & Lawson T. 2013. Chlorophyll fluorescence analysis: S guide to good practice and understanding some new applications. *Journal of Experimental Botany,* **64(13):** 3983–3998.

Nikolaou G, Neocleous D, Christou A, Kitta E & Katsoulas N. 2020. Implementing sustainable irrigation in water-scarce regions under the impact of climate change. *Agronomy,* **10(8):** 1120.

Nishiyama R, Watanabe Y, Fujita Y, Le DT, Kojima M, Werner T, et al. 2011. Analysis of cytokinin mutants and regulation of cytokinin metabolic genes reveals important regulatory roles of cytokinins in drought salt and abscisic acid responses and abscisic acid biosynthesis. *The Plant Cell,* **23(6):** 2169–2183.

Nonami H. 1998. Plant water relations and control of cell elongation at low water potentials. *Journal of Plant Research,* **111:** 373–382.

Osakabe Y, Kajita S & Osakabe K. 2011. Genetic engineering of woody plants: Current and future targets in a stressful environment. *Physiologia Plantarum,* **142(2):** 105–117.

Osakabe Y, Osakabe K, Shinozaki K & Tran LSP. 2014. Response of plants to water stress. *Frontiers in Plant Science,* **5:** 86.

Parkash V & Singh S. 2020. A review on potential plant-based water stress indicators for vegetable crops. *Sustainability,* **12(10):** 3945.

Parry MA, Andralojc PJ, Khan S, Lea PJ & Keys AJ. 2002. RuBisCO activity: Effects of drought stress. *Annals of Botany,* **89(7):** 833–839.

Parry MAJ, Delgado E, Vadell J, Keys AJ, Lawlor DW & Medrano H. 1993. Water stress and the diurnal activity of ribulose-1,5-bisphosphate carboxylase in field grown Nicotiana tabacum genotypes selected for survival at low CO_2 concentrations. *Plant Physiology and Biochemistry,* **31:** 113–113.

Pic E, de La Serve BT, Tardieu F & Turc O. 2002. Leaf senescence induced by mild water deficit follows the same sequence of macroscopic biochemical and molecular events as monocarpic senescence in pea. *Plant Physiology,* **128(1):** 236–246.

Portis Jr AR. 1995. The regulation of RuBisCO by RuBisCO activase. *Journal of Experimental Botany,* **46:** 1285–1291.

Quick WP, Schurr U, Scheibe R, Schulze ED, Rodermel SR, Bogorad L, et al. 1991. Decreased ribulose-1,5-bisphosphate carboxylase-oxygenase in transgenic tobacco transformed with "antisense" rbcS: I. Impact on photosynthesis in ambient growth conditions. *Planta,* **183:** 542–554.

Reddy AR, Chaitanya KV & Vivekanandan M. 2004. Drought-induced responses of photosynthesis and antioxidant metabolism in higher plants. *Journal of Plant Physiology,* **161(11):** 1189–1202.

Rossel JB, Walter PB, Hendrickson L, Chow WS, Poole A, Mullineaux PM, et al. 2006. A mutation affecting ascorbate peroxidase 2 gene expression reveals a link between responses to high light and drought tolerance. *Plant Cell & Environment,* **29(2):** 269–281.

Roy H & Andrews TJ. 2000. RuBisCO: Assembly and mechanism. *In:* Photosynthesis: Physiology and Metabolism, pp. 53–83.

Rucker KS, Kvien CK, Holbrook CC & Hook JE. 1995. Identification of peanut genotypes with improved drought avoidance traits. *Peanut Science,* **22(1):** 14–18.

Ruuska S, Andrews TJ, Badger MR, Hudson GS, Laisk A, Price GD, et al. 1998. The interplay between limiting processes in C3 photosynthesis studied by rapid-response gas exchange using transgenic tobacco impaired in photosynthesis. *Functional Plant Biology,* **25(8):** 859–870.

Saibo NJ, Lourenço T & Oliveira MM. 2009. Transcription factors and regulation of photosynthetic and related metabolism under environmental stresses. *Annals of Botany,* **103(4):** 609–623.

Salekdeh GH, Siopongco J, Wade LJ, Ghareyazie B & Bennett J. 2002. Proteomic analysis of rice leaves during drought stress and recovery. *Proteomics: International Edition,* **2(9):** 1131–1145.

Sarlikioti V, Driever SM & Marcelis LFM. 2010. Photochemical reflectance index as a mean of monitoring early water stress. *Annals of Alied Biology,* **157(1):** 81–89.

Servaites JC & Geiger DR. 1995. Regulation of ribulose 1,5-bisphosphate carboxylase/oxygenase by metabolites. *Journal of Experimental Botany,* **46:** 1277–1283.

Sharp RE & Le Noble ME. 2002. ABA ethylene and the control of shoot and root growth under water stress. *Journal of Experimental Botany,* **53(366):** 33–37.

Shavrukov Y, Kurishbayev A, Jatayev S, Shvidchenko V, Zotova L, Koekemoer F, et al. 2017. Early flowering as a drought escape mechanism in plants: How can it aid wheat production? *Frontiers in Plant Science,* **8:** 1950.

Silvera K, Santiago LS & Winter K. 2005. Distribution of crassulacean acid metabolism in orchids of Panama: Evidence of selection for weak and strong modes. *Functional Plant Biology,* **32(5):** 397–407.

Singh DK & Sale PW. 1998. Phosphorus suly and the growth of frequently defoliated white clover (*Trifolium repens* L.) in dry soil. *Plant and Soil,* **205:** 155–162.

Skirycz A & Inzé D. 2010. More from less: Plant growth under limited water. *Current Opinion in Biotechnology,* **21(2):** 197–203.

Specht JE, Chase K, Macrander M, Graef GL, Chung J, Markwell JP, et al. 2001. Soybean response to water: A QTL analysis of drought tolerance. *Crop Science,* **41(2):** 493–509.

Subrahmanyam D, Subash N, Haris A & Sikka AK. 2006. Influence of water stress on leaf photosynthetic characteristics in wheat cultivars differing in their susceptibility to drought. *Photosynthetica,* **44:** 125–129.

Tahir MHN, Imran M & Hussain MK. 2002. Evaluation of sunflower (*Helianthus annuus* L.) inbred lines for drought tolerance. *International Journal of Agriculture and Biology,* **3:** 398–400.

Tahkokorpi M, Taulavuori K, Laine K & Taulavuori E. 2007. After-effects of drought-related winter stress in previous and current year stems of *Vaccinium myrtillus* L. *Environmental and Experimental Botany,* **61(1):** 85–93.

Takahashi N, Yamazaki Y, Kobayashi A, Higashitani A & Takahashi H. 2003. Hydrotropism interacts with gravitropism by degrading amyloplasts in seedling roots of *Arabidopsis* and radish. *Plant Physiology,* **132(2):** 805–810.

Tezara W & Lawlor DW. 1995. Effects of water stress on the biochemistry and physiology of photosynthesis in sunflower. *Photosynthesis: From Light to Biosphere,* **4:** 625–628.

Tezara W, Mitchell VJ, Driscoll SD & Lawlor DW. 1999. Water stress inhibits plant photosynthesis by decreasing coupling factor and ATP. *Nature,* **401(6756):** 914–917.

Thornley JH & Johnson IR. 1990. Plant and Crop Modelling. Oxford: Clarendon 669.

Vu JC, Gesch RW, Allen Jr LH, Boote KH & Bowes G. 1999. CO_2 enrichment delays a rapid, drought-induced decrease in Rubisco small subunit transcript abundance. *Journal of Plant Physiology,* **155(1):** 139–142.

Wilson PB, Estavillo GM, Field KJ, Pornsiriwong W, Carroll AJ, Howell KA, et al. 2009. The nucleotidase/phosphatase SAL1 is a negative regulator of drought tolerance in *Arabidopsis*. *The Plant Journal,* **58(2):** 299–317.

Winter K & Holtum JA. 2005. The effects of salinity crassulacean acid metabolism and plant age on the carbon isotope composition of *Mesembryanthemum crystallinum* L. a halophytic C 3-CAM species. *Planta,* 222: 201–209.

Winter K & Holtum JA. 2007. Environment or development? Lifetime net CO_2 exchange and control of the expression of crassulacean acid metabolism in *Mesembryanthemum crystallinum*. *Plant Physiology,* **143(1):** 98–107.

Winter K, Garcia M & Holtum JA. 2011. Drought-stress-induced up-regulation of CAM in seedlings of a tropical cactus Opuntia elatior operating predominantly in the C3 mode. *Journal of Experimental Botany,* **62(11):** 4037–4042.

Wu QS, Xia RX & Zou YN. 2008. Improved soil structure and citrus growth after inoculation with three arbuscular mycorrhizal fungi under drought stress. *European Journal of Soil Biology*, **44(1):** 122–128.

Wullschleger SD, Yin TM, DiFazio SP, Tschaplinski TJ, Gunter LE, Davis MF, et al. 2005. Phenotypic variation in growth and biomass distribution for two advanced-generation pedigrees of hybrid poplar. *Canadian Journal of Forest Research,* **35(8):** 1779–1789.

Xu Z, Zhou G & Shimizu H. 2009. Are plant growth and photosynthesis limited by pre-drought following rewatering in grass? *Journal of Experimental Botany,* **60(13):** 3737–3749.

Xu Z, Zhou G & Shimizu H. 2010. Plant responses to drought and rewatering. *Plant Signaling & Behavior,* **5(6):** 649–654.

Yang X, Lu M, Wang Y, Wang Y, Liu Z & Chen S. 2021. Response mechanism of plants to drought stress. *Horticulturae,* **7(3):** 50.

5 | Oxidative Responses, Antioxidative System, and Redox Regulation During Water Stress

Dwaipayan Sinha[1*], Arpan Dey[2], Arun Kumar Maurya[3], Gholamreza Abdi[4], Supatra Sen[5], Satarupa Nath[6], Sherif Babatunde ADEYEMI[7], Zoofishan Kazi[8], Rimi Roy[9], Mohammad Sohail[10], Swastika Banerjee[11], Aqsa Majgaonkar[12] and Sahana Chowdhury[13]

[1] Department of Botany, Government General Degree College, Mohanpur, Paschim Medinipur, India
[2] Department of Bioscience & Bioengineering, Indian Institute of Technology Jodhpur, Rajasthan, India
[3] Department of Botany, Multanimal Modi College, Modinagar, Ghaziabad - 201204, Uttar Pradesh, India
[4] Department of Biotechnology, Persian Gulf Research Institute, Persian Gulf University, Bushehr, 75169, Iran
[5] Department of Botany, Ashutosh College, Kolkata, West Bengal, India
[6] Department of Botany, Rishi Bankim Chandra College for Women, Naihati, West Bengal, India
[7] Ethnobotany/Phytomedicine Laboratory, Department of Plant Biology, Faculty of Life Sciences, University of Ilorin, Ilorin PMB 1515, Kwara State, Nigeria
[8] Department of Botany, St. Xavier's College (Autonomous), Mumbai - 400001 Maharashtra, India
[9] Department of Botany, Jagannath Kishore College, Purulia - 723101, India
[10] Abul Wali Khan University Mardan, Khyber-Pakhtunkhwa - 23200, Pakistan
[11] Department of Botany, Sanjivani Arts, Commerce and Science College, Kopargaon, Ahmednagar, Maharashtra, India
[12] Department of Botany, St. Xavier's College (Autonomous) Mumbai - 400001 Maharashtra, India
[13] German University Bangladesh, T&T Road, Chandona Chourasta, Gazipur, Bangladesh

The profound interdependence of water stress, oxidative stress, and the complex biochemical responses elucidated in this study lays a foundation for future research endeavors to develop resilient crops adaptable to changing climates. As

*Corresponding author: dwaipayansinha@hotmail.com

a pivotal determinant of weather conditions and seasonal patterns, water plays a crucial role in sustaining biological functions and ecological stability. When water becomes scarce due to environmental changes or anthropogenic activities, it leads to water stress, negatively affecting normal plant growth and development. Plants have evolved a spectrum of morphological, anatomical, physiological, and biochemical mechanisms to counter water stress. The consequences of water stress encompass reduced photosynthesis, oxidative stress, metabolic abnormalities, stomatal closure, and disruptions in the electron transport system. Plants deploy enzymatic and non-enzymatic antioxidant defense mechanisms to mitigate stress-induced oxidative damage. Innovative approaches have shown promise in reducing damage, such as rhizobacterial inoculation, mycorrhizal fungi, and nitric oxide treatment. Genetic engineering, focusing on enhancing key enzymes, strengthens plant resilience. Exogenous application of compounds like glutathione, proline, and polyamines effectively enhances agricultural yields by bolstering the plant's ability to cope with water stress. The role of miRNAs in different plant species is highlighted, showcasing their diverse strategies for surviving drought conditions. Utilizing CRISPR/Cas9 technology to control these miRNAs is a promising avenue to improve agricultural output and enhance crop stress resistance. Omics-based research, exploring the transcriptome and genomic processes has identified genes associated with plant resilience and production. This chapter serves as a comprehensive overview of water stress-induced oxidative stress in plants, delving into the deep analysis of mitigation strategies employed to enhance plant resilience.

1. Introduction

Human endeavors and well-being depend on water, while the aquatic environment significantly influences ecological equilibrium and the global climate (Luo et al., 2022). Water, essential for sustaining all life, is also pivotal for maintaining healthy environments (Parkes & Horwitz, 2009). However, excessive utilization of water resources places strain on freshwater ecosystems, resulting in a decline in biodiversity and a reduction in ecosystem services (Vanham et al., 2021). The intricate connection between climate change and water is evident. Extreme weather events are causing water to become scarcer, less reliable, or contaminated. The consequences of these effects throughout the water cycle pose risks to sustainable development, biodiversity, and people's access to water and sanitation. Climate change exacerbates water stress, particularly in regions with meager water supplies, intensifying competition for water and potentially triggering conflicts (UNICEF, 2024).

Water stress may arise from either an excess or a deficiency of water. Water scarcity is the prevalent form of water stress, commonly called drought stress (Costa de Oliveira et al., 2014). Water stress adversely impacts various aspects of a plant's physiology, including its ability to photosynthesize (Wang et al., 2018), closure of stomata (Abdalla et al., 2022), and decreased carbon dioxide (CO_2)

availability (Flexas et al., 2006). Extended water stress significantly diminishes plant growth and productivity(Abrar et al., 2022). Furthermore, a lack of water under drought stress increases electron leakage towards O_2 during photosynthetic and respiratory activities, which increases ROS generation and oxidative stress in the plant cell (Murshed et al., 2013). The Mehler reaction models oxygen (O_2) photoreduction at PSI in plants. With the aid of superoxide dismutase (SOD), this photoreduction generates superoxide ions ($O_2^{-\bullet}$) that are disproportionate to H_2O_2 and O_2. The ascorbate peroxidase (APX) pathway rapidly converts H_2O_2 to water for detoxification. The term "water-water cycle" originated from the electrons that moved from water in PSII to water in PS I during this process (Makino et al., 2002). During photosynthesis and drought stress, the Mehler reaction increases the amount of electron leakage to O_2 (Cruz de Carvalho, 2008). The report states that wheat in drought conditions released 50% more photosynthetic electrons into the Mehler reaction than wheat under normal conditions (Biehler & Fock, 1996). Additionally, during drought stress, there was an increase in thylakoid membrane electron leakage to O_2 (Sgherri et al., 1996). Unfavorable climatic circumstances have been shown to reduce CO_2 assimilation, interfere with stomatal conductance, and perhaps cause excited triplet chlorophyll (3Chl*) to be produced (Shakirova et al., 2016). Reactive oxygen species (ROS) are produced in excess as a result of this disturbance in the photosynthetic electron transport chain (ETC), which also promotes photo-oxidation. Excess high-energy intermediates, reductants, and O_2 are the main causes of ROS generation in PSI and PSII reaction centers (Asada, 2006; Tripathy & Oelmüller, 2012; Das & Roychoudhury, 2014). During illumination, light-harvesting complexes (LHC) absorb photon energy, leading to the creation of an excited singlet chlorophyll (1Chl*). This long-lived molecule actively engages in the charge separation process, converting excitation energy into electrochemical energy. Chlorophyll triplets are created when the PSs become saturated with absorbed energy, which occurs when the amount of energy stimulated by the chlorophylls implanted in the LHC exceeds the energy that can be used by the photochemical reaction centers. When O_2 and these chlorophyll triplets interact, damaging ROS are created (Pospíšil, 2016; Lingvay et al., 2020; Levin & Schuster, 2023). When 3Chl* and O_2 interact, $1O_2$ is produced, which is very oxidizing. Besides excess light, factors such as dryness contribute to an imbalance between light absorption and utilization, resulting in the formation of singlet oxygen ($1O_2$) (Mattos & Moretti, 2016).

Because of lowered stomatal conductance brought on by abiotic stressors, the amount of CO_2 supplied to Calvin's cycle decreases. Inside the photosynthetic PSII Electron Transport Chain (ETC), this reduction results in an over-reduction of plastoquinone QA and QB. It therefore prevents P680 (the chlorophyll molecules in PSII) and pheophytin from separating their charges. Triplet chlorophyll (3P680) is formed in the PSII reaction center as a result of this series of reactions, which produces $^1O_2$ (Selmar & Kleinwächter, 2013; Sachdev et al., 2021). In situations of abiotic stress, the decreased availability of CO_2, which serves as the final electron acceptor, contributes to a reduced presence of $NADP^+$. This scarcity prompts an excess leakage of electrons from the photosynthetic electron transport system, directing them towards O_2 at the acceptor side of PSI via ferredoxin, resulting in the formation of $O_2^{-\bullet}$ (Sachdev et al., 2021).

Several plants have been shown to activate their antioxidant defense as an adaptation strategy against drought stress (Hasanuzzaman et al., 2020). The antioxidant defense system prevents protein damage, reduces lipid peroxidation in membranes, and effectively manages DNA and nucleic acid damage under stress. It does this by delaying oxidation. Thus, plants' capacities to withstand stress are improved by offering complete cellular protection (Fujita & Hasanuzzaman, 2022). Figure 1 describes the overall impact of water stress on a plant and its defensive strategies. This chapter aims to highlight the relationship that exists between oxidative stress and the antioxidant process in plants, with a focus on water stress. There has been an attempt to shed light on the complex connection between oxidative stress and water stress in plants.

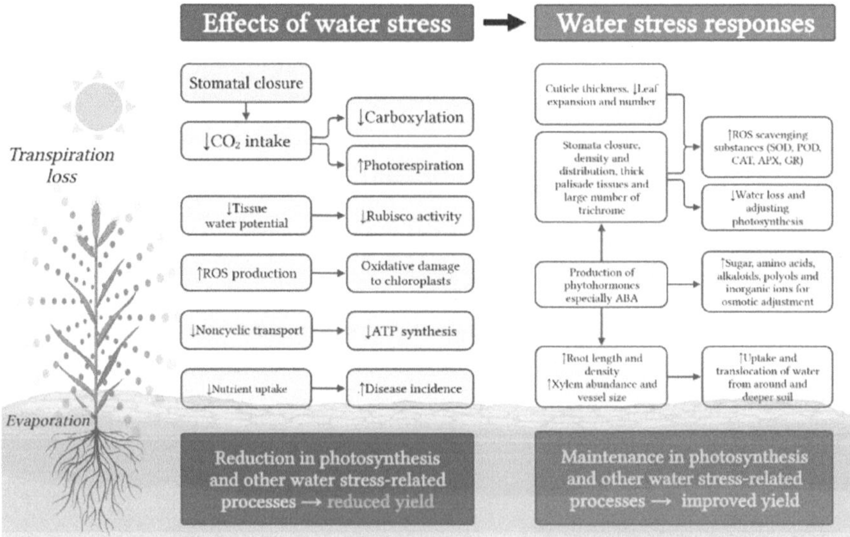

Figure 1: Impact of water stress on plants and its responses.

2. Water Stress-induced Oxidative Stress in Plants

Reactive oxygen species are forms of atmospheric O_2 that have been partially reduced. They usually happen when O_2 is excited to make singlet oxygen ($^1O_2$) or when one, two, or three electrons are transferred to O_2 to produce hydrogen peroxide (H_2O_2), $O_2^{-\bullet}$, or hydroxyl radicals (OH^-) (Scandalios, 2001). Unlike ambient O_2, ROS can oxidize diverse cellular components without restriction, which might result in the cell's oxidative destruction (Mittler et al., 2022). Plants release increased ROS as a result of environmental stressors that disturb cellular homeostasis, including salt, drought, cold, UV-B radiation, metal toxicity, and pathogen attacks (Verma & Dubey, 2003; Sharma & Dubey, 2007; Han et al., 2009). When a cell's ROS level surpasses its defensive mechanisms, it is referred to as "oxidative stress" (Sharma et al., 2012). Water stress stimulates the generation of ROS, which are byproducts of aerobic metabolism, by interfering with the electron transport system and oxidizing metabolic processes in chloroplasts, mitochondria, and microbodies (Asada, 1999; Van Breusegem et al., 2001; Ahmad et al., 2010). Both enzymatic and non-enzymatic antioxidants, in the optimum condition, control the ratio of ROS generation to consumption. Oxidative stress is caused by drought-related increases in ROS generation beyond the elimination capability of antioxidative mechanisms (Noctor & Foyer, 1998). A plant's normal physiological and biochemical processes may be disrupted by water stress, which may result in an overproduction of ROS. Because of their high reactivity, ROS have the potential to harm a variety of biological components, including membranes, protein lipids, and nucleic acids. Water stress induces oxidative stress in the plant cell, which raises the formation of ROS because of increased electron leakage towards O_2 during photosynthetic and respiratory activities (Sánchez-Rodríguez et al., 2012). When there is water stress, several processes lead to ROS production. These processes include stomata closing (Kar, 2011), reduced photosynthesis, and NADPH oxidase activation (Cruz de Carvalho, 2008; Farooq et al., 2008). The guard cells' plasma membranes include ion and water channel proteins that control the turgor pressure, which in turn controls stomatal movements (Schroeder et al., 2001; Kim et al., 2010; Engineer et al., 2016).

These movements are modulated by a combination of internal and external stimuli, including levels of ROS, calcium (Ca^{2+}), and abscisic acid (ABA), as well as environmental factors including moisture, light, and CO_2. Water stress adversely affects the gas exchange characteristics, premature leaf senescence, photosynthetic machinery, leaf expansion, oxidation of chloroplast lipids, and changes in the structure of proteins and pigments (Menconi et al., 1995). A study found that when drought conditions were applied to maize, the crop's net photosynthesis decreased by 33.22%, transpiration rate decreased by 37.84%, stomatal conductance decreased by 25.54%, water use efficiency decreased by 50.87%, intrinsic water use efficiency decreased by 11.58%, and intercellular CO_2 decreased by 5.86% (Anjum et al., 2011).

3. Water Stress-induced Nitrosative Stress in Plants

Under water stress, plants may encounter nitrosative stress when they struggle to regulate the production of reactive nitrogen species (RNS) or effectively eliminate them through detoxification or scavenging processes (Valderrama et al., 2007). Plants face water stress when they either lack sufficient water intake or experience high rates of water loss due to conditions like drought, elevated salinity, or high temperatures.

Nitric oxide (NO), a free radical, has become a vital signaling molecule in many critical physiological processes of higher plants within the past decade (Neill et al., 2008). Reactive nitrogen species (RNS), which include NO and its derivatives like N_2O_3, NO_2, $ONOO^-$, S-nitrosothiols (RSNOs), and S-nitroso glutathione (GSNO), are important components in plant biology (Molassiotis & Fotopoulos, 2011). Increased NO and S-nitrosothiol levels cause nitrosative stress, which is a factor in pathophysiological situations. It is mainly caused by the S-nitrosylation of important protein cysteine (Cys) thiols and metal cofactors (Foster et al., 2009). Under the circumstances of water stress, reactive nitrogen species are produced by several methods. These processes include interactions with ROS, nitrate reductase activity, and NO synthase enzyme-like activity. As with ROS, NO has been demonstrated to be important for many different developmental processes, including floral transition, senescence, xylogenesis, leaf expansion, germination and dormancy of seeds, root growth, root gravitropism, stomatal closure, and programmed cell death (Wendehenne et al., 2004; Besson-Bard et al., 2008; Neill et al., 2008).

4. Water Stress and the Role of Nitric Oxide

In many different circumstances, plants experience drought stress, which can be made worse by intense sunlight and warm temperatures. These conditions hinder photosynthesis, which causes oxidative stress and the production of ROS, both of which lower agricultural plant yields (Ederli et al., 2019). Reactive nitrogen species (RNS), also known as NO, is an essential signaling molecule which is involved in many important biological processes, such as stomatal closure (Sun et al., 2019; Hancock, 2020). Numerous studies indicate that NO may play a part in how the ecosystem reacts to abiotic stressors such as ozone, salt, heat shock, drought, and UV radiation (Cassia et al., 2019; Iqbal et al., 2021; Shang et al., 2022). Plants can biosynthesize NO through both oxidative and reductive processes. A polyamine-mediated mechanism, a hydroxylamine pathway, and an enzyme similar to NOS are included in the former, whereas NR and xanthine oxidoreductase (XOR) are included in the latter. L-arginine can be the substrate for an enzyme that resembles NOS, which then produces NO and L-citrulline. Numerous cofactors, including BH4, CaM (calmodulin), flavin adenine dinucleotide, flavin mononucleotide, Ca^{2+}, and O_2, are required for this activity (Gupta et al., 2022). The most extensively researched method of NO production in plants is the reduction of nitrite, either enzymatically or non-enzymatically. According to León and Costa-Broseta (2020), nitrate reductase is an enzyme that catalyzes the conversion of nitrites into NO (León & Costa-Broseta, 2020). Furthermore, the primary enzyme systems responsible for this reductive NO synthesis in plants include the mitochondrial electron transport chain and the unique complex between NR and NOFNiR, which was discovered in the *Chlamydomonas reinhardtii* (Astier et al., 2018). Drought stress has been shown to reduce photosynthesis and increase ROS formation, leading to photooxidative stress. This stress is frequently exacerbated by high air temperatures and light irradiation (Wang et al., 2018; Villadangos et al., 2023). When drought severely hinders photosynthesis, secondary metabolites may be able to boost the functional functions of antioxidant enzymes. To avoid permanent oxidative damage caused by various environmental stimuli, plants organize certain components of their antioxidant machinery, which is further enhanced by the formation of isoprenoid and phenylpropanoid (Akula & Ravishankar, 2011; Tattini et al., 2015; Talbi et al., 2020).

By activating phenylalanine ammonia-lyase (PAL), NO can thereby stimulate the production of secondary metabolites under various abiotic stresses (Hao et al., 2009; Kováčik et al., 2009). An early NO burst in the roots was seen in water-stressed rice seedlings treated with NH_4^+ ions, but not in seedlings treated with NO_3^-, according to a rice study. For treatments with NO_3, as opposed to NH_4^+, the generation of ROS in the former, led to significantly more severe root oxidative damage induced by water stress. Inducing NO production three hours after NO_3^- treatment reduced oxidative damage. However, blocking the early NO burst with the NO scavenger 2-(4-carboxyphenyl)-4,4,5,5-tetramethylimidazoline-1-oxyl-3-oxide (c-PTIO) exacerbated root oxidative damage in NH_4^+ treatment. The NOS inhibitor N(G)-nitro-L-arginine methyl ester (L-NAME) inhibited NO production in roots three hours after NH_4^+ administration and intensified oxidative damage caused by water stress (Cao et al., 2019). In another sugarcane experiment, it was shown that when plants were given 100% NO_3^- during a period of drought, they exhibited increased root NO_3^- and NO_2^-. Additionally, the same plants showed enhanced root NO production and root NR function. Additionally, it was shown that plants receiving higher NO_3^- had higher rates of photosynthetic activity and stomatal conductance, which were linked to greater development of roots. Increased catalase (CAT) activity in the leaves and APX and SOD activity in the roots of plants treated with 100% NO_3 under water deficiency circumstances decreased the formation of ROS. When the plants were given a NO scavenger, these positive responses to water constraint were eliminated, suggesting that NO increased leaf gas exchange and plant growth (Pissolato et al., 2020). Water-challenged plants treated with 1 M SNP (as a NO donor) were shown to have a greater relative water content than plants treated with 0 M SNP in another sunflower experiment. Drought did not impact SOD activity, but it did increase proline and malondialdehyde (MDA) levels and decrease pyrogallol peroxidase (PG-POD) activity. Proline content and PG-POD activity rose while MDA levels decreased in plants under water stress treated with 10 M SNP (Cechin et al., 2015). In apricots, water stress increased leaf temperature, proline, MDA content, and membrane permeability, but decreased leaf area, stomatal conductivity, chlorophyll content, relative water content, dry and fresh weight, and nutritional components. Under water stress conditions, silicon (Si), NO, and significantly fewer negative effects of water stress were seen on leaf temperature, stomata conductivity, dry and fresh weight, membrane permeability, leaf area, L-proline, relative water content, and MDA content when Si + NO combination treatments were applied. Shoot dry weight, chlorophyll content, stomata conductivity, and leaf relative water content rose by 59%, 55%, 12%, and 8%, respectively, in Si + NO treated apricot saplings (Bakır et al., 2022). In one research, the effects of NO and salicylic acid (SA) on wheat plants under osmotically stressed conditions caused by polyethylene glycol were examined. Osmotic stress was found to increase MDA, lipoxygenase (LOX) activity, and H_2O_2 generation while decreasing chlorophyll levels, shoot fresh weight (SFW), and membrane stability index (MSI). Additionally, enzymatic antioxidant activities such as glutathione reductase, guaiacol peroxidase, and SOD were elevated in response to osmotic stress. Conversely, SA or SNP pretreatment increased antioxidant activity, which led to increased Chl, and MSI and lowered LOX activity and MDA level, therefore mitigating the negative effects of osmotic stress. However, the protective benefits of SA and SNP were reversed or lessened in plants with MB after pretreatment, suggesting that the protective effects were most likely related to NO signaling (Naser Alavi et al., 2014).

Application of NO and ABA partially closed stomata in water-stressed grapevines, increasing water potential and restricting vine development. Additionally, NO and ABA reduced the accumulation of amino acids, which was linked to decreased protein degradation, and increased the activity of guaiacol peroxidase as well as the amount of specific sugars and anthocyanins in leaves. Water stress impacts in grapevine leaves may be overcome by NO and ABA through different pathways, as demonstrated by their differential responses to NO in terms of boosting tricarboxylic acid cycle intermediates, terpenes that stabilize membranes, and activating APX activity (Pontin et al., 2021). Investigations were conducted on the impact of sodium nitroprusside (SNP) on osmotic stress in banana roots caused by polyethylene glycol (PEG). Both plant growth and the quantity of yellow leaves were reduced by SNP treatment. Comparing PEG and PEG+SNP against the control, it was discovered that PEG+SNP varied in how thirty and ninety proteins were expressed, respectively. Energy and carbohydrate metabolisms were linked to most of the protein discrepancies. A report states that banana roots treated with SNP showed lower activity levels of SOD and APX when compared to those treated with PEG (Amnan et al., 2021). When plants are subjected to water constraint, SNP treatment increases NO, proline, chloroplastidic pigments, osmotic potential, and allows for a constant level of gas exchange, chlorophyll fluorescence characteristics, and antioxidant enzyme activities. When SNP was applied to plants experiencing water stress, the levels of ROS and MDA were considerably lowered. As a result, even under stressful circumstances, SNP preserved physiological and biochemical responses at reference levels, delaying the start of water shortage symptoms in Crambe plants (Batista et al., 2018). A different research examined how ABA and NO protected Brassica genotypes from drought stress caused by PEG. The response of enzymatic and non-enzymatic antioxidants to reducing oxidative stress in the leaves and roots of PEG-stressed cultivars varied depending on whether they were fed NO or/and ABA. Increasing both enzymatic and non-enzymatic antioxidant activities allowed Varina to develop tolerance, while Pusa Jagannath only showed antioxidant protection through the accumulation of non-enzymatic antioxidants (Sahay et al., 2019).

5. Effect of Water Stress-induced Oxidative Stress in Plants

5.1 Drought Stress and Membrane Functions

Drought stress negatively affects several physiological systems, and plants react to adjust to adverse circumstances. Biological membranes are the initial target of many abiotic stresses. It is well acknowledged that maintaining membrane integrity and stability under water stress plays a critical role in a plant's capacity to resist drought (Dastborhan & Ghassemi-Golezani, 2015). The membrane stability index (MSI) is a physiological metric that has also been widely used to evaluate heat and drought tolerance. It has been reported that the membrane stability index measures the membrane's resistance to drought stress and determines the extent of damage (Sakya et al., 2018). Drought and heat stress have been shown to increase electrolyte leakage, suggesting that lipid-protein structural alterations and cellular function loss are the causes of membrane instability (Alhaithloul, 2019). In particular, drought stress damages the cell membrane and causes a loss of membrane integrity. Conversely, mild water stress-induced osmotic stress may alter the physical characteristics of membrane lipids that cells can sense through sensory proteins placed on the PM, including mechanosensitive ion channels or receptor kinases (Couchoud et al., 2019). Drought significantly decreased the biomass of roots and shoots in an experiment on *Brassica juncea*; however, rhizobacterial inoculation increased plant biomass compared to uninoculated plants. Additionally, the total chlorophyll concentration, photosynthetic characteristics, membrane stability index, and relative water content were all improved by inoculation. Similarly, inoculation boosted the activity of antioxidative enzymes including APX, CAT, and SOD during the vegetative and reproductive phases of development, perhaps increasing stress tolerance by preserving the equilibrium of ROS. Drought significantly reduced proline, H_2O_2, and total soluble sugar accumulation in plants treated with rhizobacteria (Asha et al., 2021). According to different research, proline concentration rose during drought stress, but RWC, MSI, protein content, leaf area, plant height, and yield characteristics all decreased (Bangar et al., 2019). According to a study on chickpeas, the effects of drought stress were found to be negative in terms of relative water content (RWC) and membrane stability index (MSI). However, under normal conditions, treatment with biochar and arbuscular mycorrhizal fungus, either separately or in combination, significantly increased RWC and MSI (Hashem et al., 2019).

5.2 Water Stress and Oxidation of Membranes

ROS production in cells is a sign of oxidative stress brought on by water limitation. Due to increased electron leakage towards O_2 during photosynthesis, plant cells undergo oxidative stress during droughts (Cruz de Carvalho, 2008).

On the PSII electron acceptor side, electrons escape to molecular O_2, leading to the formation of superoxide ($O_2^{-\bullet}$). This superoxide subsequently dismutates into hydrogen peroxide (H_2O_2), which is then reduced to hydroxyl radicals by non-heme iron. On the PSII electron donor side, incomplete water oxidation generates H_2O_2, which is converted by manganese into hydroxyl radicals. At elevated temperatures, lipoxygenase-induced lipid peroxidation results in $^1O_2$; nevertheless, partial water oxidation generates H_2O_2, which manganese subsequently converts to hydroxyl radical (Pospíšil, 2016). The oxidation of membrane lipids is caused by all of these ROS. The primary byproducts of lipid peroxidation are lipid hydroperoxides (LOOH). Reactive oxygen species, which induce oxidative stress, come from them in a variety of forms. The main byproducts of lipid peroxidation are called lipid hydroperoxides (LOOH). They produce a range of ROS that lead to oxidative damage (Nam, 2011). Furthermore, several reactive aldehydes are produced when membrane lipids undergo peroxidation. As aldehydes produced by lipid peroxidation, malonaldidehyde (MDA) and 4-hydroxynonenal (HNE) have attracted a lot of attention. Because of its significant synthesis during lipid peroxidation, MDA is especially notable and is frequently used as a marker for oxidative stress evaluation. Conversely, HNE is well known for having remarkable biological potency (Barrera et al., 2018). Many investigations have been conducted to determine the impact of oxidative stress caused by drought on plants. Lipid peroxidation was one of the measures of oxidative stress in a few investigations on drought-induced oxidative stress, as shown in Table 1.

5.3 Water Stress and Alteration of Proteins

Plants that experience water stress go through a series of cellular processes that lead to an excess of ROS being produced (Osakabe et al., 2014). Oxidative damage is caused by ROS reacting with proteins (Schieber & Chandel, 2014). Reactive oxygen species, often known as the protein oxidation process, directly encourage the covalent alteration of a protein entity. Oxidation may occur in both the side chains of amino acids and the backbones of proteins, resulting in the fragmentation of proteins and the creation of cross-links between them. Oxidative alterations can alter a protein's shape, structure, solubility, susceptibility to proteolysis, and enzyme activity, among other physical and chemical characteristics (Zhang et al., 2013). As we have discussed earlier, oxidative stress can induce lipid peroxidation, leading to disruption of the integrity of the lipid bilayer, and compromising the fluidity and permeability of the membrane (Su et al., 2019). Membrane proteins, which are embedded within the lipid bilayer, can be directly affected by these changes, leading to altered protein function, stability, and conformation.

The oxidized secondary product of lipid peroxidation, 4-HNE adds aldehyde substances with –SH groups of cysteine or amino group of lysine or imidazole group of histidine, causing protein damage. This oxidative modification can result

Table 1: Selected studies on the effect of drought-induced lipid peroxidation on plants

Sr. No	Name of the plant	Type of experiment	Mode of stress applied	Effect on membrane lipids	Reference
1.	*Lycium ruthenicum*	Pot experiment	Not watered until the end of the experiment	Significant increase in MDA content of roots and leaves as compared to control	Guo et al., 2018
2.	*Maclura pomifera*	Randomized block design/Pot experiment	Pots watered to 75%, 50%, and 30% of field capacity.	13 nmol g^{-1} DW of MDA were observed in plants grown at 30% of field capacity for 22 days	Khaleghi et al., 2019
3.	*Oryza sativa*	Seedlings were grown in Conviron plant growth chamber	Treatment with PEG to mimic drought stress	50% increase in MDA content	Dalal & Tripathy, 2018
4.	*Triticum aestivum*	Field experiment	$NaClO_3$ in a water solution	Increase in MDA content in U1 and Njivka cultivars	Dvojković et al., 2023
5.	*Glycine max*	Pot experiment	Variation of irrigation regimes	The MDA levels of soybean plants subjected to water deficit for eight days exhibited a 45% increase in comparison to the MDA levels of plants exposed to water deficit for six, four, and two days	Schneider et al., 2020
6.	*Brassica napus*	Plot study	Variation of irrigation regimes	Increase in MDA content under severe water stress	Lotfi et al., 2015
7.	*Hordeum vulgare*	Field study	Water deficit by less irrigation	Significant increase in MDA content in water-stressed plants	Abdelaal et al., 2018
8.	*Avena sativa*	Sand culture study	Application of 20% PEG-6000 to mimic water stress	Significant increase in MDA content due to drought stress	Xie et al., 2021

in changes in protein structure and function. Oxidized membrane proteins may lose their enzymatic activity, receptor-ligand interactions, or transport functions, compromising vital cellular processes such as signaling, nutrient uptake, and ion homeostasis (Zhang et al., 2013). Another secondary product MDA promotes cross-linking between proteins, causing alteration of biochemical properties of protein (Pizzimenti et al., 2013; Zhu et al., 2022). This cross-linking can induce the development of protein conglomerates, impairing protein functionality and causing cellular dysfunction. Moreover, aggregated proteins can also trigger proteolytic degradation pathways, further exacerbating the detrimental effects on cellular membranes (Balasubramaniam et al., 2019).

6. Effect on Antioxidant Enzymes

6.1 Superoxide Dismutase

Plant SODs activity is essential for the response to drought stress because it shields cells from ROS-induced oxidative damage. Dismutation of $O_2^{-\bullet}$into H_2O_2 and O_2 is catalyzed by SODs. SOD activity shows complexity as it drops, or initially decreases, and subsequently increases, or stays the same as stress intensity increases. The above variations may be because the degree of water scarcity that plants perceive rather than the actual water shortage itself triggers their reaction to it (Yang et al., 2021). Numerous studies have investigated the effects of drought stress on SOD activity in plants. For instance, in maize seedlings, SOD activity increased in response to drought stress, indicating its participation in the antioxidant defense mechanism. By scavenging $O_2^{-\bullet}$, the increased SOD activity helps to lessen their harmful effects and the oxidative damage brought on by drought stress (Ru et al., 2023). In another study, it was observed that drought stress induced the upregulation of SOD gene expression and increased SOD activity in rice plants (Wang et al., 2019). A study with *Maclura pomifera* also showed similar responses to increased SOD activity under drought stress (Khaleghi et al., 2019).

One study assessed the increased tolerance to drought in rice plants that were transgenically altered to express a pea-specific form of manganese superoxide dismutase (MnSOD). The study examined the expression of the pea MnSOD gene in transgenic rice plants subjected to polyethylene glycol (PEG) 6000-induced drought stress. Even when subjected to methyl viologen (MV), the transgenic rice leaf slices showed less electrolyte leakage than leaf slices from the wild-type plants following PEG treatment. According to Wang et al. (2005), this suggests that the transgenic plants were more resilient to oxidative stress caused by MV or PEG (Wang et al., 2005). Another study examined the subcellular location and reactions of different isoforms of SOD in local wheat cultivars that were exposed to persistent soil drought. Mn-SOD was detected in the mitochondrial fractions, whereas Fe-SOD was identified in the chloroplast fraction. Copper/zinc superoxide dismutase (Cu/Zn-SOD) is distributed across all subcellular fractions.

Drought stress specifically increased the expression of Mn-SOD (Huseynova et al., 2014). The oxidative stress tolerance of different genotypes of maize that differed in drought sensitivity was also investigated. It was observed that the activity of SOD was significantly reduced in the sensitive genotypes while it remained unchanged in the genotypes that were tolerant to drought conditions (Chugh et al., 2011). In a separate study, drought and heat stress increased the activity of total SOD in maize as compared to the control (Hussain et al., 2019). In another study, it was found that the drought-tolerant species of *Amaranthus tricolour* was found to contain higher activities of SOD and CAT and a significant increase in ascorbate-glutathione content (Sarker & Oba, 2018). Another study aimed to assess the activity of SOD in *Haloxylon aphyllum* and *Haloxylon persicum*. The findings demonstrated a considerable ($p < 0.01$) increase in SOD activity in two *Haloxylon* species as a result of drought stress. Severe tensions caused a rise in enzyme activity during the drought. When two species of *Haloxylon* were subjected to mild water deprivation for 7 and 14 days, their enzyme activity was higher than that observed in the control treatment but lower than the activity shown in the severe water deprivation treatments of 21 and 28 days (Arabzadeh & Khavari-Ne, 2013). Drought stress also increases the SOD activity and proline content in the upper and bottom foliage of *Glycine max* (Tian et al., 2023). In another study, the activity of SOD in response to drought was investigated in transgenic Populus expressing pine cytosolic glutamine synthetase. The results of the investigation indicated that the occurrence of drought stress caused a decrease in the activity of the copper-containing superoxide dismutases (CSDs) in the plastids of GS poplar plants, while simultaneously increasing the activity of the iron-containing superoxide dismutases (FSDs) in the plastids, compared to the wild type (Molina-Rueda et al., 2013).

6.2 Peroxidase

Peroxidases (PODs), a group of isozymes found universally in plants, are monomeric glycoproteins containing heme, capable of using either H_2O_2 or O_2 to oxidize a wide range of molecules (Yoshida et al., 2003). These POD in plants play a crucial role in safeguarding cells against oxidative harm induced by ROS. Numerous studies have investigated POD activity under drought stress conditions in plants. In an experiment involving two *Brassica napus canola* cultivars subjected to water stress, both cultivars exhibited heightened total phenolic contents and increased POD activities, indicating their response to the stress (Akram et al., 2018).

Another study demonstrated increased POD activity in drought-tolerant genotypes of Cicer arietinum in comparison to the drought-susceptible ones. The enhanced POD activity helps to scavenge excess H_2O_2 and minimize oxidative damage induced by drought stress (Patel & Hemantaran, 2012). ERF194, the AP2/ERF transcription factor, is essential for poplar development and tolerance under drought stress. A study was done on ERF194-overexpressed (OX), suppressed-expression (RNA interference, RNAi), and non-transgenic (WT) poplar clone

717 to find out how the ERF194 transcription factor functions physiologically in poplar growth and in resistance to drought. The results showed that OX contained higher quantities of starch, NSC, soluble sugar, chlorophyll, and POD, as well as a higher water potential and higher levels of SOD, CAT, and POD activities. Based on these findings, it appears that ERF194 helps poplar become more drought-tolerant (Huan et al., 2023). In *Carex duriuscula*, mild drought stress increased the activities of glutathione peroxidase (GPX), POD, and CAT in the leaves, while only POD and CAT increased in the roots. On the other hand, the plant's roots and leaves displayed elevated levels of POD, CAT, SOD, and GPX amid severe drought stress (Hou et al., 2021). When *Dahlia pinnata* was stressed by drought, the activities of antioxidant enzymes, such as POD, first increased and then reduced (Fan et al., 2011).

Additionally, it has been shown that soybean plants treated with *Bradyrhizobium japonicum*, an arbuscular mycorrhizal fungus, showed improved CAT and POD activity under drought. Furthermore, it was found that the application of biofertilizer treatments during drought stress circumstances led to an increase in the relative expression of the CAT and POD genes (Sheteiwy et al., 2021). In C5 and Zafar 1 genotypes of *Vicia faba*, the activities of antioxidant enzymes namely CAT, POD, and SOD are better thereby enabling the cultivars to perform better and resist the deteriorating effects of growth performance under drought conditions (Siddiqui et al., 2015). In a separate experiment, it was found that application of silicon dioxide nanoparticles to Rosa damascene enhanced the tolerance of water deficit by increasing the effectiveness of antioxidant enzymes, including POD, CAT, guaiacol peroxidase (GPX), and SOD. It also reduced lipid peroxidation and H_2O_2 concentration (Hajizadeh et al., 2022). In some instances, overexpression of transcription factors of one plant in a separate plant results in the upregulation of POD activity. One such example is the HvbZIP21 from wild barley, when overexpressed in *Arabidopsis* resulted in increased activity of POD and other antioxidant enzymes, including SOD and, CAT, thereby increasing the resilience of the plant towards drought (Pan et al., 2022).

6.3 Catalase

Catalase is a significant enzyme responsible for metabolizing H_2O_2, a non-radical ROS. Its role involves the breakdown of H_2O_2, ensuring the cell maintains an appropriate level of this molecule critical for cellular signaling activities (Nandi et al., 2019). Catalase exhibits dual enzymatic activity, which is contingent upon the concentration of H_2O_2. When the quantity of H_2O_2 is elevated, CAT exhibits catalytic activity by effectively decomposing H_2O_2 into H_2O and O_2 through a catalytic reaction. Nevertheless, when H_2O_2 is present in a small amount, and there is a suitable hydrogen donor such as ethanol, methanol, phenol, or other substances, CAT functions as a POD, eliminating H_2O_2 while oxidizing its substrate (Scibior & Czeczot, 2006). Reactive oxygen species are generated in plants when they encounter environmental challenges, such as dryness and high temperatures. Reactive oxygen species significantly influence plant growth and development during drought conditions by causing oxidative stress (Lee & Park, 2012). The function of CAT in plants under drought and water scarcity will be covered in this section. In one research, it was shown that potato plants under drought stress had elevated levels of the enzymes CAT, POD, and SOD. Proline, soluble proteins, and total soluble sugars all increased in quantity. Additionally, compared to plants that were not treated with the PGPR, plants treated with the PGPR-*Bacillus subtilis* HAS31 showed increased levels of chlorophyll, soluble proteins, total soluble sugars, and photosynthetic activity, as well as higher enzymatic activities of CAT, POD, and SOD (Lee & Park, 2012). The *E. coli* KatE gene—which produces CAT—was inserted into cotton plants using Agrobacterium-mediated transformation in a different study. The study results showed that the transgenic lines expressed EcKatE in response to drought, and that total CAT activity was only increased compared to the wild type in response to salt or drought stress. Furthermore, the drought-induced expression of EcKatE resulted in elevated activities of SOD and POD, a reduction in H_2O_2 levels, enhanced photosystem II performance, and an increase in relative water content (Liu et al., 2022).

In another study, The activity of CAT, the concentration of H_2O_2, soluble protein, and soluble sugar in drought-stressed rice plants rose as the duration and severity of drought stress increased. The drought-resistant cultivar exhibited a reduced photosynthetic impacted area, an extended period of CAT enzyme activity, and a decreased accumulation of H_2O_2 compared to the drought-sensitive cultivar. While POD and CAT enzymes retain their capacity to remove H_2O_2 across extended periods of drought, APX enzymes appear to function as a quick reaction mechanism for scavenging H_2O_2 under drought-induced stress (Wang et al., 2019). In a study using two cultivars of *Triticum durum* subjected to drought conditions, H_2O_2 production was seen during the initial phases of stress in the tolerant Barakatli-95 variety. However, in the susceptible Garagylchyg-2 variety, substantial levels of H_2O_2 were only accumulated during later stages. The leaves and roots of the drought-tolerant cultivar Barakatli-95 had the highest CAT activity, which is crucial for neutralizing H_2O_2 (Huseynova et al., 2015). In maize, drought stress increases the generation of MDA. In addition, higher activity of CAT along with other antioxidant enzymes, namely SOD, POD, and glutathione reductase, was also observed (Anjum et al., 2016). Similar results were also observed in another study using maize in which drought stress increased the production of H_2O_2, $O_2^{-\bullet}$, and MDA. Furthermore, the membrane permeability and accumulation of osmolytes in the form of proline and soluble sugars also increased along with an increase in the activity of CAT and other antioxidant enzymes (Yao et al., 2016). In a recent study, it was found that Moroccan barley cultivars subjected to drought stress exhibited higher MDA content as well as an increase in CAT activity (Ferioun et al., 2023). In addition, in a comparative study of the effect of drought stress on *Cleome spinosa* (C3) and *Cleome gynandra* (C4), it was observed that drought stress increased the activity of CAT and other antioxidant enzymes in both species. Furthermore, the expression of the CAT1 gene in *C. spinosa* was shown to be linked with CAT activity. However, in *C. gynandra*, no correlation was observed between CAT1 gene expression and CAT activity. In addition to the elevated levels of MDA, the concentration of H_2O_2 was much higher in *C. spinosa* compared to *C. gynandra* when subjected to drought stress. These findings indicate that in *C. spinosa*, the antioxidant defense mechanism was inadequate to inhibit the escalating production of ROS under stressful conditions (Uzilday et al., 2012).

6.4 Ascorbate Peroxidase

The antioxidant enzyme APX is widely distributed. It works by removing or reducing ROS, such as H_2O_2, to water, using reduced ascorbate as the electron source, in contrast to CAT and other PODs. When it comes to efficiently removing ROS, ascorbate outperforms other antioxidant enzymes. This is because the compounds $O_2^{-\circ}$, hydroxyl radical (OH^-), $^1O_2$, and H_2O_2 may all react with ascorbate (Kumar, 2022). In response to abiotic and biotic stresses, as well as at different stages of plant development, the expression of the APX gene is modulated. Protecting plant cells from adverse environmental conditions is directly attributed to the APX responses. Additionally, mutant plants' growth, physiology, and antioxidant metabolism are altered, suggesting that APX genes are involved in the normal development of plants (Caverzan et al., 2012). The functions of APX in plants under drought stress will be covered in this section. It has been observed that the damage brought on by water stress circumstances can be lessened in transgenic tobacco plants by overexpressing Cu/Zn sod and apx, or at least apx, in the cytosol. It appears that cyt apx overexpression shields tobacco membranes from water stress since oxidative stress indicators such as electrolyte leakage, lipid peroxidation, and H_2O_2 levels were often greater in non-transformed plants than in transgenic lines (Faize et al., 2011). Additionally, it has been discovered that in *Vigna unguiculata*, variations in the APX gene transcript buildup between the drought-tolerant cultivar 'EPACE-1' and the drought-sensitive cultivar '1183' were brought about by a lack of water. The ability of chloroplastic APX genes to efficiently detoxify ROS at the site of their formation was demonstrated by their early response to a gradual loss of water in the tolerant plant (D'Arcy-Lameta et al., 2006). Additionally, the more sensitive '1183' was also capable of responding to drought by activating all of its APX genes (D'Arcy-Lameta et al., 2006). In another study, when two wheat genotypes—the drought-tolerant Plainsman V and the drought-sensitive Cappelle Desprez—were exposed to a slight water shortage, the relative APX transcript levels in each genotype revealed distinct changes. Both genotypes showed an increase in cytosolic APX1 expression, but only the drought-tolerant genotype showed an increase in cytosolic APX2 expression. The drought-tolerant genotype showed an increase in thylakoid APX transcript levels, whereas the cultivar that was more vulnerable to drought showed an increase in stromal APX2 expression levels (Sečenji et al., 2010). Pearl millet was treated with *Rhizobacteria Bacillus amyloliquefaciens* (MMR04), which resulted in enhanced proline, APX, and SOD activity. In addition, the MDA level was reduced by up to 2.3 times when compared to the untreated plants. The relative expression of the marker genes for ethylene-responsive factor (ERF-1B) and drought-responsive factor (DREB-1E) was significantly reduced, as demonstrated by gene expression analysis, by 1.8 and 3.3 times, respectively. In contrast, compared to the control group, plants treated with MMR04 and exposed to severe drought conditions showed an increase in the relative expression of APX1 and SOD1 of 2.2 and 2.9 times, respectively (Murali et al., 2021). Plants that have undergone genetic modification to produce higher levels of antioxidant enzymes in various species have exhibited improved resilience to a variety of stresses, including drought. A research looked at transgenic tobacco plants' elevated expression of the Populus peroxisomal ascorbate peroxidase (PpAPX) gene. The 35S promoter of the cauliflower mosaic virus or the rd29 promoter regulated the gene. 35S-PpAPX transgenic plants showed a 50% increase in APX enzyme activity as a consequence of a significant overexpression of the PpAPX gene. When transgenic plants were exposed to salt or drought, their leaves showed an 80% increase in APX activity. Moreover, during the vegetative phase, the transgenic tobacco showed significant improvements in salt tolerance and drought resistance. According to RNA blot research, PpAPX gene expression in rd29Ap-PpAPX plants was considerably lower under normal growth conditions but showed a discernible rise during drought stress (Li et al., 2009).

6.5 Glutathione Reductase

The plants are susceptible to experiencing internal strain caused by various harmful chemical species grouped as ROS/RNS. These elements are effectively countered and eliminated by the active function of antioxidant mechanisms in resilient species (Mittler et al., 2022). Within the antioxidant pathway, the Ascorbate-glutathione (AsA-GSH) pathway serves as a constituent where an imperative enzyme called Glutathione reductase (GR; EC 1.8.1.7) plays a significant role in converting oxidized glutathione (GSSG) into its reduced form (GSH) (Yousuf et al., 2012). The transcriptional profile and activity of this enzyme are modified under various stressful conditions such as abiotic stress (Anderson & Davis, 2004), photooxidative stress in *Chlamydomonas* cells (Lin et al., 2018), drought stress in *Triticum aestivum* (Gill et al., 2013), *J. curcas* (Mudalkar et al., 2017), and water stress in tomatoes (Murshed et al., 2013). Glutathione (L-ɣ-glutamyl-L-cysteinylglycine, GSH) stands as a crucial biomolecule within the cellular antioxidant system, demonstrating high efficacy in scavenging ROS and RNS to maintain cellular equilibrium (Gill et al., 2013). Glutathione reductase, also known as GSR or GR, belongs to the NADPH-dependent oxidoreductase family and is present in both prokaryotic and eukaryotic organisms (Gill et al., 2013). This enzyme's high conservation across nature likely stems from its vital role.

Glutathione reductase, an oxidoreductase NAD(P)H-dependent cellular enzyme, exists as a flavor-protein-bound homodimer (Wang et al., 2020a). Its active site houses a cysteine residue (Cys) crucial for catalyzing the NADPH-dependent reduction of GSSG to GSH (Vašková et al., 2023). The coenzyme FAD initiates the disulfide reduction reaction through a nucleophilic attack on GSSR by a transient FADH-anion (Fagan & Palfey, 2010). This conversion aids in maintaining a high GSH/GSSG ratio during various abiotic stresses. While primarily located in chloroplasts (70-80%), GR is also found in mitochondria and the cytosol. Chloroplasts' oxidative conditions and the availability of NADPH contribute to this specific localization (Edwards et al., 1990; Creissen et al., 1994).

Studies on the purification and characterization of GR have revealed its homodimeric nature, with a molecular mass ranging from 100 to 150 kDa, containing one FAD per monomer in species like *Spinacia oleracea, Pisum sativum*, and *Zea mays* (Lascano et al., 2001). However, GR is reported as a monomer in *Chlamydomonas reinhardtii* and a heterodimer in *Pisum sativum* and *Zea mays* (Kalt-Torres et al., 1984; Mahan & Burke, 1987). Notably, a peroxisomal glutathione reductase (GR; EC 1.6.4.2) has been identified and purified in response to abiotic stress in pea (*Pisum sativum*) leaves using immune gold microscopy (Romero-Puertas et al., 2006). Two homoeologous groups consisting of seven glutathione reductase (GR) genes have been found in *Triticum aestivum*. The same group also discovered three GR genes in *Aegilops tauschii, Brachypodium distachyon, Triticum urartu*, and *Sorghum bicolor*. GR proteins were classified into two classes, Class I and Class II, by comparative phylogenetic analysis (Madhu et al., 2022). Class I and Class II are expected to be located in the cytoplasm and chloroplasts, respectively, and each group exhibits conserved exon–intron and motif patterns. TaGR gene differential expression and elevated GR enzyme activity reveal their functions in salt, heat, drought, and arsenic stress (Gill et al., 2013). Moreover, GR operates in conjunction with Glyoxalase (GLYI and GLYII) to confer drought tolerance in *J. curcas* (Mudalkar et al., 2017). During seed imbibition in oat plants, GR responses to drought treatments have been instrumental in improving stress tolerance and seed vigor (Sun et al., 2022). Drought-induced oxidative stress leads to elevated MDA production, mitigated by heightened activity of enzyme groups such as CAT, SOD, and GR (Anjum et al., 2016). Consequently, the abundance of research on GR underscores its protective role in plants against oxidative stress.

6.6 Dehydroascorbate Reductase (DHAR) and Monodehydroascorbate Reductase (MDHAR)

Aerobic metabolism, though very effective and efficient in terms of energetics, side effects are seen as cellular ROS generation during stress conditions (Hasanuzzaman et al., 2012; Sharma et al., 2012; Zandi & Schnug, 2022). These harmful ROS are removed through the operation of an antioxidant system that includes both enzymatic as well as non-enzymatic molecules (Ahmad et al., 2010). Ascorbic acid (AsA), which is widely recognized for its antioxidant properties, efficiently prevents ROS/RNS from being neutralized (Devasagayam et al., 2004). The active AsA is generated through the AsA-glutathione (GSH) cycle with the help of a twin enzyme system known as dehydroascorbate reductase (DHAR; EC 1.8.5.1) and mono-dehydroascorbate reductase (MDHAR; EC 1.6.5.4). The ascorbate–glutathione cycle up-regulated in response to drought stress in apple leaves (Ma et al., 2011), *Prunus* hybrids (Sofo et al., 2005) and found that hydrogen sulfide (H_2S) and H_2O_2 participated in the up-regulation of cycle in wheat leaves (Shan & Ou, 2018; Shan et al., 2018), Jasmonic acid (JA) for acquisition of water stress tolerance in *Agropyron cristatum* (Shan & Liang, 2010).

As AsA acts as an antioxidant, it gets oxidized after reacting with ROS/RNS and forms an unstable radical as monodehydroascorbate molecule that further breaks down into ascorbate and dehydroascorbate. The enzymes monodehydroascorbate reductase (MDHAR) and dehydroascorbate reductase (DHAR) help in reducing these molecules and bringing them again into active form. Although both enzymes help regenerate ascorbate for maintenance of ROS scavenging DHAR, play a pivotal role in glutathione homeostasis in plants (Ding et al., 2020), as seen in tobacco and tomato where high expression of only MDAHR conferred osmotic tolerance. Experimentally it was found that overexpression of GR genes in transgenic tobacco plants conferred protection against drought stress (Chen et al., 2016). In tomato (*Solanum lycopersicum* L. cv. Micro-Tom), water stress affects the expression of relative transcript levels of the ascorbate-glutathione cycle enzymes, including MDHAR, DHAR, and GR. This expression is dependent on the plant's developmental stage (Makino et al., 2002). However, it was discovered that in tomatoes, constitutively expressed DHAR and MDHAR affect the fruit but not foliar ascorbate levels (Haroldsen et al., 2011). In growing *Poa pratensis* L. (Bian & Jiang, 2009) and rice seedlings (Sharma & Dubey, 2005), dryness also increased the activity of these antioxidant enzymes (DHAR, MDAHR). These important genes were inducted in white clover in response to abiotic challenges such as drought and exogenous abscisic acid (ABA) or spermidine (Spd) therapy (Zhang et al., 2015). These datasets point to its potential function in reducing the oxidative damage brought on by abiotic stress.

7. Induction of Water Stress Metabolites

7.1 Glutathione

Glutathione (GSH) is a crucial antioxidant and redox regulator that plays a significant role in plant responses to water stress. Glutathione, as a vital component of the antioxidant defense system, helps plants combat the damaging effects of ROS and maintain cellular redox homeostasis (Fahad et al., 2017). Drought commonly leads to restricted growth and developmental processes, diminished growth, and decreased final crop output (Fahad et al., 2017). The lack of water during a drought reduces the ability of stomata to function properly, as well as the pace at which electrons are transported across the membrane, the diffusion of CO_2, the effectiveness of carboxylation, and the process of photosynthesis. This results in the production of ROS, which in turn may cause oxidative damage (Mishra et al., 2023). The metabolic enzymes glutathione and its products interact with plant growth regulators, which are thought to be critical for plant development under both favorable and unfavorable conditions (Hasanuzzaman et al., 2017). The role of glutathione in water stress response includes several key functions. Firstly, it acts as a primary antioxidant by directly scavenging ROS or participating in enzymatic reactions mediated by GPX and glutathione-S-transferase (Hasanuzzaman et al., 2017). This allows glutathione to convert harmful ROS, such as H_2O_2 and lipid peroxides, into less toxic forms, preventing oxidative damage to cellular components (Foyer & Noctor, 2011; Corpas et al., 2013).

Secondly, glutathione is involved in maintaining cellular redox balance. It functions as a crucial redox buffer in plants, existing in two forms: reduced glutathione (GSH) and oxidized glutathione (GSSG) (Dorion et al., 2021). Under normal conditions, the ratio of GSH to GSSG is high, indicating a reduced state that supports optimal cellular function (Zitka et al., 2012). However, water stress disrupts this balance, increasing GSSG levels. Glutathione reductase enzyme helps regenerate GSH from GSSG, restoring the redox equilibrium and protecting cellular integrity (Koffler et al., 2014). Moreover, glutathione acts as an important signaling molecule in response to water stress (Koffler et al., 2014). Changes in the GSH/GSSG ratio can modulate the activity of redox-sensitive proteins, transcription factors, and enzymes involved in stress signaling pathways (Alkazemi et al., 2021; Kalinina & Novichkova, 2021). Glutathione participates in the regulation of gene expression, influencing the activation of stress-responsive genes involved in adaptive responses (Foyer & Noctor, 2011).

Under conditions of water stress, plants produce and regulate glutathione through intricate mechanisms. Two enzymatic steps requiring ATP are needed to synthesize GSH from its amino acids: first, γ-glutamylcysteine must be created from glutamate and cysteine, and then GSH must be formed from γ-glutamylcysteine and glycine (Lu, 2013). γ-glutamylcysteine synthetase (γ-ECS) is the enzyme responsible for catalyzing the rate-limiting step (Chen et al., 2005). Water stress can raise the expression and activity of γ-ECS and other enzymes involved in glutathione biosynthesis, which increases glutathione production (Shah & Dubey, 1997; Lu et al., 2021). Moreover, the existence of the precursor amino acid cysteine and the activity of the rate-limiting enzyme glutamate cysteine ligase (GCL) are the main variables affecting the synthesis of GSH. The two subunits that comprise GCL are the modifier subunit (GCLM) and the catalytic subunit (GCLC) (Lu, 2013). Under situations of water stress, glutathione metabolism is also modulated by transcriptional control and post-translational changes (Foyer & Noctor, 2011; Kaur & Asthir, 2015).

Numerous studies have demonstrated that elevated endogenous GSH levels can confer resilience against drought stress and that GSH mitigates ROS-induced damage. Drought conditions in *Vigna radiata* caused the GSH/GSSG ratio to decrease while endogenous GSH and glutathione disulfide (GSSG) levels increased. Moreover, APX and glutathione S-transferase activities increased whereas those of CAT, dehydroascorbate reductase (DHAR), and monodehydroascorbate reductase (MDHAR) decreased. In response to drought stress, the activities of the enzymes glyoxalase I (Gly I) and glyoxalase II (Gly II) rose (Nahar et al., 2015). Exogenous GSH supplementation significantly increased the various components of the antioxidant and glyoxalase systems in drought-stricken *Vigna radiata* seedlings, as opposed to drought stress alone (Nahar et al., 2015). Drought stress caused a significant rise in H_2O_2 and lipid peroxidation levels, suggesting that *B. napus* was under severe oxidative stress. The impacts of drought stress caused the GSSG content of the seedlings to noticeably increase. By reducing oxidative stress and improving resistance to drought conditions by raising the amounts of naturally produced GSH, the use of selenium from an external source is shown to be beneficial. It was found that there was an increase in the GSH/GSSG ratio as well as the activity of the AsA-GSH cycle enzymes (Hasanuzzaman & Fujita, 2011). Water scarcity in rice caused Ascorbic Acid (AsA) and Glutathione (GSH) levels to drop, along with their corresponding redox ratios (AsA/DHA and GSH/GSSG). In contrast to the sensitive seedlings, the decrease was less noticeable in the seedlings that were more resilient to the water shortage. As the water deficit increased in the young plants of both rice cultivars, the expressions of glutathione transferase, dehydroascorbate reductase, APX, and monodehydroascorbate reductase increased. Nonetheless, the juvenile plants of the cultivar resistant to drought exhibited higher levels of activity (Pyngrope et al., 2013).

7.2 Proline

Proline is a non-essential amino acid that plays diverse roles in plant physiology, particularly in response to environmental stresses, including water stress (Hayat et al., 2012). It accumulates in plant cells under adverse conditions and acts as an osmoprotectant and compatible solute (Yoshiba et al., 1997). Proline helps maintain cellular water potential and prevents dehydration by acting as an osmolyte (Lehmann et al., 2010). It plays a crucial role in stabilizing proteins, membranes, and cellular structures, protecting them from damage caused by water stress (Verslues & Sharma, 2010). During water stress, proline accumulation helps maintain cellular water potential and prevent water loss by acting as an osmolyte. It balances osmotic pressure within cells, reducing water stress-induced damage. Additionally, proline's presence supports protein stability and membrane integrity, safeguarding cellular structures under water stress (Szabados & Savouré, 2010).

Furthermore, proline acts as a signaling molecule in response to water stress. It triggers the activation of stress-responsive genes, including those involved in antioxidant defense, stress tolerance, and the synthesis of protective compounds (Khedr, 2003; Banu et al., 2009; Nounjan et al., 2012). Proline signaling pathways help coordinate adaptive responses, promoting plant survival and resilience during water stress (Verslues & Sharma, 2010). By regulating osmotic balance, proline enables plants to withstand water deficits and maintain cellular integrity. By modulating gene expression, proline helps plants adapt to water stress and enhance their tolerance to adverse environmental conditions (Szabados & Savouré, 2010). There are two primary mechanisms via which proline is biosynthesized: the glutamate pathway and the ornithine pathway. Proline is produced in the glutamate pathway by the enzymes Δ1-pyrroline-5-carboxylate reductase (P5CR) and Δ1-pyrroline-5-carboxylate synthetase (P5CS) from glutamic acid (Yan et al., 2021). The ornithine pathway involves the conversion of ornithine to proline using ornithine aminotransferase (OAT) and Δ(1)-pyrroline-5-carboxylate reductase (P5CR) (Da Rocha et al., 2012; Fichman et al., 2015). These pathways are upregulated under water stress conditions, leading to increased proline accumulation in plants (Verslues & Sharma, 2010).

Numerous investigations have been conducted to examine the impact of water stress on plants. Both resistant and sensitive cultivars of peanuts showed increased expression of proline biosynthesis-related genes (P5CS1, P5CS2a, P5CS2b, and P5CR) when they were exposed to water stress. However, during stress, the relative expression of the proline catabolism genes (ProDH1, ProDH2) was solely upregulated in the sensitive cultivar (Furlan et al., 2020). Drought caused Barley's water content to steadily decrease while proline and ABA levels increased in the plant's roots and leaves of both genotypes. Additionally, a robust positive correlation was seen between proline accumulation and Δ1 pyrroline-5-carboxylate synthase activity. The rapid rise in Δ1 pyrroline-5-carboxylate synthetase activity and proline buildup was preceded by an increase in the transcriptional up-regulation of Δ1 pyrroline-5-carboxylate synthetase. The correlation between variations in Δ1 pyrroline-5-carboxylate synthetase expression, Δ1 pyrroline-5-carboxylate synthetase activity, and ABA levels indicates that ABA plays a role in the transcriptional and enzymatic regulation of proline synthesis during drought conditions (Bandurska et al., 2017). In one research, when watermelon drought-tolerant germplasm M08 and drought-susceptible line Y34 were exposed to drought stress, significant accumulations of free proline were found in the roots, stems, and leaves of the former. The ClP5CS1 gene, in contrast to its paralog ClP5CS2, largely contributes to proline synthesis in leaves via the Glu pathway, according to an investigation of gene expression and enzyme activity (Wang et al., 2022b). A study on rice revealed that proline exhibited significant accumulation during drought stress, but saw a quick decline upon re-watering. Moreover, the drought-tolerant cultivar exhibited a greater capacity for proline accumulation compared to sensitive types (Dien et al., 2019). A recent study on rice discovered that transgenic rice plants, which were genetically modified to express OsOLP1, a newly identified osmotin-like protein, exhibited significant drought tolerance. These plants had a leaf water content of up to 65% and a survival rate of 53.1%. This was achieved by regulating stomatal closure, resulting in a 96% closure rate, and promoting proline content by more than 2.5 times. Additionally, the plants accumulated 1.5 times more endogenous ABA, a hormone involved in stress response, and enhanced lignin synthesis by approximately 50%.

Nevertheless, the OsOLP1 deletion lines exhibited a significant decrease in ABA content, a reduction in lignin deposition, and a diminished ability to withstand drought (Yan et al., 2023).

In another study it was observed that the overexpression of TaERF87 (ethylene-responsive factors) improves the ability of wheat to withstand drought conditions, but silencing TaERF87 makes wheat more susceptible to drought. Through the use of biochemical experiments, RNA sequencing demonstrated that TaERF87 directly binds to GCC-box elements, resulting in the activation of the proline biosynthesis genes TaP5CS1 and TaP5CR1. Furthermore, the collaboration between TaERF87 and the bHLH transcription factor TaAKS1 significantly amplifies the activation of TaP5CS1 and TaP5CR1 gene expression. TaAKS1 OE enhances wheat's ability to tolerate drought by stimulating the accumulation of proline (Du et al., 2023). In another study using wheat, it was found that the tolerant cultivar exhibited considerably elevated P5CS expressions in comparison to the sensitive cultivar when subjected to drought stress. The expression of P5CS was nearly 3.1 times more in the tolerant cultivar compared to the sensitive one, 48 hours after stress induction. The plants subjected to stress and treated with a combination of silicon and salicylic acid exhibited a greater level of gene expression. The proline content exhibited a gradual rise upon treatment with silicon and salicylic acid. The highest proline content was seen when silicon and salicylic acid were applied simultaneously (Maghsoudi et al., 2018).

7.3 Glycine-betaine

Glycine betaine is a compatible osmolyte that is anticipated to play an essential function in osmoregulation in plants exposed to harsh environmental stimuli such as high-temperature stress (Zulfiqar et al., 2022). The Osmo protectant glycine betaine (GB) production is an adaptive metabolic characteristic that enables plants to cope with osmotic stress resulting from drought, salinity, or cold temperatures (Muñoz-Clares et al., 2014). Glycine betaine is a tiny zwitterionic molecule produced by a variety of species. It was first extracted from the juice of *Beta vulgaris* (sugar beetroot), therefore the name betaine. Although betaine, trimethylglycine (TMG), betaine anhydrous, oxyneurine, and N, N, N-trimethylglycine are the most often used terms for glycine betaine, the IUPAC nomenclature for it is 2-trimethylammonioacetate (Kar et al., 2015). To preserve chloroplasts through osmoregulation, GB mostly accumulates in situations of water stress. So, thylakoid membrane damage is avoided, and photosynthetic production is sustained (Dikilitas et al., 2020). High quantities of proline and/or other osmolytes may be a useful tactic to recover from the unfolding of proteins caused by stress exposure since both substances can stabilize membranes and proteins (Ortbauer, 2013). To deal with stressful situations, GB takes part in several plant responses, such as controlling gene expression (Einset & Connolly, 2009), regulating enzyme concentration and activity (Figueroa-Soto & Valenzuela-Soto, 2018), and ensuring appropriate protein folding and association (Figueroa-Soto & Valenzuela-Soto, 2018). The thylakoid membrane's lipids and proteins are preserved by GB, which also increases photosynthetic capacity (Niu et al., 2023). Moreover, GB increases the antioxidant and vital enzyme concentrations and activities of the Calvin cycle (Hernandez-Leon & Valenzuela-Soto, 2023). Numerous studies have shown that glycine betaine contributes to resistance to drought stress. According to reports, glycine betaine given exogenously to rice enhanced its development in both irrigated and drought situations (Farooq et al., 2008). Exogenous administration of glycine betaine and/or abscisic acid enhanced drought resistance in *Axonopus compressus* (Nawaz & Wang, 2020). The application of glycine betaine externally to drought-stressed maize improved plant development and yield characteristics, as well as the concentrations of chlorophyll pigments, osmolytes, total phenolics, ascorbic acid, and the activities of enzymes that neutralize ROS, according to a similar study (Shafiq et al., 2021). Glycine betaine was shown to lessen oxidative stress in drought-prone maize in a related experiment (Shemi et al., 2021). The codA gene from *Arthrobacter globiformis*, which codes for choline oxidase and facilitates the conversion of choline to GB, was effectively incorporated into the plastid genome of the potato (*Solanum tuberosum*) using plastid genetic engineering. It was shown that in codA-transplastomic potato plants, GB accumulated in identical amounts in both the tubers and the leaves. The codA-transplastomic potato plants showed marginally greater levels of GB than wild-type plants, and the plant's resilience to drought stress was significantly increased by the localization of GB in plastids (You et al., 2019).

7.4 Polyamines

Polyamines are naturally occurring organic compounds found in every living organism (Sagar et al., 2021). Aliphatic nitrogenous bases with two or more amino groups and a low molecular weight are known as polyamines. In plants, the primary polyamines are putrescine (put), spermidine (spm), and spermine (spd) (Chen et al., 2019). They are connected with the regulation of cell division, differentiation, and morphogenesis (Rao & Wang, 2010). Additionally, polyamines participate in maintaining cellular homeostasis, controlling gene expression, and modulating responses to various stresses (Seo et al., 2019).

Water stress triggers a complex cascade of molecular events in plants, leading to the activation of various stress-responsive genes (Osakabe et al., 2014). Under conditions of water stress, it has been observed that polyamines tend to accumulate within plants (Chen et al., 2019). By upregulating the expression of genes encoding important enzymes involved in polyamine biosynthesis, such as arginine decarboxylase (ADC) and ornithine decarboxylase (ODC), water stress encourages the production of polyamines (Fortes et al., 2011).

Polyamines play a crucial role in plant adaptation to water stress by mediating physiological, biochemical, and molecular changes. By exerting their influence, polyamines possess the ability to modulate the dimensions of potassium channels and pores within the plasma membrane of guard cells. This intricate regulation significantly impacts the opening and closing of these pores, thus enabling them to effectively manage the loss of water in plants (Chen et al., 2019). At an optimal level, putrescine (put) has the potential to initiate physiological processes and stimulate the production of osmotic adjustment substances, including free amino acids, soluble sugars, and proline. This compensatory response helps counteract the detrimental effects of drought stress (Sánchez-Rodríguez et al., 2016; Mohammadi et al., 2018). Consequently, polyamines act as osmoprotectants by upholding cellular water balance and safeguarding against dehydration-induced damage (Alcázar et al., 2020). Polyamines also contribute to the stabilization of membranes and proteins under water stress conditions, thereby enhancing plant tolerance (Seo et al., 2019).

Reactive oxygen species accumulate under water deficit stress conditions, leading to oxidative damage in plants (Nxele et al., 2017). Polyamines have been found to enhance the scavenging of ROS and minimize oxidative stress (Murray Stewart et al., 2018). They interact with ROS directly and indirectly by modulating the activities of antioxidant enzymes, such as CAT, superoxide SOD, and POD (Tanou et al., 2014). Polyamines possess the capability to reduce the rate of membrane lipid peroxidation, thereby effectively enhancing tolerance to hypoxia stress (Wu et al., 2018). The presence of polyamines helps to maintain redox homeostasis and minimize the harmful effects of ROS (Murray Stewart et al., 2018). Moreover, polyamines regulate the stress-responsive gene expression and activate the signaling pathways that trigger the protective molecules synthesis, including antioxidants and stress-related proteins (Liu et al., 2015). Harnessing the potential of polyamines through genetic engineering or exogenous application may provide avenues for improving crop productivity under water-limited conditions and contribute to sustainable agriculture in the face of climate change (Shao et al., 2022).

7.5 Alpha-tocopherol

Under water stress conditions, plants initiate a series of internal changes leading to an excessive production of ROS. The presence of ROS under such stress can prompt oxidative stress, causing harm to crucial cellular elements like proteins, lipids, and nucleic acids (Schieber & Chandel, 2014). To counteract this oxidative harm, plants trigger various defense mechanisms, with alpha-tocopherol playing a pivotal role in alleviating the detrimental effects of ROS (Hasanuzzaman et al., 2021).

Alpha-tocopherol, also known as Vitamin E, is a lipid-soluble compound naturally synthesized by plants and is present in multiple plant tissues, including leaves, seeds, and fruits (Fritsche et al., 2017). Its crucial function lies in shielding cells from oxidative damage by neutralizing harmful free radicals, preventing lipid peroxidation, and preserving cell membrane integrity (Galli et al., 2022; Ponnampalam et al., 2022). The key roles of alpha tocopherols in maintaining plant resilience under water stress conditions are:

- **Antioxidant Activity:** Alpha-tocopherol acts as a potent antioxidant by scavenging and neutralizing ROS, preventing oxidative damage to cell membranes and other cellular structures (Sharifi-Rad et al., 2020). They effectively suppress or chemically eliminate $^1O_2$ with spin-paired valence electrons. Through the mechanism of resonance energy transfer, a single molecule of alpha-tocopherol can inactivate up to 120 molecules of $^1O_2$ (Szarka et al., 2012). It also helps maintain the integrity and functionality of chloroplasts, mitochondria, and other subcellular organelles under water-stress conditions (Zandi & Schnug, 2022).

- **Enhancing Membrane Fluidity:** Water stress can cause changes in the fluidity and permeability of cellular membranes (Wang et al., 2020b). Alpha-tocopherol plays a vital role in preserving membrane fluidity by impeding lipid peroxidation through its capacity to donate its phenolic hydrogen to lipid free radicals (Raederstorff et al., 2015). This ensures efficient nutrient uptake and transport across membranes, facilitating the overall adaptation and survival of plants under water stress (Ali et al., 2019).
- **Stabilizing Photosynthetic Machinery:** During water stress, photosynthesis is often compromised due to reduced water availability (S. Lisar et al., 2012). Alpha-tocopherol protects the photosynthetic machinery by minimizing lipid peroxidation and maintaining the integrity of chloroplast membranes(Shah et al., 2021). This ensures the efficient capture and utilization of light energy, reducing the negative impact of water stress on photosynthesis (Ali et al., 2019).
- **Modulating Plant Hormone Signaling:** Alpha-tocopherol is involved in the regulation of phytohormone signaling pathways, including abscisic acid (ABA) (Zhao et al., 2023). ABA serves as a pivotal hormone governing the regulatory mechanism of plants in response to water stress (Aslam et al., 2022).

7.6 Ascorbate and Monodehydroascorbate

Monodehydroascorbate and ascorbate are important plant compounds with diverse roles in cellular functions, particularly in response to environmental challenges such as water stress. Monodehydroascorbate, an oxidized form of ascorbate, functions as a significant antioxidant (Foyer & Noctor, 2011). Ascorbate, a crucial antioxidant in plants, relies on the essential roles of monodehydroascorbate and dehydroascorbate reductase for its regeneration, maintaining the ability to ROS scavenge (Shin et al., 2013). The ascorbate-glutathione cycle, which is made up of these elements, is a vital redox mechanism that keeps cellular redox equilibrium intact and protects plants from oxidative damage brought on by water stress. ROS are natural byproducts that are involved in signaling inside cells (Snezhkina et al., 2019). In plants, ascorbate acts as a potent antioxidant, directly scavenging ROS generated during water stress and minimizing oxidative damage to cellular components (Potters et al., 2010). Ascorbate also serves as a cofactor for enzymes like APX, involved in ROS detoxification (Talla et al., 2011). Ascorbate peroxidases efficiently eliminate ROS by converting ascorbate to monodehydroascorbate (MDHA), which can then be further transformed into dehydroascorbate (DHA) and ASC through disproportionation. MDHA reductases (MDHARs) and DHA reductases (DHARs) play pivotal roles in converting MDHA and DHA back into ascorbate, contributing to the preservation of cellular redox balance and reducing oxidative damage caused by abiotic stress (Chen et al., 2003; Smirnoff, 2018). For example, ascorbate enhances maize's drought resistance by promoting abscisic acid (ABA) synthesis, reducing water loss, and increasing resistance to osmotic stress (Terzi et al., 2015). Ascorbate is also believed to influence abiotic stress responses through epigenetic mechanisms, acting as a cofactor for ten-eleven-translocation (TET) dioxygenase, ascorbate increases the production of 5-hydroxymethylcytosine (5hmC). TET dioxygenase, a type of 2-oxoglutarate-dependent dioxygenase (2-ODD), facilitates the conversion of 5-methylcytosine (5mC) to 5hmC through oxidation (Dickson et al., 2013). Evidence suggests that TET dioxygenase plays a crucial role in the demethylation process of genomic 5mC in green algae, implying a potential occurrence of DNA demethylation catalyzed by TET enzymes in plants (Xue et al., 2019). Ascorbate's involvement in abiotic stress responses extends to altering plant cell composition, coordinating hormone production, and regulating gene expression through epigenetic mechanisms.

8. Mitigation Strategies

Major and obvious markers of drought-induced oxidative damage encompass LPO (MDA) and H_2O_2 build-up, resulting in the impairment of numerous cellular and biological activities, which include photosynthesis, respiration, and transpiration, stomatal conductance, membrane operations, efficiency in using water, and the carboxylation performance (Hasanuzzaman et al., 2012). To deal with all these abiotic stresses which produce ROS and lead to oxidative damage, applications of biotechnological tools, omics technology, breeding technologies, and genetic engineering will be the best suited methodologies that will allow agricultural researchers to produce better versions of crop varieties with all the stress-resistant capabilities (Wani et al., 2016; Anwar & Kim, 2020). Application of these modern technologies have been marked to have significant roles in the upgradation of conventional means of agricultural practices (Bera et al., 2022). Implementation of technologies like utilization of miRNA, Crispr/Cas9 gene editing system, practices of Omics technology are the most promising tools to improve plant stress tolerance mechanisms to vanquish abiotic stress.

8.1 Application of MicroRNAs to Mitigate Drought Stress in Plants

miRNAs or microRNAs are non-coding, endogenous, small sized single-stranded riboregulators of the order of 20-24 nt (Shriram et al., 2016). They are found in large numbers in higher organisms. The miRNAs are associated with every possible biological and metabolic pathway (Sunkar et al., 2012). As a consequence, miRNAs are now considered a promising avenue for improving crops, increasing plant productivity, and preventing abiotic stresses like drought, salinity, and temperature changes, among others (Pandita, 2019).

One of the most catastrophic elements that adversely affects the yield of crops is drought, which has an extensive effect on plant growth and development (Pagano et al., 2021; Zhang et al., 2022). By screening an RNA library derived from sugar cane, *Oryza*, and *Arabidopsis*, an abundance of drought-responsive miRNAs have been identified (Shriram et al., 2016). To find candidates that could enhance rice drought tolerance, genome-wide analyses of six genotypes of *Oryza* were conducted (Xia et al., 2020) and it came to light that miRNAs like miR169, miR398, and miR5505 had the potential to achieve drought resistance. Apart from that, drought dramatically upregulates and momentarily influences miR169g and miR393 in *Oryza* (Zhao et al., 2007). In sensitive soybean cultivars, miRNAs such as miR166-5P, miR1513c, miR169-3P, miR397ab, and miRSeq13 were shown to be upregulated, whereas their tolerant counterparts showed downregulation (Kulcheski et al., 2011). Additionally, miR169 was identified to have different impacts across different species, for example being slightly elevated in *Triticum aestivum* (Akdogan et al., 2016) and downregulated in the plant *Arabidopsis thaliana* (Li et al., 2016). Through the use of the short tandem target mimic approach (STTM), research on miR160 and miR165/166 has demonstrated that they impact plant development and drought tolerance in *Arabidopsis* (Zhang et al., 2022). Zea Mays drought-tolerant inbreed has been displayed to consist of 3 conserved miRNAs, namely miR156, miR159, and miR319 (Li et al., 2013).

8.2 Application of Crispr/Cas9 to Mitigate Oxidative Stress Tolerance

To safeguard the worldwide supply of food amid increasing demand for food, we must devote more attention to ongoing research that seeks to improve crop yields under severe environmental stresses done to abiotic factors (Chowdhury et al., 2022). One of the most widely used techniques for manipulating crops is genome engineering, which consists of gene displacement, gene disruptions, and elimination of genes (Zaidi et al., 2017). The umbrella term of molecular biology has gone through a significant transformation as a consequence of the discovery of programmable nucleases that may produce double-stranded breaks (Zafar et al., 2020). Some of these are transcription affector-like effector nucleases (TALENs) and zinc finger nucleases (ZFNs). However, the discovery of Crispr/Cas9 quickly piqued the fascination of scientists researching both animals and plants owing to its readily apparent advantages over ZFNs and TALENs (Mao et al., 2013). This is because the target design is straightforward. The Crispr/Cas9 system has been successfully implemented in a broad spectrum of organisms, spanning bacteria, animals, and plants. Multiplex genome editing utilizing miR482b and miR482c concurrently is employed with Crispr/Cas9 gene editing tools to elevate *Phytopthora infestans* (Biswas et al., 2021). When Crispr/Cas9 produced ARGOS8 variations in the plant *Zea mays*, drought-resistant ability was enhanced (Shi et al., 2017). Nevertheless, it is essential to pay attention to selecting an appropriate promoter for Cas9 expression, designing guide RNA, and establishing an effective gene delivery mechanism when employing Crispr/Cas9 in plants to enhance their ability to withstand adverse environmental conditions caused by abiotic factors (Osakabe & Osakabe, 2017). Rice PARAQUAT TOLERANCE 3 knockout mutants OsPQT3 created with the aid of Crispr/Cas9 technology exhibit increased resistance to both salt and oxidative stresses (Alfatih et al., 2020). To improve drought tolerance, Crispr/Cas9 was implemented in *Arabidopsis* for the introduction of OsT2 utilizing the simultaneous expression of Cas9 and GFP (Osakabe et al., 2016). In *Arabidopsis*, this third-generation gene editing approach was carried out to boost expression of the genes for AVP1 and AREB1, which ultimately assisted the plant tolerate extreme drought conditions (Park et al., 2017; Roca Paixão et al., 2019). With the assistance of Crispr/Cas9, OsNCED3 overexpression in rice has been shown to have higher ABA accumulation and improvement against drought as well as salinity stress (Huang et al., 2018).

Crispr/Cas9 generated the OsPYL mutant, which simultaneously raised rice's tolerance level and its total manufacturing capacity (Chowdhury et al., 2022). In polyploid wheat, Crispr/Cas9 has been successfully utilized to stimulate genes such as TaERF3 and TaDREB2 that respond to abiotic stress, enhancing wheat's capability for dealing with drought-related conditions (Kim et al., 2018). By leveraging Crispr/Cas9 technology, researchers have described an enhanced stress response in *Arabidopsis* plants under drought conditions. Researchers accomplished this by employing genuine guide RNAs that targeted the OsT2/AHA1 gene and a tissue-specific AtEF1 promoter for Cas9 production (Biswas et al., 2021). Through this approach they suppressed ROS and promoted the biosynthesis of HSFs and HSPs ultimately optimizing heat tolerance in tomato plants (Li et al., 2022). Researchers have leveraged Crispr/Cas9 technology to demonstrate that the rice annexin gene OsAnn5 plays a crucial role in cold stress tolerance at the seedling stage (Shen et al., 2017). OsNAC006 may act in heat tolerance by lowering the oxidative response, as evidenced by the unambiguous proof in rice that OsNAC006 knockout with the help of Crispr/Cas9 demonstrated in the increase of $O_2^{-\bullet}$ and H_2O_2 level. By using a Cas9-OsRR22-gRNA expression vector in rice, Crispr/Cas9-focused mutagenesis of the OsRR22 gene resulted in salt tolerance as well (Zhang et al., 2019a). The capitalization of the OsPRP gene by employing Crispr/Cas9 can be adopted to boost the ability to tolerate cold in rice as studies have demonstrated that wiping out Proline-rich Properties (PRPs) may enhance cold responsiveness (Nawaz et al., 2019). The parameters governing the growth of fruit during severe heat stress have benefited as a consequence of using the Crispr/Cas9 technology on the tomato S gene, SlAGAMOUS-LIKE 6 (SlAGL6) (Klap et al., 2017). In addition, the OsMIR528 gene was discovered to function as an advantageous regulator against salt-induced stress, and Crispr/Cas9 genome editing was executed to engineer the multipurpose gene OsBBS1, which offers resilience against saline tolerance (Khan et al., 2022). Through the implementation of Crispr/Cas9 machinery to enhance the expression of the TaPEPK2 gene, it was established that this gene can tolerate extreme temperatures in both wheat and *Arabidopsis* (Zhang et al., 2019b). Plants like soybeans suffer significantly from stress caused by abiotic components, but intriguingly, integration of Crispr/Cas9 conciliated alteration in the ABA-induced transcriptional repressors (AITRs) gene has been shown to have the ability to defend against stress caused by salinity and resist against drought stress in *Arabidopsis* (Li et al., 2022). A key protein that modulates cajal body function is coilin. The ability of coilin to withstand salt stress has been illustrated by the genetic modification of just a portion of the potato coilin gene that encodes the CTD through the Crispr/Cas9 system and the horrifying loss of development of roots and fall of leaves that resulted in the WT under salinity overload (Makhotenko et al., 2019). According to Janni M et al. (2020), Crispr/Cas9 is a highly potential practice that intentionally transformed the genetic material in more than twenty distinct crops, comprising rice, potato, cotton, maize, soybean, and sorghum (Janni et al., 2020). It is the most sophisticated machinery that is used to ameliorate intricate characteristics in plants that ultimately satisfy the stringent requirements of today's agricultural industry (Ahmed et al., 2021). In light of this, Crispr technology will soon play a key role in assisting researchers studying plants to create varieties of crops that are capable of withstanding stress and thriving in an environment that is always changing (Shelake et al., 2022).

8.3 Application of Omics to Mitigate Oxidative Stress Tolerance

Omics is a rapidly expanding and developing area of research that offers the methodology, technological expertise, inter- and transdisciplinary advancement, and furtherment needed to improve awareness and identification of the genomic and transcriptomic activities of the plant as a whole (Ali et al., 2022). As a result, several researchers are employing various kinds of omics methodologies to gain insight into how plants respond to several abiotic challenges and stresses (Gupta et al., 2018; Parida et al., 2018). The broadening of omics technology through the integration of novel technologies, from genomics and proteomics to lipidomics and phenomics. Bioinformatics techniques have already been employed to incorporate the data gathered using omics technology (Ozturk Gokce et al., 2020). Therefore, to increase both productivity and quality under a variety of abiotic challenges, plant biologists will be able to identify favorable and potential candidate genes through the successful integration of diverse omics techniques (Kumar et al., 2022).

8.3.1 Genomic Approach

Genomics is an important arm of omics technology that unlocks precious data from a given genome of an organism (Gilliham et al., 2017). The main emphasis is given to the physical coherence of the genome to acknowledge and identify the important genomic features for the whole chromosome. According to Akpinar et al. (2013), genomics has been mainly categorized into three classes: structural, functional, and comparative (Akpınar et al., 2013). Each of these categories has a distinct objective and methodology. Since it can depict the functional properties of genes and how they interact with other genes in regulatory networks, functional genomics has been successfully employed to identify many genes that are involved in tolerating abiotic stresses in plants (Zhang et al., 2017; Joshi et al., 2019). Comparative genomics can compare more than one genome to reveal the shared and distinct characteristics between them, whereas structural genomics assembles extremely detailed genetic and physical maps (Kamali & Singh, 2022). The swift growth of sequencing technologies that consist of next-generation sequencing (NGS), Roche, SOLiD applied biosystem, and Illumina (Solexa) platforms, which can perform whole plant genome sequencing and can identify conceivable genes associated with abiotic stress tolerance, has tremendously sophisticated genomics technology (Unamba et al., 2015; Mehta et al., 2019). According to Clemente et al. (2013), genomic tools are beneficial in unraveling the mutagenicity of plant stress responses through genome sequence, protein and metabolite profiles and variations in them, interactions between proteins, and mutant screens (Pérez-Clemente et al., 2013).

8.3.2 Transcriptomics Approaches

The word transcriptome implies the total population of mRNAs or messenger RNAs inside a cell or tissue organism (Lata, 2015). The phrase "transcriptome" implies the procedure of sequencing, characterizing, and quantifying every RNA molecule from the genome, thereby allowing a substantial amount of quantitative investigation into gene expression (Ozturk Gokce et al., 2020). Northern blotting, a low-output technology, was used for providing insights into the transcriptome of plants. However, as the decades progressed, new technology advancements also happened, which in the end culminated in the introduction of technologies with high throughput like serial analysis of gene expression (SAGE), expressed sequence tags (EST) sequencing, and massively parallel signature sequences (MPSS) for determining the level of transcripts (Bagati et al., 2018). The most promising method of comprehending how plants adapt to stress caused by abiotic factors, according to approximately 25000 research articles in Google Scholar, is transcriptome investigations or studies (IMADI et al., 2015). The overwhelming majority of transcriptome research has been undertaken by using the model crop *Arabidopsis thaliana*, with the remaining studies being carried out on more difficult and complicated species, which include wheat (Kayıhan & Eyidoğan, 2019). Microarrays have been used at the global level to generate expression profiles. For example, the transcriptome alterations caused by salinity stress have been analyzed using AtH1 in the Arabidopsis GeneChip from Affymetrix. This analysis has revealed that 35 percent of the genome, or over 8000 genes, had changed expression in response to salt stress (Ouakfaoui & Miki, 2005). The rice heat stress transcription element OsHsfA2e, which enhanced salt tolerance, was further identified by microarray studies on transgenic *Arabidopsis*. Two rice cultivars' leaf and root tissues were subjected to transcriptome analysis under salinity stress, and it turned out that transporter proteins including OsCHX11, OsNGC1, and OsTPC1 were expressed differently in both salt-tolerant and salt-sensitive cultivars (Senadheera et al., 2009). In comparison with microarrays, RNA sequencing is an innovative technique that can more accurately detect the transcriptome and generate more effective quantitation (Jain, 2012). This approach has already been employed in plants which include rice, sorghum, and maize to examine transcriptome and genome-wide gene expression responses in drought stress (Kayıhan & Eyidoğan, 2019).

Transcriptomics advancements have contributed to the discovery of numerous novel regulators that are capable of helping elicit many kinds of abiotic stresses, examples of which include GLYCOALKALOID METABOLISM (GAME) 9, an AP2/ERF transcription factor which has been uncovered in tomato as the primary controlling factor of the SGA pathway (Chaudhary et al., 2019). To figure out how challenges like temperatures, salt, dryness, and toxic metals impact various crops with examples including *Arabidopsis*, *Chrysanthemum*, and cassava, substantial transcriptomics research has been undertaken so far (Sharma & Negi, 2021).

8.3.3 Proteomic Approaches

The proteomic investigation delivers a novel strategy for addressing issues associated with protein verification, transcriptional profiles, post-translational modifications or PTMs, and interactions between proteins throughout stress-related and non-stressful circumstances (Hashiguchi et al., 2010; Nam et al., 2012). Proteomic studies are well-documented in their reaction to various pressures such as salt, drought, and extreme temperatures. They are used to explore both the reactions of plants to conditions as well as the intricate nature of the processes involved in biochemistry (Zhu et al., 2012; Aghaei & Komatsu, 2013; Cramer et al., 2013; Xuan et al., 2013). Several proteomic types of machinery, particularly 2-DE, DIGE, iTRAQ, and LC-MS/MS, have been successfully employed in relative proteomics examinations to recognize different protein types during circumstances of drought (Singh et al., 2022). At the time of drought, administering 2D-E to a barley crop using a MALDI-TOF-TOF mass spectrometer revealed 38 differentially abundant proteins involved in the generation of energy, the process of photosynthesis amino acid metabolism, and the stress response (Wang et al., 2015). Utilizing *Oryza sativa*, identical investigations yielded 102 proteins: 29% of such proteins have applications in bioenergy production and metabolic processes; 22% in cell protection and rescues; 21% in the synthesis of protein and storage class; 9% in cell signaling; and 19% in proteins with undetermined activities (Agrawal et al., 2016). Sunflower proteomics was employed to examine how defense/disease and energy-associated proteins declined substantially in the considerably less resilient sunflower variety and rose in the more accommodating one upon being subjected to drought-related stress. Enhanced conveyance of water, conservation of energy, and an antioxidant defense mechanism have been suggested as significant approaches to regulating the development of plants in water-limited circumstances (Ghaffari et al., 2013). In grapevine (*Vitis vinifera* L.), it was found that proteome responses to water stress often entailed an abundance of proteins for translation, as well as steroid and antioxidant metabolic processes (Cramer et al., 2013). It was reported in a study that 34 significantly expressed proteins that are associated with the process of photosynthesis reaction to stress, protein manufacturing and deterioration, transcription of genes, and signaling are present in chickpea via 2D-E with (LC-MS/MS) (Vessal et al., 2020). In chickpeas, the most prevalent proteins were those linked to metabolism. These proteins came from a variety of pathways, including the process of glycolysis known as the TCA cycle, gluconeogenesis, and the pentose phosphate pathway, which includes subunits of the ATP synthase enzyme in addition to 6-phosphogluconate dehydrogenase, phosphoglycerate mutase, fructose-bisphosphate aldolase, isocitrate dehydrogenase, transaldolase, malate dehydrogenase, and glyceraldehyde-3-phosphate dehydrogenase (Gupta et al., 2020). In a study, it was noteworthy to observe an elevation of ethylene-responsive transcription factor (ERF) and GTP-binding protein in drought-stressed leaves of cotton (*Gossypium herbaceum* L) (Deeba et al., 2012). Thus, it would be reasonable to conclude that the proteomic methodology is crucial since it enables plant physiological researchers to gain insight into what happens within a cell when exposed to an outside stimulus. Considering the methods for proteome analysis are simple to use and the results are precise, proteomics has attracted significant interest from all corners of the globe.

9. Conclusion

In summary, this article has delved into the intricate relationship between water stress and oxidative stress in plants, illuminating the adverse effects of water stress on cellular equilibrium. Essential to this defense mechanism are antioxidant enzymes, which play a critical role in shielding against the detrimental impact of reactive oxygen species (ROS), thereby augmenting our comprehension of plant innate defense systems. Leveraging antioxidant enzymes and cutting-edge biotechnological methods presents an opportunity to cultivate an eco-friendlier and more adaptable agricultural milieu. Furthermore, the exploration of innovative approaches to alleviate oxidative stress has unveiled promising avenues for plant biologists. MicroRNA, CRISPR/Cas9, and omics methodologies have emerged as influential tools, affording precision in identifying specific genetic components and signaling pathways implicated in the response to oxidative stress. The integration of these technologies not only holds intriguing potential for bolstering plants' resilience to stress but also introduces fresh possibilities for sustainable agriculture in response to evolving environmental conditions. The insights gleaned from this review make a substantial contribution to our collective understanding as we grapple with challenges posed by climate change and resource constraints. The interplay between water stress, oxidative stress, and the intricate biochemical responses elucidated in this study lays the groundwork for future research endeavors focused on developing robust, climate-resilient crops. This amalgamation of scientific breakthroughs holds promise for ensuring food security and environmental sustainability in the future.

To conclude, the responses of plants to water stress involve a complex interplay of diverse biochemical and molecular mechanisms, with key contributors fostering stress tolerance. The antioxidant system, featuring enzymes like superoxide dismutase (SOD), catalase (CAT), and ascorbate peroxidase (APX), assumes a central role in eliminating reactive oxygen species (ROS), thereby averting oxidative damage to cellular components. Proline, with its dual role as an osmoprotectant and signaling molecule, aids in preserving cellular water potential and shielding proteins and membranes from the adverse effects of water stress. Additionally, glutathione exhibits a multifaceted involvement in the water stress response. Serving as a primary antioxidant, it directly scavenges ROS and engages in enzymatic reactions that transform harmful ROS into less toxic forms. Moreover, glutathione plays a pivotal role in sustaining cellular redox balance, acting as a redox buffer and regulating the activity of redox-sensitive proteins, transcription factors, and enzymes associated with stress signaling pathways. Glycine betaine, functioning as a compatible osmolyte, contributes to osmoregulation in plants confronting harsh environmental conditions. Its accumulation during water stress aids in the preservation of chloroplasts, membrane stabilization, and the augmentation of antioxidant and vital enzyme concentrations. Proline, categorized as a non-essential amino acid, assumes various roles as an osmoprotectant, signaling molecule, and stabilizer of proteins and membranes. The upregulation of proline biosynthesis under water stress conditions contributes to heightened stress tolerance. Polyamines, encompassing putrescine, spermidine, and spermine, accumulate in response to water stress, facilitating adaptive responses. Polyamines modulate potassium channels, instigate physiological processes, and enhance ROS scavenging, thereby mitigating oxidative stress. Alpha-tocopherol, also known as Vitamin E, plays a pivotal role in counteracting the detrimental effects of ROS by functioning as a potent antioxidant, improving membrane fluidity, stabilizing the photosynthetic machinery, and modulating plant hormone signaling. Lastly, ascorbate and monodehydroascorbate, integral components of the ascorbate-glutathione cycle, contribute to ROS scavenging and the preservation of cellular redox balance. Ascorbate's involvement in abiotic stress responses extends to epigenetic mechanisms, emphasizing its diverse role in plant adaptation to water stress. Comprehending the intricate network of these biochemical and molecular responses provides valuable insights for devising strategies to enhance crop resilience to water stress, ultimately contributing to sustainable agriculture in the face of evolving climatic conditions.

References

Abdalla M, Ahmed MA, Cai G, Wankmüller F, Schwartz N, Litig O, et al. 2022. Stomatal closure during water deficit is controlled by below-ground hydraulics. *Ann Bot,* **129:** 161–170.

Abdelaal KAA, Hafez YM, El-Afry MM, Tantawy DS & Alshaal T. 2018. Effect of some osmoregulators on photosynthesis, lipid peroxidation, antioxidative capacity, and productivity of barley (Hordeum vulgare L.) under water deficit stress. *Environmental Science and Pollution Research,* **25:** 30199–30211.

Abrar MM, Sohail M, Saqib M, Akhtar J, Abbas G, Wahab HA, et al. 2022. Interactive salinity and water stress severely reduced the growth, stress tolerance, and physiological responses of guava (Psidium Guajava L.). *Sci Rep,* **12:** 18952.

Aghaei K & Komatsu S. 2013. Crop and medicinal plants proteomics in response to salt stress. *Front Plant Sci,* **4:** 1–9.

Agrawal L, Gupta S, Mishra SK, Pandey G, Kumar S, Chauhan PS, et al. 2016. Elucidation of complex nature of PEG induced drought-stress response in rice root using comparative proteomics approach. *Front Plant Sci,* **7:** 1–18.

Ahmad P, Jaleel CA, Salem MA, Nabi G & Sharma S. 2010. Roles of enzymatic and nonenzymatic antioxidants in plants during abiotic stress. *Crit Rev Biotechnol,* **30:** 161–175.

Ahmed T, Noman M, Shahid M, Muhammad S, Tahir Ul Qamar M, Ali MdA, et al. 2021. Potential application of CRISPR/Cas9 system to engineer abiotic stress tolerance in plants. *Protein Pept Lett,* **28:** 861–877.

Akdogan G, Tufekci ED, Uranbey S & Unver T. 2016. miRNA-based drought regulation in wheat. *Funct Integr Genomics,* **16:** 221–233.

Akpınar BA, Lucas SJ & Budak H. 2013. Genomics approaches for crop improvement against abiotic stress. *The Scientific World Journal,* **2013:** e361921.

Akram NA, Iqbal M, Muhammad A, Ashraf M, Al-Qurainy F & Shafiq S. 2018. Aminolevulinic acid and nitric oxide regulate oxidative defense and secondary metabolisms in canola (Brassica napus L.) under drought stress. *Protoplasma,* **255:** 163–174.

Akula R & Ravishankar GA. 2011. Influence of abiotic stress signals on secondary metabolites in plants. *Plant Signal Behav,* **6:** 1720–1731.

Alcázar R, Bueno M & Tiburcio AF. 2020. Polyamines: Small amines with large effects on plant abiotic stress tolerance. *Cells,* **9**.

Alfatih A, Wu J, Jan SU, Zhang Z, Xia J & Xiang C. 2020. Loss of rice PARAQUAT TOLERANCE 3 confers enhanced resistance to abiotic stresses and increases grain yield in field. *Plant Cell Environ,* **43:** 2743–2754.

Alhaithloul HAS. 2019. Impact of combined heat and drought stress on the potential growth responses of the desert grass Artemisia sieberi alba: Relation to biochemical and molecular adaptation. *Plants,* **8:** 416.

Ali A, Altaf MT, Nadeem MA, Karaköy T, Shah AN, Azeem H, et al. 2022. Recent advancement in OMICS approaches to enhance abiotic stress tolerance in legumes. *Front Plant Sci,* **13:** 952759.

Ali Q, Ali S, Iqbal N, Javed MT, Rizwan M, Khaliq R, et al. 2019. Alpha-tocopherol fertigation confers growth physio-biochemical and qualitative yield enhancement in field grown water deficit wheat (Triticum aestivum L.). *Sci Rep,* **9:** 12924.

Alkazemi D, Rahman A & Habra B. 2021. Alterations in glutathione redox homeostasis among adolescents with obesity and anemia. *Sci Rep,* **11:** 3034.

Amnan MAM, Pua T-L, Lau S-E, Tan BC, Yamaguchi H, Hitachi K, et al. 2021. Osmotic stress in banana is relieved by exogenous nitric oxide. *Peer J,* **9:** e10879.

Anderson JV & Davis DG. 2004. Abiotic stress alters transcript profiles and activity of glutathione S-transferase, glutathione peroxidase, and glutathione reductase in Euphorbia esula. *Physiol Plant,* **120:** 421–433.

Anjum SA, Tanveer M, Ashraf U, Hussain S, Shahzad B, Khan I, et al. 2016. Effect of progressive drought stress on growth, leaf gas exchange, and antioxidant production in two maize cultivars. *Environmental Science and Pollution Research,* **23:** 17132–17141.

Anjum SA, Wang LC, Farooq M, Hussain M, Xue LL & Zou CM. 2011. Brassinolide application improves the drought tolerance in maize through modulation of enzymatic antioxidants and leaf gas exchange. *J Agron Crop Sci,* **197:** 177–185.

Anwar A & Kim J-K. 2020. Transgenic breeding approaches for improving abiotic stress tolerance: Recent progress and future perspectives. *Int J Mol Sci,* **21:** 2695.

Arabzadeh N & Khavari-Ne RA. 2013. Effect of drought stress on superoxide dismutase activity in two species of Haloxylon aphyllum and Haloxylon persicum. *Pakistan Journal of Biological Sciences,* **16:** 351–361.

Asada K. 1999. The water-water cycle in chloroplasts: Scavenging of active oxygens and dissipation of excess photons. *Annu Rev Plant Physiol Plant Mol Biol,* **50:** 601–639.

Asada K. 2006. Production and scavenging of reactive oxygen species in chloroplasts and their functions. *Plant Physiol,* **141:** 391–396.

Asha AD, Nivetha N, Krishna GK, Thakur JK, Rathi MS, Manjunatha BS, et al. 2021. Amelioration of short-term drought stress during different growth stages in Brassica juncea by rhizobacteria mediated maintenance of ROS homeostasis. *Physiol Plant,* **172:** 1880–1893.

Aslam L, Kaur R, Hussain S, Kapoor N & Mahajan R. 2022. LC-MS/MS identification and structural characterization of isolated cyclotides from precursor sequences of Viola odorata L. petiole tissue using computational approach. *J Biosci,* **47:** 1–18.

Astier J, Gross I & Durner J. 2018. Nitric oxide production in plants: An update. *J Exp Bot,* **69:** 3401–3411.

Bagati S, Mahajan R, Nazir M, Dar AA & Zargar SM. 2018. "Omics": A gateway towards abiotic stress tolerance. pp. 1–45. *In:* Zargar SM & Zargar MY (eds), Abiotic Stress-Mediated Sensing and Signaling in Plants: An Omics Perspective, Springer, Singapore.

Bakır AG, Bolat I, Korkmaz K, Hasan MM & Kaya O. 2022. Exogenous nitric oxide and silicon applications alleviate water stress in apricots. *Life,* **12:** 1454.

Balasubramaniam M, Ayyadevara S, Ganne A, Kakraba S, Penthala NR, Du X, et al. 2019. Aggregate interactome based on protein cross-linking interfaces predicts drug targets to limit aggregation in neurodegenerative diseases. *iScience,* **20:** 248–264.

Bandurska H, Niedziela J, Pietrowska-Borek M, Nuc K, Chadzinikolau T & Radzikowska D. 2017. Regulation of proline biosynthesis and resistance to drought stress in two barley (Hordeum vulgare L.) genotypes of different origin. *Plant Physiology and Biochemistry,* **118:** 427–437.

Bangar P, Chaudhury A, Tiwari B, Kumar S, Kumari R & Bhat K. 2019. Morphophysiological and biochemical response of mungbean [Vigna radiata (L.) Wilczek] varieties at different developmental stages under drought stress. *Turkish Journal of Biology,* **43:** 58–69.

Banu MstNA, Hoque MdA, Watanabe-Sugimoto M, Matsuoka K, Nakamura Y, Shimoishi Y, et al. 2009. Proline and glycinebetaine induce antioxidant defense gene expression and suppress cell death in cultured tobacco cells under salt stress. *J Plant Physiol,* **166:** 146–156.

Barrera G, Pizzimenti S, Daga M, Dianzani C, Arcaro A, Cetrangolo GP, et al. 2018. Lipid peroxidation-derived aldehydes, 4-hydroxynonenal and malondialdehyde in aging-related disorders. *Antioxidants,* **7:** 102.

Batista PF, Costa AC, Müller C, Silva-Filho R de O, Barbosa da Silva F, Merchant A, et al. 2018. Nitric oxide mitigates the effect of water deficit in Crambe abyssinica. *Plant Physiology and Biochemistry,* **129:** 310–322.

Bera K, Dutta P & Sadhukhan S. 2022. Plant responses under abiotic stress and mitigation options towards agricultural sustainability. pp. 3–28. *In:* Roy S, Mathur P, Chakraborty AP & Saha SP (eds), Plant Stress: Challenges and Management in the New Decade, Springer International Publishing, Cham.

Besson-Bard A, Pugin A & Wendehenne D. 2008. New insights into nitric oxide signaling in plants. *Annu Rev Plant Biol,* **59:** 21–39.

Bian S & Jiang Y. 2009. Reactive oxygen species, antioxidant enzyme activities and gene expression patterns in leaves and roots of Kentucky bluegrass in response to drought stress and recovery. *Sci Hortic,* **120:** 264–270.

Biehler K & Fock H. 1996. Evidence for the contribution of the mehler-peroxidase reaction in dissipating excess electrons in drought-stressed wheat. *Plant Physiol,* **112:** 265–272.

Biswas D, Saha SC & Dey A. 2021. CRISPR-Cas genome-editing tool in plant abiotic stress-tolerance. *Plant Gene,* **26:** 100286.

Cao X, Zhu C, Zhong C, Zhang J, Wu L, Jin Q, et al. 2019. Nitric oxide synthase-mediated early nitric oxide burst alleviates water stress-induced oxidative damage in ammonium-supplied rice roots. *BMC Plant Biol,* **19:** 108.

Cassia R, Amenta M, Fernández MB, Nocioni M & Dávila V. 2019. The role of nitric oxide in the antioxidant defense of plants exposed to UV-B radiation. pp. 555–572. *In:* Reactive Oxygen, Nitrogen and Sulfur Species in Plants, John Wiley & Sons, Ltd.

Caverzan A, Passaia G, Rosa SB, Ribeiro CW, Lazzarotto F & Margis-Pinheiro M. 2012. Plant responses to stresses: Role of ascorbate peroxidase in the antioxidant protection. *Genet Mol Biol,* **35:** 1011–1019.

Cechin I, Cardoso GS, Fumis T de F & Corniani N. 2015. Nitric oxide reduces oxidative damage induced by water stress in sunflower plants. *Bragantia,* **74:** 200–206.

Chaudhary J, Khatri P, Singla P, Kumawat S, Kumari A, et al. 2019. Advances in omics approaches for abiotic stress tolerance in tomato. *Biology (Basel),* **8:** 90.

Chen D, Shao Q, Yin L, Younis A & Zheng B. 2019. Polyamine function in plants: Metabolism, regulation on development, and roles in abiotic stress responses. *Front Plant Sci,* **9:** 1945.

Chen Y, Han Y, Zhang M, Zhou S, Kong X & Wang W. 2016. Overexpression of the wheat expansin gene TaEXPA2 improved seed production and drought tolerance in transgenic tobacco plants. *PLoS One,* **11:** e0153494.

Chen Y, Shertzer HG, Schneider SN, Nebert DW & Dalton TP. 2005. Glutamate cysteine ligase catalysis: Dependence on ATP and modifier subunit for regulation of tissue glutathione levels. *J Biol Chem,* **280:** 33766–33774.

Chen Z, Young TE, Ling J, Chang S-C & Gallie DR. 2003. Increasing vitamin C content of plants through enhanced ascorbate recycling. *Proceedings of the National Academy of Sciences,* **100:** 3525–3530.

Chowdhury N, Das D, Sarki YN, Sharma M, Singha DL & Chikkaputtaiah C. (2022). Genome editing and CRISPR-Cas technology for enhancing abiotic stress tolerance in cereals. pp. 259–294. *In:* Roychoudhury A, Aftab, T & Acharya, K (eds), Omics Approach to Manage Abiotic Stress in Cereals. Springer Nature Singapore. https://doi.org/10.1007/978-981-19-0140-9_11.

Chugh V, Kaur N & Gupta AK. 2011. Evaluation of oxidative stress tolerance in maize (Zea mays L.) seedlings in response to drought. *Indian J Biochem Biophys,* **48:** 47–53.

Corpas F, Palma J, del Río L & Barroso J. 2013. Protein tyrosine nitration in higher plants grown under natural and stress conditions. *Front Plant Sci,* **4.**

Costa de Oliveira A, Marini N & Farias DR. 2014. Climate change: New breeding pressures and goals. pp. 284–293. *In:* Van Alfen N.K. (eds), Encyclopedia of Agriculture and Food Systems, Academic Press, Oxford.

Couchoud M, Der C, Girodet S, Vernoud V, Prudent M & Leborgne-Castel N. 2019. Drought stress stimulates endocytosis and modifies membrane lipid order of rhizodermal cells of Medicago truncatula in a genotype-dependent manner. *BMC Plant Biol,* **19:** 221.

Cramer GR, Van Sluyter SC, Hopper DW, Pascovici D, Keighley T & Haynes PA. 2013. Proteomic analysis indicates massive changes in metabolism prior to the inhibition of growth and photosynthesis of grapevine (Vitis vinifera L.) in response to water deficit. *BMC Plant Biol,* 13: 49.

Creissen GP, Broadbent P, Kular B, Reynolds H, Wellburn AR & Mullineaux PM. 1994. Manipulation of glutathione reductase in transgenic plants: Implications for plants' responses to environmental stress. *Proceedings of the Royal Society of Edinburgh Section B Biological Sciences,* **102:** 167–175.

Cruz de Carvalho MH. 2008. Drought stress and reactive oxygen species. *Plant Signal Behav,* **3:** 156–165.

Dalal VK & Tripathy BC. 2018. Water-stress induced downsizing of light-harvesting antenna complex protects developing rice seedlings from photo-oxidative damage. *Sci Rep,* **8:** 5955.

D'Arcy-Lameta A, Ferrari-Iliou R, Contour-Ansel D, Pham-Thi A-T & Zuily-Fodil Y. 2006. Isolation and characterization of four ascorbate peroxidase cDNAs responsive to water deficit in cowpea leaves. *Ann Bot,* **97:** 133–140.

Das K & Roychoudhury A. 2014. Reactive oxygen species (ROS) and response of antioxidants as ROS-scavengers during environmental stress in plants. *Front Environ Sci,* **2.**

Dastborhan S & Ghassemi-Golezani K. 2015. Influence of seed priming and water stress on selected physiological traits of borage. *Folia Horticulturae,* **27:** 151–159.

Deeba F, Pandey AK, Ranjan S, Mishra A, Singh R, Sharma YK, et al. 2012. Physiological and proteomic responses of cotton (Gossypium herbaceum L.) to drought stress. *Plant Physiology and Biochemistry,* **53:** 6–18.

Devasagayam TPA, Tilak JC, Boloor KK, Sane KS, Ghaskadbi SS & Lele RD. 2004. Free radicals and antioxidants in human health: Current status and future prospects. *J Assoc Physicians India,* **52:** 794–804.

Dickson KM, Gustafson CB, Young JI, Züchner S & Wang G. 2013. Ascorbate-induced generation of 5-hydroxymethylcytosine is unaffected by varying levels of iron and 2-oxoglutarate. *Biochem Biophys Res Commun,* **439:** 522–527.

Dien DC, Mochizuki T & Yamakawa T. 2019. Effect of various drought stresses and subsequent recovery on proline, total soluble sugar and starch metabolisms in Rice (Oryza sativa L.) varieties. *Plant Prod Sci,* **22:** 530–545.

Dikilitas M, Simsek E & Roychoudhury A. 2020. Role of proline and glycine betaine in overcoming abiotic stresses. pp. 1–23. *In:* Roychoudhury A & Tripathi DK (eds), Protective Chemical Agents in the Amelioration of Plant Abiotic Stress, Wiley.

Ding H, Wang B, Han Y & Li S. 2020. The pivotal function of dehydroascorbate reductase in glutathione homeostasis in plants. *J Exp Bot,* **71:** 3405–3416.

Dorion S, Ouellet JC & Rivoal J. 2021. Glutathione metabolism in plants under stress: Beyond reactive oxygen species detoxification. *Metabolites,* **11:** 641.

Du L, Huang X, Ding L, Wang Z, Tang D, Chen B, et al. 2023. TaERF87 and TaAKS1 synergistically regulate TaP5CS1/TaP5CR1-mediated proline biosynthesis to enhance drought tolerance in wheat. *New Phytologist,* **237:** 232–250.

Dvojković K, Plavšin I, Novoselović D, Šimić G, Lalić A, Čupić T, et al. 2023. Early antioxidative response to desiccant-stimulated drought stress in field-grown traditional wheat varieties. *Plants,* **12:** 249.

Ederli L, Bianchet C, Paolocci F, Alqurashi M, Gehring C & Pasqualini S. 2019. Drought stress induces a biphasic NO accumulation in Arabidopsis thaliana. *Plant Signal Behav,* **14:** e1573098.

Edwards EA, Rawsthorne S & Mullineaux Philip M. 1990. Subcellular distribution of multiple forms of glutathione reductase in leaves of pea (Pisum sativum L.). *Planta,* **180**.

Einset J & Connolly EL. 2009. Glycine betaine enhances extracellular processes blocking ROS signaling during stress. *Plant Signal Behav,* **4:** 197–199.

Engineer CB, Hashimoto-Sugimoto M, Negi J, Israelsson-Nordström M, Azoulay-Shemer T, Rappel W-J, et al. 2016. CO_2 sensing and CO_2 regulation of stomatal conductance: Advances and open questions. *Trends Plant Sci,* **21:** 16–30.

Fagan RL & Palfey BA. 2010. Flavin-dependent enzymes. pp. 37–113. *In:* Comprehensive Natural Products II, Elsevier.

Fahad S, Bajwa AA, Nazir U, Anjum SA, Farooq A, Zohaib A, et al. 2017. Crop production under drought and heat stress: Plant responses and management options. *Front Plant Sci,* **8:** 1147.

Faize M, Burgos L, Faize L, Piqueras A, Nicolas E, Barba-Espin G, et al. 2011. Involvement of cytosolic ascorbate peroxidase and Cu/Zn-superoxide dismutase for improved tolerance against drought stress. *J Exp Bot,* **62:** 2599–2613.

Fan S, Yuan Z, Feng L, Wang X, Ding X & Zhen H. 2011. Effects of drought stress on physiological and biochemical parameters of Dahlia pinnata. *Ying Yong Sheng Tai Xue Bao,* **22:** 651–657.

Farooq M, Basra SMA, Wahid A, Cheema ZA, Cheema MA & Khaliq A. 2008. Physiological role of exogenously applied glycinebetaine to improve drought tolerance in fine grain aromatic rice (Oryza sativa L.). *J Agron Crop Sci,* **194:** 325–333.

Ferioun M, Srhiouar N, Bouhraoua S, El Ghachtouli N & Louahlia S. 2023. Physiological and biochemical changes in Moroccan barley (Hordeum vulgare L.) cultivars submitted to drought stress. *Heliyon,* **9:** e13643.

Fichman Y, Gerdes SY, Kovács H, Szabados L, Zilberstein A & Csonka LN. 2015. Evolution of proline biosynthesis: Enzymology, bioinformatics, genetics, and transcriptional regulation. *Biological Reviews,* **90:** 1065–1099.

Figueroa-Soto CG & Valenzuela-Soto EM. 2018. Glycine betaine rather than acting only as an osmolyte also plays a role as regulator in cellular metabolism. *Biochimie,* **147:** 89–97.

Flexas J, Ribas-Carbó M, Bota J, Galmés J, Henkle M, Martínez-Cañellas S, et al. 2006. Decreased Rubisco activity during water stress is not induced by decreased relative water content but related to conditions of low stomatal conductance and chloroplast CO_2 concentration. *New Phytologist,* **172:** 73–82.

Fortes AM, Costa J, Santos F, Seguí-Simarro JM, Palme K, Altabella T, et al. 2011. Arginine decarboxylase expression, polyamines biosynthesis and reactive oxygen species during organogenic nodule formation in hop. *Plant Signal Behav,* **6:** 258–269.

Foster MW, Liu L, Zeng M, Hess DT & Stamler JS. 2009. A genetic analysis of nitrosative stress. *Biochemistry,* **48:** 792–799.

Foyer CH & Noctor G. 2011. Ascorbate and glutathione: The heart of the redox hub. *Plant Physiol,* **155:** 2–18.

Fritsche S, Wang X & Jung C. 2017. Recent advances in our understanding of tocopherol biosynthesis in plants: An overview of key genes, functions, and breeding of vitamin E improved crops. *Antioxidants (Basel),* **6.**

Fujita M & Hasanuzzaman M. 2022. Approaches to enhancing antioxidant defense in plants. *Antioxidants,* **11:** 925.

Furlan AL, Bianucci E, Giordano W, Castro S & Becker DF. 2020. Proline metabolic dynamics and implications in drought tolerance of peanut plants. *Plant Physiology and Biochemistry,* **151:** 566–578.

Galli F, Bonomini M, Bartolini D, Zatini L, Reboldi G, Marcantonini G, et al. 2022. Vitamin E (Alpha-Tocopherol) metabolism and nutrition in chronic kidney disease. *Antioxidants,* **11:** 989.

Ghaffari M, Toorchi M, Valizadeh M & Komatsu S. 2013. Differential response of root proteome to drought stress in drought sensitive and tolerant sunflower inbred lines. *Functional Plant Biology,* **40:** 609.

Gill SS, Anjum NA, Hasanuzzaman M, Gill R, Trivedi DK, Ahmad I, et al. 2013. Glutathione and glutathione reductase: A boon in disguise for plant abiotic stress defense operations. *Plant Physiology and Biochemistry,* **70:** 204–212.

Gilliham M, Able JA & Roy SJ. 2017. Translating knowledge about abiotic stress tolerance to breeding programmes. *The Plant Journal,* **90:** 898–917.

Guo YY, Yu HY, Yang MM, Kong DS & Zhang YJ. 2018. Effect of drought stress on lipid peroxidation, osmotic adjustment and antioxidant enzyme activity of leaves and roots of Lycium ruthenicum Murr. seedling. *Russian Journal of Plant Physiology,* **65:** 244–250.

Gupta KJ, Kaladhar VC, Fitzpatrick TB, Fernie AR, Møller IM & Loake GJ. 2022. Nitric oxide regulation of plant metabolism. *Mol Plant,* **15:** 228–242.

Gupta N, Singh A, Zahra S & Kumar S. 2018. PtRFdb: A database for plant transfer RNA-derived fragments. *Database,* 2018.

Gupta S, Mishra SK, Misra S, Pandey V, Agrawal L, Nautiyal CS, et al. 2020. Revealing the complexity of protein abundance in chickpea root under drought-stress using a comparative proteomics approach. *Plant Physiology and Biochemistry,* **151:** 88–102.

Hajizadeh HS, Azizi S, Rasouli F & Okatan V. 2022. Modulation of physiological and biochemical traits of two genotypes of Rosa damascena Mill. by SiO_2-NPs under in vitro drought stress. *BMC Plant Biol,* **22:** 538.

Han C, Liu Q & Yang Y. 2009. Short-term effects of experimental warming and enhanced ultraviolet-B radiation on photosynthesis and antioxidant defense of Picea asperata seedlings. *Plant Growth Regul,* **58:** 153–162.

Hancock JT. 2020. Nitric oxide signaling in plants. *Plants,* **9:** 1550.

Hao G, Du X, Zhao F, Shi R & Wang J. 2009. Role of nitric oxide in UV-B-induced activation of PAL and stimulation of flavonoid biosynthesis in Ginkgo biloba callus. *Plant Cell, Tissue and Organ Culture (PCTOC),* **97:** 175–185.

Haroldsen VM, Chi-Ham CL, Kulkarni S, Lorence A & Bennett AB. 2011. Constitutively expressed DHAR and MDHAR influence fruit, but not foliar ascorbate levels in tomato. *Plant Physiology and Biochemistry,* **49:** 1244–1249.

Hasanuzzaman M, Bhuyan MHM, Zulfiqar F, Raza A, Mohsin S, Mahmud J, et al. 2020. Reactive oxygen species and antioxidant defense in plants under abiotic stress: Revisiting the crucial role of a universal defense regulator. *Antioxidants,* **9:** 681.

Hasanuzzaman M & Fujita M. 2011. Selenium pretreatment upregulates the antioxidant defense and methylglyoxal detoxification system and confers enhanced tolerance to drought stress in rapeseed seedlings. *Biol Trace Elem Res,* **143:** 1758–1776.

Hasanuzzaman M, Hossain MA, da Silva JAT & Fujita M. 2012. Plant response and tolerance to abiotic oxidative stress: Antioxidant defense is a key factor. pp. 261–315. Crop Stress and its Management: Perspectives and Strategies, Springer Netherlands, Dordrecht.

Hasanuzzaman M, Nahar K, Anee TI & Fujita M. 2017. Glutathione in plants: Biosynthesis and physiological role in environmental stress tolerance. *Physiology and Molecular Biology of Plants,* **23:** 249–268.

Hasanuzzaman M, Raihan MdRH, Masud AAC, Rahman K, Nowroz F, Rahman M, et al. 2021. Regulation of reactive oxygen species and antioxidant defense in plants under salinity. *Int J Mol Sci,* **22:** 9326.

Hashem A, Kumar A, Al-Dbass AM, Alqarawi AA, Al-Arjani A-BF, Singh G, et al. 2019. Arbuscular mycorrhizal fungi and biochar improves drought tolerance in chickpea. *Saudi J Biol Sci,* **26:** 614–624.

Hashiguchi A, Ahsan N & Komatsu S. 2010. Proteomics application of crops in the context of climatic changes. *Food Research International,* **43:** 1803–1813.

Hayat S, Hayat Q, Alyemeni MN, Wani AS, Pichtel J & Ahmad A. 2012. Role of proline under changing environments: A review. *Plant Signal Behav,* **7:** 1456–1466.

Hernandez-Leon SG & Valenzuela-Soto EM. 2023. Glycine betaine is a phytohormone-like plant growth and development regulator under stress conditions. *J Plant Growth Regul,* **42:** 5029–5040.

Hou P, Wang F, Luo B, Li A, Wang C, Shabala L, et al. 2021. Antioxidant enzymatic activity and osmotic adjustment as components of the drought tolerance mechanism in Carex duriuscula. *Plants,* **10:** 436.

Huan X, Wang X, Zou S, Zhao K, Han Y & Wang S. 2023. Transcription factor ERF194 modulates the stress-related physiology to enhance drought tolerance of poplar. *Int J Mol Sci,* **24:** 788.

Huang Y, Guo Y, Liu Y, Zhang F, Wang Z, Wang H, et al. 2018. 9-cis-epoxycarotenoid dioxygenase 3 regulates plant growth and enhances multi-abiotic stress tolerance in rice. *Front Plant Sci,* **9:** 162.

Huseynova IM, Aliyeva DR & Aliyev JA. 2014. Subcellular localization and responses of superoxide dismutase isoforms in local wheat varieties subjected to continuous soil drought. *Plant Physiology and Biochemistry,* **81:** 54–60.

Huseynova IM, Aliyeva DR, Mammadov AC & Aliyev JA. 2015. Hydrogen peroxide generation and antioxidant enzyme activities in the leaves and roots of wheat cultivars subjected to long-term soil drought stress. *Photosynth Res,* **125:** 279–289.

Hussain HA, Men S, Hussain S, Chen Y, Ali S, Zhang S, et al. 2019. Interactive effects of drought and heat stresses on morpho-physiological attributes, yield, nutrient uptake and oxidative status in maize hybrids. *Sci Rep,* **9:** 3890.

Imadi SR, Kazi AG, Ahanger MA, Gucel S & Ahmad P. 2015. Plant transcriptomics and responses to environmental stress: An overview. *Journal of Genetics,* **94(3):** 525–537. https://doi.org/10.1007/s12041-015-0545-6.

Iqbal Y, Raouf Malik A, Iqbal T, Hammad Aziz M, Ahmed F, Abolaban FA, et al. 2021. Green synthesis of ZnO and Ag-doped ZnO nanoparticles using Azadirachta indica leaves: Characterization and their potential antibacterial, antidiabetic, and wound-healing activities. *Mater Lett,* **305:** 130671.

Jain M. 2012. Next-generation sequencing technologies for gene expression profiling in plants. *Brief Funct Genomics,* **11:** 63–70.

Janni M, Gullì M, Maestri E, Marmiroli M, Valliyodan B, Nguyen HT, et al. 2020. Molecular and genetic bases of heat stress responses in crop plants and breeding for increased resilience and productivity. *J Exp Bot,* **71:** 3780–3802.

Joshi R, Gupta BK, Pareek A, Singh MB & Singla-Pareek SL. 2019. Functional genomics approach towards dissecting out abiotic stress tolerance trait in plants. pp. 1–24. *In:* Rajpal VR, Sehgal D, Kumar A & Raina SN (eds), Genetic Enhancement of Crops for Tolerance to Abiotic Stress: Mechanisms and Approaches, Vol. I, Springer International Publishing, Cham.

Kalinina E & Novichkova M. 2021. Glutathione in protein redox modulation through S-glutathionylation and S-nitrosylation. *Molecules,* **26:** 435.

Kalt-Torres W, Burke JJ & Anderson JM. 1984. Chloroplast glutathione reductase: Purification and properties. *Physiol Plant,* **61:** 271–278.

Kamali S & Singh A. 2022. Genomics and transcriptomics approaches to understand abiotic stress response in rice. pp. 405–433. *In:* Roychoudhury A, Aftab T & Acharya K (eds), Omics Approach to Manage Abiotic Stress in Cereals, Springer Nature, Singapore.

Kar JR, Hallsworth JE & Singhal RS. 2015. Fermentative production of glycine betaine and trehalose from acid whey using Actinopolyspora halophila (MTCC 263). *Environ Technol Innov,* **3:** 68–76.

Kar RK. 2011. Plant responses to water stress: Role of reactive oxygen species. *Plant Signal Behav,* **6:** 1741–1745.

Kaur G & Asthir B. 2015. Proline: A key player in plant abiotic stress tolerance. *Biol Plant,* **59:** 609–619.

Kayıhan C & Eyidoğan F. 2019. Omics in oxidative stress tolerance in crops. pp. 195–224. *In:* Hasanuzzaman M, Fotopoulos V, Nahar K & Fujita M (eds), Reactive Oxygen, Nitrogen and Sulfur Species in Plants, Wiley.

Khaleghi A, Naderi R, Brunetti C, Maserti BE, Salami SA & Babalar M. 2019. Morphological, physicochemical and antioxidant responses of Maclura pomifera to drought stress. *Sci Rep,* **9:** 19250.

Khan I, Zhang Y, Akbar F & Khan J. 2022. Abiotic stress tolerance in cereals through genome editing. pp. 295–319. *In:* Roychoudhury A, Aftab T & Acharya K (eds), Omics Approach to Manage Abiotic Stress in Cereals, Springer Nature, Singapore.

Khedr AHA. 2003. Proline induces the expression of salt-stress-responsive proteins and may improve the adaptation of Pancratium maritimum L. to salt-stress. *J Exp Bot,* **54:** 2553–2562.

Kim D, Alptekin B & Budak H. 2018. CRISPR/Cas9 genome editing in wheat. *Funct Integr Genomics,* **18:** 31–41.

Kim T-H, Böhmer M, Hu H, Nishimura N & Schroeder JI. 2010. Guard cell signal transduction network: Advances in understanding abscisic acid, CO_2, and Ca^{2+} signaling. *Annu Rev Plant Biol,* **61:** 561–591.

Klap C, Yeshayahou E, Bolger AM, Arazi T, Gupta SK, Shabtai S, et al. 2017. Tomato facultative parthenocarpy results from Sl AGAMOUS-LIKE 6 loss of function. *Plant Biotechnol J,* **15:** 634–647.

Koffler BE, Luschin-Ebengreuth N, Stabentheiner E, Müller M & Zechmann B. 2014. Compartment specific response of antioxidants to drought stress in Arabidopsis. *Plant Science,* **227:** 133–144.

Kováčik J, Klejdus B & Bačkor M. 2009. Nitric oxide signals ROS scavenger-mediated enhancement of PAL activity in nitrogen-deficient Matricaria chamomilla roots: Side effects of scavengers. *Free Radic Biol Med,* **46:** 1686–1693.

Kulcheski FR, De Oliveira LF, Molina LG, Almerão MP, Rodrigues FA, Marcolino J, et al. 2011. Identification of novel soybean microRNAs involved in abiotic and biotic stresses. *BMC Genomics,* **12:** 307.

Kumar P. 2022. Measurement of ascorbate peroxidase activity in sorghum. *Bio Protoc,* **12**.

Kumar P, Choudhary M, Halder T, Prakash NR, Singh V, et al. 2022. Salinity stress tolerance and omics approaches: Revisiting the progress and achievements in major cereal crops. *Heredity (Edinb),* **128:** 497–518.

Lascano HR, Casano LM, Melchiorre MN & Trippi VS. 2001. Biochemical and molecular characterisation of wheat chloroplastic glutathione reductase. *Biol Plant,* **44:** 509–516.

Lata MC. 2015. Advances in omics for enhancing abiotic stress tolerance in millets. *Proceedings of the Indian National Science Academy,* **81**.

Lee S & Park C-M. 2012. Regulation of reactive oxygen species generation under drought conditions in Arabidopsis. *Plant Signal Behav,* **7:** 599–601.

Lehmann S, Funck D, Szabados L & Rentsch D. 2010. Proline metabolism and transport in plant development. *Amino Acids,* **39:** 949–962.

León J & Costa-Broseta Á. 2020. Present knowledge and controversies, deficiencies, and misconceptions on nitric oxide synthesis, sensing, and signaling in plants. *Plant Cell Environ,* **43:** 1–15.

Levin G & Schuster G. 2023. LHC-like proteins: The guardians of photosynthesis. *Int J Mol Sci,* **24:** 2503.

Li J, Fu F, An M, Zhou S, She Y & Li W. 2013. Differential expression of microRNAs in response to drought stress in maize. *J Integr Agric,* **12:** 1414–1422.

Li S-B, Xie Z-Z, Hu C-G & Zhang J-Z. 2016. A review of auxin response factors (ARFs) in plants. *Front Plant Sci,* **7:** 47.

Li X, Xu S, Fuhrmann-Aoyagi MB, Yuan S, Iwama T, Kobayashi M, et al. 2022. CRISPR/Cas9 technique for temperature, drought, and salinity stress responses. *Curr Issues Mol Biol,* **44:** 2664–2682.

Li Y, Hai R, Du X, Jiang X & Lu H. 2009. Over-expression of a Populus peroxisomal ascorbate peroxidase (PpAPX) gene in tobacco plants enhances stress tolerance. *Plant Breeding,* **128:** 404–410.

Lin T, Rao M, Lu H, Chiou C, Lin S, Chao H, et al. 2018. A role for glutathione reductase and glutathione in the tolerance of <scp> *Chlamydomonas reinhardtii* </scp> to photo-oxidative stress. *Physiol Plant,* **162:** 35–48.

Lingvay M, Akhtar P, Sebők-Nagy K, Páli T & Lambrev PH. 2020. Photobleaching of chlorophyll in light-harvesting complex II increases in lipid environment. *Front Plant Sci,* **11**.

Lisar SYS, Motafakkerazad R, Hossain MM & Rahman IMM. 2012. Water stress in plants: Causes, effects and responses. pp. 1-14. *In:* Rahman, I.M.M. & Hasegawa, H. (eds), Water Stress, InTech, Croatia.

Liu J-H, Wang W, Wu H, Gong X & Moriguchi T. 2015. Polyamines function in stress tolerance: From synthesis to regulation. *Front Plant Sci,* **6**.

Liu R, Jiao T, Li J, Wang A, Li Y, Wu S, et al. 2022. Drought-induced increase in catalase activity improves cotton yield when grown under water-limiting field conditions. *J Agron Crop Sci,* **208:** 853–867.

Lotfi R, Pessarakli M, Gharavi-Kouchebagh P & Khoshvaghti H. 2015. Physiological responses of Brassica napus to fulvic acid under water stress: Chlorophyll a fluorescence and antioxidant enzyme activity. *Crop J,* **3:** 434–439.

Lu B, Luo X, Gong C & Bai J. 2021. Overexpression of γ-glutamylcysteine synthetase gene from Caragana korshinskii decreases stomatal density and enhances drought tolerance. *BMC Plant Biol,* **21:** 444.

Lu SC. 2013. Glutathione synthesis. *Biochimica et Biophysica Acta (BBA) - General Subjects,* **1830:** 3143–3153.

Luo P, Mu Y, Wang S, Zhu W, Mishra BK, Huo A, et al. 2022. Exploring sustainable solutions for the water environment in Chinese and Southeast Asian cities. *Ambio,* **51:** 1199–1218.

Ma Y-H, Ma F-W, Wang Y-H & Zhang J-K. 2011. The responses of the enzymes related with ascorbate–glutathione cycle during drought stress in apple leaves. *Acta Physiol Plant,* **33**.

Madhu Null, Kaur A, Tyagi S, Shumayla Null, Singh K & Upadhyay SK. 2022. Exploration of glutathione reductase for abiotic stress response in bread wheat (Triticum aestivum L.). *Plant Cell Rep,* **41:** 639–654.

Maghsoudi K, Emam Y, Niazi A, Pessarakli M & Arvin MJ. 2018. P5CS expression level and proline accumulation in the sensitive and tolerant wheat cultivars under control and drought stress conditions in the presence/absence of silicon and salicylic acid. *J Plant Interact,* **13:** 461–471.

Mahan JR & Burke JJ. 1987. Purification and characterization of glutathione reductase from corn mesophyll chloroplasts. *Physiol Plant,* **71:** 352–358.

Makhotenko AV, Khromov AV, Snigir EA, Makarova SS, Makarov VV, Suprunova TP, et al. 2019. Functional analysis of coilin in virus resistance and stress tolerance of potato Solanum tuberosum using CRISPR-Cas9 editing. *Dokl Biochem Biophys,* **484:** 88–91.

Makino A, Miyake C & Yokota A. 2002. Physiological functions of the water–water cycle (Mehler reaction) and the cyclic electron flow around PSI in rice leaves. *Plant Cell Physiol,* **43:** 1017–1026.

Mao Y, Zhang H, Xu N, Zhang B, Gou F & Zhu J-K. 2013. Application of the CRISPR–Cas system for efficient genome engineering in plants. *Mol Plant,* **6:** 2008–2011.

Mattos LM & Moretti CL. 2016. Oxidative stress in plants under drought conditions and the role of different enzymes. *Enzyme Engineering,* **5**.

Mehta S, James D & Reddy MK. 2019. Omics technologies for abiotic stress tolerance in plants: Current status and prospects. pp. 1–34. Recent Approaches in Omics for Plant Resilience to Climate Change, Springer International Publishing, Cham.

Menconi M, Sgherri CLM, Pinzino C & Navari-lzzo F. 1995. Activated oxygen production and detoxification in wheat plants subjected to a water deficit programme. *J Exp Bot,* **46:** 1123–1130.

Mishra N, Jiang C, Chen L, Paul A, Chatterjee A & Shen G. 2023. Achieving abiotic stress tolerance in plants through antioxidative defense mechanisms. *Frontiers in Plant Science,* **14:** 1110622. https://doi.org/10.3389/fpls.2023.1110622.

Mittler R, Zandalinas SI, Fichman Y & Van Breusegem F. 2022. Reactive oxygen species signalling in plant stress responses. *Nat Rev Mol Cell Biol,* **23:** 663–679.

Mohammadi H, Ghorbanpour M & Brestic M. 2018. Exogenous putrescine changes redox regulations and essential oil constituents in field-grown Thymus vulgaris L. under well-watered and drought stress conditions. *Ind Crops Prod,* **122:** 119–132.

Molassiotis A & Fotopoulos V. 2011. Oxidative and nitrosative signaling in plants. *Plant Signal Behav,* **6:** 210–214.

Molina-Rueda JJ, Tsai CJ & Kirby EG. 2013. The populus superoxide dismutase gene family and its responses to drought stress in transgenic poplar overexpressing a pine cytosolic glutamine synthetase (GS1a). *PLoS One,* **8:** e56421.

Mudalkar S, Sreeharsha RV & Reddy AR. 2017. Involvement of glyoxalases and glutathione reductase in conferring abiotic stress tolerance to Jatropha curcas L. *Environ Exp Bot,* **134:** 141–150.

Muñoz-Clares RA, Riveros-Rosas H, Garza-Ramos G, González-Segura L, Mújica-Jiménez C & Julián-Sánchez A. 2014. Exploring the evolutionary route of the acquisition of betaine aldehyde dehydrogenase activity by plant ALDH10 enzymes: Implications for the synthesis of the osmoprotectant glycine betaine. *BMC Plant Biol,* **14:** 149.

Murali M, Singh SB, Gowtham HG, Shilpa N, Prasad M, Aiyaz M, et al. 2021. Induction of drought tolerance in Pennisetum glaucum by ACC deaminase producing PGPR-Bacillus amyloliquefaciens through antioxidant defense system. *Microbiol Res,* **253:** 126891.

Murray Stewart T, Dunston TT, Woster PM & Casero RA. 2018. Polyamine catabolism and oxidative damage. *Journal of Biological Chemistry,* **293:** 18736–18745.

Murshed R, Lopez-Lauri F & Sallanon H. 2013. Effect of water stress on antioxidant systems and oxidative parameters in fruits of tomato (Solanum lycopersicon L, cv. Micro-tom). *Physiology and Molecular Biology of Plants,* **19:** 363–378.

Nahar K, Hasanuzzaman M, Alam MdM & Fujita M. 2015. Glutathione-induced drought stress tolerance in mung bean: Coordinated roles of the antioxidant defence and methylglyoxal detoxification systems. *AoB Plants,* **7:** plv069.

Nam MH, Huh SM, Kim KM, Park WW, Seo JB, Cho K, et al. 2012. Comparative proteomic analysis of early salt stress-responsive proteins in roots of SnRK2 transgenic rice. *Proteome Sci,* **10:** 25.

Nam T. 2011. Lipid peroxidation and its toxicological implications. *Toxicol Res,* **27:** 1–6.

Nandi A, Yan L-J, Jana CK & Das N. 2019. Role of catalase in oxidative stress- and age-associated degenerative diseases. *Oxid Med Cell Longev,* **2019:** 1–19.

Naser Alavi SM, Arvin MJ & Manoochehri Kalantari K. 2014. Salicylic acid and nitric oxide alleviate osmotic stress in wheat (Triticum aestivum L.) seedlings. *J Plant Interact,* **9:** 683–688.

Nawaz G, Han Y, Usman B, Liu F, Qin B & Li R. 2019. Knockout of OsPRP1, a gene encoding proline-rich protein, confers enhanced cold sensitivity in rice (Oryza sativa L.) at the seedling stage. *3 Biotech,* **9:** 254.

Nawaz M & Wang Z. 2020. Abscisic acid and glycine betaine mediated tolerance mechanisms under drought stress and recovery in Axonopus compressus: A new insight. *Sci Rep,* **10:** 6942.

Neill S, Bright J, Desikan R, Hancock J, Harrison J & Wilson I. 2008. Nitric oxide evolution and perception. *J Exp Bot,* **59:** 25–35.

Niu T, Zhang J, Li J, Gao X, Ma H, Gao Y, et al. 2023. Effects of exogenous glycine betaine and cycloleucine on photosynthetic capacity, amino acid composition, and hormone metabolism in Solanum melongena L. *Sci Rep,* **13:** 7626.

Noctor G & Foyer CH. 1998. Ascorbate and glutathione: Keeping active oxygen under control. *Annu Rev Plant Physiol Plant Mol Biol,* **49:** 249–279.

Nounjan N, Nghia PT & Theerakulpisut P. 2012. Exogenous proline and trehalose promote recovery of rice seedlings from salt-stress and differentially modulate antioxidant enzymes and expression of related genes. *J Plant Physiol,* **169:** 596–604.

Nxele X, Klein A & Ndimba BK. 2017. Drought and salinity stress alters ROS accumulation, water retention, and osmolyte content in sorghum plants. *South African Journal of Botany,* **108:** 261–266.

Ortbauer M. 2013. Abiotic stress adaptation: Protein folding stability and dynamics. pp. 1–21. *In:* Vahdati K (eds), Abiotic Stress – Plant Responses and Applications in Agriculture, InTech.

Osakabe Y, Osakabe K, Shinozaki K & Tran L-SP. 2014. Response of plants to water stress. *Front Plant Sci,* **5**.

Osakabe Y, Watanabe T, Sugano SS, Ueta R, Ishihara R, Shinozaki K, et al. 2016. Optimization of CRISPR/Cas9 genome editing to modify abiotic stress responses in plants. *Sci Rep,* **6:** 26685.

Osakabe Y & Osakabe K. 2017. Genome editing to improve abiotic stress responses in plants. *Prog Mol Biol Transl Sci.*, **149:** 99–109. doi: 10.1016/bs.pmbts.2017.03.007. Epub 2017 May 3. PMID: 28712503.

Ouakfaoui S El & Miki B. 2005. The stability of the Arabidopsis transcriptome in transgenic plants expressing the marker genes nptII and uidA. *The Plant Journal,* **41:** 791–800.

Ozturk Gokce ZN, Akbas S, Ayten S, Azimi MH, Das R, Guven SB, et al. 2020. Abiotic stress tolerance in field crops: Integration of omics approaches. pp. 503–526. *In:* Hasanuzzaman M (eds), Agronomic Crops, Springer Singapore.

Pagano L, Rossi R, Paesano L, Marmiroli N & Marmiroli M. 2021. miRNA regulation and stress adaptation in plants. *Environ Exp Bot,* **184:** 104369.

Pan R, Buitrago S, Feng Z, Abou-Elwafa SF, Xu L, Li C, et al. 2022. HvbZIP21, a novel transcription factor from wild barley confers drought tolerance by modulating ROS scavenging. *Front Plant Sci,* **13:** 878459.

Pandita D. 2019. Plant MIRnome: miRNA biogenesis and abiotic stress response. pp. 449–474. *In:* Hasanuzzaman M, Hakeem KR, Nahar K & Alharby HF (eds), Plant Abiotic Stress Tolerance, Springer International Publishing, Cham.

Parida AK, Panda A & Rangani J. 2018. Metabolomics-guided elucidation of abiotic stress tolerance mechanisms in plants. pp. 89–131. *In:* David B. Teplow (eds), Plant Metabolites and Regulation Under Environmental Stress, Elsevier.

Park J-J, Dempewolf E, Zhang W & Wang Z-Y. 2017. RNA-guided transcriptional activation via CRISPR/dCas9 mimics overexpression phenotypes in Arabidopsis. *PLoS One,* 12: e0179410.

Parkes MW & Horwitz P. 2009. Water, ecology and health: Ecosystems as settings for promoting health and sustainability. *Health Promot Int,* **24:** 94–102.

Patel PK & Hemantaran A. 2012. Antioxidant defence system in chickpea (Cicer arietinum L.): Influence by drought stress implemented at pre- and post-anthesis stage. *American Journal of Plant Physiology,* **7:** 164–173.

Pérez-Clemente RM, Vives V, Zandalinas SI, López-Climent MF, Muñoz V & Gómez-Cadenas A. 2013. Biotechnological approaches to study plant responses to stress. *Biomed Res Int,* **2013:** 1–10.

Pissolato MD, Silveira NM, Prataviera PJC, Machado EC, Seabra AB, Pelegrino MT, et al. 2020. Enhanced nitric oxide synthesis through nitrate supply improves drought tolerance of sugarcane plants. *Front Plant Sci,* **11**.

Pizzimenti S, Ciamporcero E, Daga M, Pettazzoni P, Arcaro A, Cetrangolo G, et al. 2013. Interaction of aldehydes derived from lipid peroxidation and membrane proteins. *Front Physiol,* **4**.

Ponnampalam EN, Kiani A, Santhiravel S, Holman BWB, Lauridsen C & Dunshea FR. 2022. The importance of dietary antioxidants on oxidative stress, meat and milk production, and their preservative aspects in farm animals: Antioxidant action, animal health, and product quality-invited review. *Animals (Basel),* **12**.

Pontin M, Murcia G, Bottini R, Fontana A, Bolcato L & Piccoli P. 2021. Nitric oxide and abscisic acid regulate osmoprotective and antioxidative mechanisms related to water stress tolerance of grapevines. *Aust J Grape Wine Res,* **27:** 392–405.

Pospíšil P. 2016. Production of reactive oxygen species by photosystem II as a response to light and temperature stress. *Front Plant Sci,* **7**.

Potters G, Horemans N & Jansen MAK. 2010. The cellular redox state in plant stress biology – A charging concept. *Plant Physiology and Biochemistry,* **48:** 292–300.

Pyngrope S, Bhoomika K & Dubey RS. 2013. Reactive oxygen species, ascorbate–glutathione pool, and enzymes of their metabolism in drought-sensitive and tolerant indica rice (Oryza sativa L.) seedlings subjected to progressing levels of water deficit. *Protoplasma,* **250:** 585–600.

Raederstorff D, Wyss A, Calder PC, Weber P & Eggersdorfer M. 2015. Vitamin E function and requirements in relation to PUFA. *British Journal of Nutrition,* **114:** 1113–1122.

Rao JN & Wang J-Y. 2010. Regulation of Gastrointestinal Mucosal Growth. Morgan & Claypool Life Sciences, San Rafael (CA).

Roca Paixão JF, Gillet F-X, Ribeiro TP, Bournaud C, Lourenço-Tessutti IT, Noriega DD, et al. 2019. Improved drought stress tolerance in Arabidopsis by CRISPR/dCas9 fusion with a Histone AcetylTransferase. *Sci Rep,* **9:** 8080.

Da Rocha IMA, Vitorello VA, Silva JS, Ferreira-Silva SL, Viégas RA, Silva EN, et al. 2012. Exogenous ornithine is an effective precursor and the δ-ornithine amino transferase pathway contributes to proline accumulation under high N recycling in salt-stressed cashew leaves. *J Plant Physiol,* **169:** 41–49.

Romero-Puertas MC, Corpas FJ, Sandalio LM, Leterrier M, Rodríguez-Serrano M, Del Río LA, et al. 2006. Glutathione reductase from pea leaves: Response to abiotic stress and characterization of the peroxisomal isozyme. *New Phytol,* **170:** 43–52.

Ru C, Hu X, Chen D & Wang W. 2023. Droughts and thermo-priming enhance acclimation to later drought and heat stress in maize seedlings by improving leaf physiological activity. *Agronomy,* **13:** 1124.

Sachdev S, Ansari SA, Ansari MI, Fujita M & Hasanuzzaman M. 2021. Abiotic stress and reactive oxygen species: Generation, signaling, and defense mechanisms. *Antioxidants,* **10:** 277.

Sagar NA, Tarafdar S, Agarwal S, Tarafdar A & Sharma S. 2021. Polyamines: Functions, metabolism, and role in human disease management. *Medical Sciences,* **9:** 44.

Sahay S, Khan E & Gupta M. 2019. Nitric oxide and abscisic acid protects against PEG-induced drought stress differentially in Brassica genotypes by combining the role of stress modulators, markers and antioxidants. *Nitric Oxide,* **89:** 81–92.

Sakya AT, Sulistyaningsih E, Indradewa D & Purwanto BH. 2018. Physiological characters and tomato yield under drought stress. *IOP Conf Ser Earth Environ Sci,* **200:** 012043.

Sánchez-Rodríguez E, Rubio-Wilhelmi M del M, Blasco B, Leyva R, Romero L & Ruiz JM. 2012. Antioxidant response resides in the shoot in reciprocal grafts of drought-tolerant and drought-sensitive cultivars in tomato under water stress. *Plant Sci,* **188–189:** 89–96.

Sánchez-Rodríguez E, Romero L & Ruiz JM. 2016. Accumulation of free polyamines enhances the antioxidant response in fruits of grafted tomato plants under water stress. *J Plant Physiol,* **190:** 72–78.

Sarker U & Oba S. 2018. Catalase, superoxide dismutase and ascorbate-glutathione cycle enzymes confer drought tolerance of Amaranthus tricolor. *Sci Rep,* **8:** 16496.

Scandalios JG. 2001. Molecular responses to oxidative stress. pp. 181–208. *In:* Hawkesford MJ & Buchner P (eds), Molecular Analysis of Plant Adaptation to the Environment, Springer Netherlands, Dordrecht.

Schieber M & Chandel NS. 2014. ROS function in redox signaling and oxidative stress. *Current Biology,* **24(10):** R453–R462. https://doi.org/10.1016/j.cub.2014.03.034.

Schneider JR, Müller M, Klein VA, Rossato-Grando LG, Barcelos RP, Dalmago GA, et al. 2020. Soybean plant metabolism under water deficit and xenobiotic and antioxidant agent application. *Biology (Basel),* **9:** 266.

Schroeder JI, Kwak JM & Allen GJ. 2001. Guard cell abscisic acid signalling and engineering drought hardiness in plants. *Nature,* **410:** 327–330.

Scibior D & Czeczot H. 2006. Catalase: Structure, properties, functions. *Postepy Hig Med Dosw (Online),* **60:** 170–180.

Sečenji M, Hideg É, Bebes A & Györgyey J. 2010. Transcriptional differences in gene families of the ascorbate–glutathione cycle in wheat during mild water deficit. *Plant Cell Rep,* **29:** 37–50.

Selmar D & Kleinwächter M. 2013. Stress enhances the synthesis of secondary plant products: The impact of stress-related over-reduction on the accumulation of natural products. *Plant Cell Physiol,* **54:** 817–826.

Senadheera P, Singh RK & Maathuis FJM. 2009. Differentially expressed membrane transporters in rice roots may contribute to cultivar dependent salt tolerance. *J Exp Bot,* **60:** 2553–2563.

Seo SY, Kim YJ & Park KY. 2019. Increasing polyamine contents enhances the stress tolerance via reinforcement of antioxidative properties. *Front Plant Sci,* **10:** 1331.

Sgherri CLM, Pinzincp C & Navari-Izzo F. 1996. Sunflower seedlings subjected to increasing stress by water deficit: Changes in O_2^- production Aated to the composition of thylakoid membranes. *Physiol Plant,* **96:** 446–452.

Shafiq S, Akram NA, Ashraf M, García-Caparrós P, Ali OM & Latef AAHA. 2021. Influence of glycine betaine (natural and synthetic) on growth, metabolism and yield production of drought-stressed maize (Zea mays L.) plants. *Plants,* **10:** 2540.

Shah K & Dubey RS. 1997. Effect of cadmium on proline accumulation and ribonuclease activity in rice seedlings: Role of proline as a possible enzyme protectant. *Biol Plant,* **39:** 121–130.

Shah W, Ullah S, Ali S, Idrees M, Khan MN, Ali K, et al. 2021. Effect of exogenous alpha-tocopherol on physio-biochemical attributes and agronomic performance of lentil (Lens culinaris Medik.) under drought stress. *PLoS One,* **16:** e0248200.

Shakirova FM, Allagulova ChR, Maslennikova DR, Klyuchnikova EO, Avalbaev AM & Bezrukova MV. 2016. Salicylic acid-induced protection against cadmium toxicity in wheat plants. *Environ Exp Bot,* **122:** 19–28.

Shan C & Liang Z. 2010. Jasmonic acid regulates ascorbate and glutathione metabolism in Agropyron cristatum leaves under water stress. *Plant Science,* **178:** 130–139.

Shan C & Ou X. 2018. Hydrogen peroxide is involved in the regulation of ascorbate and glutathione metabolism in wheat leaves under water stress. *Cereal Res Commun,* **46:** 21–30.

Shan C, Zhang S & Ou X. 2018. The roles of H_2S and H_2O_2 in regulating AsA-GSH cycle in the leaves of wheat seedlings under drought stress. *Protoplasma,* **255:** 1257–1262.

Shang J-X, Li X, Li C & Zhao L. 2022. The role of nitric oxide in plant responses to salt stress. *Int J Mol Sci,* **23:** 6167.

Shao J, Huang K, Batool M, Idrees F, Afzal R, Haroon M, et al. 2022. Versatile roles of polyamines in improving abiotic stress tolerance of plants. *Front Plant Sci,* **13:** 1003155.

Sharifi-Rad M, Anil Kumar NV., Zucca P, Varoni EM, Dini L, Panzarini E, et al. 2020. Lifestyle, oxidative stress, and antioxidants: Back and forth in the pathophysiology of chronic diseases. *Front Physiol,* **11**.

Sharma P & Dubey RS. 2005. Drought induces oxidative stress and enhances the activities of antioxidant enzymes in growing rice seedlings. *Plant Growth Regul,* **46:** 209–221.

Sharma P & Dubey RS. 2007. Involvement of oxidative stress and role of antioxidative defense system in growing rice seedlings exposed to toxic concentrations of aluminum. *Plant Cell Rep,* **26:** 2027–2038.

Sharma P, Jha AB, Dubey RS & Pessarakli M. 2012. Reactive oxygen species, oxidative damage, and antioxidative defense mechanism in plants under stressful conditions. *J Bot,* **2012:** e217037.

Sharma S & Negi N. 2021. Omics technology for abiotic stress response in plants. *European Journal of Molecular Biotechnology,* **07:** 2020.

Shelake RM, Kadam US, Kumar R, Pramanik D, Singh AK & Kim J-Y. 2022. Engineering drought and salinity tolerance traits in crops through CRISPR-mediated genome editing: Targets, tools, challenges, and perspectives. *Plant Commun,* **3:** 100417.

Shemi R, Wang R, Gheith E-SMS, Hussain HA, Hussain S, Irfan M, et al. 2021. Effects of salicylic acid, zinc and glycine betaine on morpho-physiological growth and yield of maize under drought stress. *Sci Rep,* **11:** 3195.

Shen C, Que Z, Xia Y, Tang N, Li D, He R, et al. 2017. Knock out of the annexin gene OsAnn3 via CRISPR/Cas9-mediated genome editing decreased cold tolerance in rice. *Journal of Plant Biology,* **60:** 539–547.

Sheteiwy MS, Ali DFI, Xiong Y-C, Brestic M, Skalicky M, Hamoud YA, et al. 2021. Physiological and biochemical responses of soybean plants inoculated with Arbuscular mycorrhizal fungi and Bradyrhizobium under drought stress. *BMC Plant Biol,* **21:** 195.

Shi J, Gao H, Wang H, Lafitte HR, Archibald RL, Yang M, et al. 2017. ARGOS 8 variants generated by CRISPR-Cas9 improve maize grain yield under field drought stress conditions. *Plant Biotechnol J,* **15:** 207–216.

Shin S-Y, Kim M-H, Kim Y-H, Park H-M & Yoon H-S. 2013. Co-expression of monodehydroascorbate reductase and dehydroascorbate reductase from Brassica rapa effectively confers tolerance to freezing-induced oxidative stress. *Mol Cells,* **36:** 304–315.

Shriram V, Kumar V, Devarumath RM, Khare TS & Wani SH. 2016. MicroRNAs as potential targets for abiotic stress tolerance in plants. *Front Plant Sci,* **7**.

Siddiqui M, Al-Khaishany M, Al-Qutami M, Al-Whaibi M, Grover A, Ali H, et al. 2015. Response of different genotypes of Faba Bean plant to drought stress. *Int J Mol Sci,* **16:** 10214–10227.

Singh PK, Indoliya Y, Agrawal L, Awasthi S, Deeba F, Dwivedi S, et al. 2022. Genomic and proteomic responses to drought stress and biotechnological interventions for enhanced drought tolerance in plants. *Curr Plant Biol,* **29:** 100239.

Smirnoff N. 2018. Ascorbic acid metabolism and functions: A comparison of plants and mammals. *Free Radic Biol Med,* **122:** 116–129.

Snezhkina AV, Kudryavtseva AV, Kardymon OL, Savvateeva MV, Melnikova NV, Krasnov GS, et al. 2019. ROS generation and antioxidant defense systems in normal and malignant cells. *Oxid Med Cell Longev,* **2019:** 6175804.

Sofo A, Tuzio AC, Dichio B & Xiloyannis C. 2005. Influence of water deficit and rewatering on the components of the ascorbate–glutathione cycle in four interspecific Prunus hybrids. *Plant Science,* **169:** 403–412.

Su L-J, Zhang J-H, Gomez H, Murugan R, Hong X, Xu D, et al. 2019. Reactive oxygen species-induced lipid peroxidation in Apoptosis, Autophagy, and Ferroptosis. *Oxid Med Cell Longev,* **2019:** 1–13.

Sun LR, Yue CM & Hao FS. 2019. Update on roles of nitric oxide in regulating stomatal closure. *Plant Signal Behav,* **14:** e1649569.

Sun M, Sun S, Jia Z, Ma W, Mao C, Ou C, et al. 2022. Genome-wide analysis and expression profiling of glutathione reductase gene family in oat (Avena sativa) indicate their responses to abiotic stress during seed imbibition. *Int J Mol Sci,* **23:** 11650.

Sunkar R, Li Y-F & Jagadeeswaran G. 2012. Functions of microRNAs in plant stress responses. *Trends Plant Sci,* **17:** 196–203.

Szabados L & Savouré A. 2010. Proline: A multifunctional amino acid. *Trends Plant Sci,* **15:** 89–97.

Szarka A, Tomasskovics B & Bánhegyi G. 2012. The ascorbate-glutathione-α-tocopherol triad in abiotic stress response. *Int J Mol Sci,* **13:** 4458–4483.

Talbi S, Rojas JA, Sahrawy M, Rodríguez-Serrano M, Cárdenas KE, Debouba M, et al. 2020. Effect of drought on growth, photosynthesis and total antioxidant capacity of the saharan plant Oudeneya africana. *Environ Exp Bot,* **176:** 104099.

Talla S, Riazunnisa K, Padmavathi L, Sunil B, Rajsheel P & Raghavendra AS. 2011. Ascorbic acid is a key participant during the interactions between chloroplasts and mitochondria to optimize photosynthesis and protect against photoinhibition. *J Biosci,* **36:** 163–173.

Tanou G, Ziogas V, Belghazi M, Christou A, Filippou P, Job D, et al. 2014. Polyamines reprogram oxidative and nitrosative status and the proteome of citrus plants exposed to salinity stress. *Plant Cell Environ,* **37:** 864–885.

Tattini M, Loreto F, Fini A, Guidi L, Brunetti C, Velikova V, et al. 2015. Isoprenoids and phenylpropanoids are part of the antioxidant defense orchestrated daily by drought-stressed Platanus × acerifolia plants during Mediterranean summers. *New Phytologist,* **207:** 613–626.

Terzi R, Kalaycıoglu E, Demiralay M, Saglam A & Kadioglu A. 2015. Exogenous ascorbic acid mitigates accumulation of abscisic acid, proline and polyamine under osmotic stress in maize leaves. *Acta Physiol Plant,* **37:** 43.

Tian Y, Li X, Zhou X, Qu Z, Wang X & Dong S. 2023. Effects of drought stress on SOD activity and pro content in different parts of soybean leaves. *Legume Research – An International Journal,* **46(8):** 995–1000.

Tripathy BC & Oelmüller R. 2012. Reactive oxygen species generation and signaling in plants. *Plant Signal Behav,* **7:** 1621–1633.

Unamba CIN, Nag A & Sharma RK. 2015. Next generation sequencing technologies: the doorway to the unexplored genomics of non-model plants. *Front Plant Sci,* **6**.

Uzilday B, Turkan I, Sekmen AH, Ozgur R & Karakaya HC. 2012. Comparison of ROS formation and antioxidant enzymes in Cleome gynandra (C4) and Cleome spinosa (C3) under drought stress. *Plant Science,* **182:** 59–70.

Valderrama R, Corpas FJ, Carreras A, Fernández-Ocaña A, Chaki M, Luque F, et al. 2007. Nitrosative stress in plants. *FEBS Lett,* **581:** 453–461.

Van Breusegem F, Vranová E, Dat JF & Inzé D. 2001. The role of active oxygen species in plant signal transduction. *Plant Science,* **161:** 405–414.

Vanham D, Alfieri L, Flörke M, Grimaldi S, Lorini V, de Roo A, et al. 2021. The number of people exposed to water stress in relation to how much water is reserved for the environment: A global modelling study. *Lancet Planet Health,* **5:** e766–e774.

Vašková J, Kočan L, Vaško L & Perjési P. 2023. Glutathione-related enzymes and proteins: A review. *Molecules,* **28:** 1447.

Verma S & Dubey RS. 2003. Lead toxicity induces lipid peroxidation and alters the activities of antioxidant enzymes in growing rice plants. *Plant Science,* **164:** 645–655.

Verslues PE & Sharma S. 2010. Proline metabolism and its implications for plant-environment interaction. *The Arabidopsis Book,* American Society of Plant Biologists, **8:** 1–23. e0140.

Vessal S, Arefian M & Siddique KHM. 2020. Proteomic responses to progressive dehydration stress in leaves of chickpea seedlings. *BMC Genomics,* **21:** 523.

Villadangos S, González C & Munné-Bosch S. 2023. Photo-oxidative stress leading to oxylipin accumulation may trigger drought-induced leaf abscission in olive trees. *Physiol Plant,* **175:** e13941.

Wang F-Z, Wang Q-B, Kwon S-Y, Kwak S-S & Su W-A. 2005. Enhanced drought tolerance of transgenic rice plants expressing a pea manganese superoxide dismutase. *J Plant Physiol,* **162:** 465–472.

Wang N, Zhao J, He X, Sun H, Zhang G & Wu F. 2015. Comparative proteomic analysis of drought tolerance in the two contrasting Tibetan wild genotypes and cultivated genotype. *BMC Genomics,* **16:** 432.

Wang X, Liu H, Yu F, Hu B, Jia Y, Sha H, et al. 2019. Differential activity of the antioxidant defence system and alterations in the accumulation of osmolyte and reactive oxygen species under drought stress and recovery in rice (Oryza sativa L.) tillering. *Sci Rep,* **9:** 8543.

Wang Y, Wang Q & Hou Y. 2020a. A new cold-adapted and salt-tolerant glutathione reductase from Antarctic Psychrophilic Bacterium Psychrobacter sp. and its resistance to oxidation. *Int J Mol Sci,* **21:** 420.

Wang Y, Zhang X, Huang G, Feng F, Liu X, Guo R, et al. 2020b. Dynamic changes in membrane lipid composition of leaves of winter wheat seedlings in response to PEG-induced water stress. *BMC Plant Biol,* **20:** 84.

Wang Z, Li G, Sun H, Ma L, Guo Y, Zhao Z, et al. 2018. Effects of drought stress on photosynthesis and photosynthetic electron transport chain in young apple tree leaves. *Biol Open,* **7:** bio035279.

Wang Z, Yang Y, Yadav V, Zhao W, He Y, Zhang X, et al. 2022. Drought-induced proline is mainly synthesized in leaves and transported to roots in watermelon under water deficit. *Hortic Plant J,* **8:** 615–626.

Wani SH, Sah SK, Hossain MA, Kumar V & Balachandran SM. 2016. Transgenic approaches for abiotic stress tolerance in crop plants. pp. 345–396. *In:* Al-Khayri JM, Jain SM, & Johnson D V (eds), Advances in Plant Breeding Strategies: Agronomic, Abiotic and Biotic Stress Traits, Springer International Publishing, Cham.

Wendehenne D, Durner J & Klessig DF. 2004. Nitric oxide: A new player in plant signalling and defence responses. *Curr Opin Plant Biol,* **7:** 449–455.

Wu J, Shu S, Li C, Sun J & Guo S. 2018. Spermidine-mediated hydrogen peroxide signaling enhances the antioxidant capacity of salt-stressed cucumber roots. *Plant Physiology and Biochemistry,* **128:** 152–162.

Xia H, Yu S, Kong D, Xiong J, Ma X, Chen L, et al. 2020. Temporal responses of conserved miRNAs to drought and their associations with drought tolerance and productivity in rice. *BMC Genomics,* **21:** 232.

Xie H, Li M, Chen Y, Zhou Q, Liu W, Liang G, et al. 2021. Important physiological changes due to drought stress on oat. *Front Ecol Evol,* **9**.

Xuan J, Song Y, Zhang H, Liu J, Guo Z & Hua Y. 2013. Comparative proteomic analysis of the stolon cold stress response between the C4 perennial grass species Zoysia japonica and Zoysia metrella. *PLoS One,* **8:** e75705.

Xue J-H, Chen G-D, Hao F, Chen H, Fang Z, Chen F-F, et al. 2019. A vitamin-C-derived DNA modification catalysed by an algal TET homologue. *Nature,* **569:** 581–585.

Yan J, Ninkuu V, Fu Z, Yang T, Ren J, Li G, et al. 2023. OsOLP1 contributes to drought tolerance in rice by regulating ABA biosynthesis and lignin accumulation. *Front Plant Sci,* **14:** 1163939.

Yan L, Li S, Riaz M & Jiang C. 2021. Proline metabolism and biosynthesis behave differently in response to boron-deficiency and toxicity in Brassica napus. *Plant Physiology and Biochemistry,* **167:** 529–540.

Yang X, Lu M, Wang Y, Wang Y, Liu Z & Chen S. 2021. Response mechanism of plants to drought stress. *Horticulturae,* **7:** 50.

Yao HM, Li YS, Zhang TZ, Zhao J, Wang C, Wang HN, et al. 2016. Effects of combined drought and salinity stress on germination and physiological characteristics of maize (Zea mays). *Ying Yong Sheng Tai Xue Bao,* **27:** 2301–2307.

Yoshiba Y, Kiyosue T, Nakashima K, Yamaguchi-Shinozaki K & Shinozaki K. 1997. Regulation of levels of proline as an osmolyte in plants under water stress. *Plant Cell Physiol,* **38:** 1095–1102.

Yoshida K, Kaothien P, Matsui T, Kawaoka A & Shinmyo A. 2003. Molecular biology and application of plant peroxidase genes. *Appl Microbiol Biotechnol,* **60:** 665–670.

You L, Song Q, Wu Y, Li S, Jiang C, Chang L, et al. 2019. Accumulation of glycine betaine in transplastomic potato plants expressing choline oxidase confers improved drought tolerance. *Planta,* **249:** 1963–1975.

Yousuf PY, Hakeem KUR, Chandna R & Ahmad P. 2012. Role of glutathione reductase in plant abiotic stress. pp. 149–158. *In:* Abiotic Stress Responses in Plants, Springer New York, New York, NY.

Zafar SA, Zaidi SS-A, Gaba Y, Singla-Pareek SL, Dhankher OP, Li X, et al. 2020. Engineering abiotic stress tolerance via CRISPR/Cas-mediated genome editing. *J Exp Bot,* **71:** 470–479.

Zaidi SS-A, Mahfouz MM & Mansoor S. 2017. CRISPR-Cpf1: A new tool for plant genome editing. *Trends Plant Sci,* **22:** 550–553.

Zandi P & Schnug E. 2022. Reactive oxygen species, antioxidant responses and implications from a microbial modulation perspective. *Biology (Basel),* **11:** 155.

Zhang A, Liu Y, Wang F, Li T, Chen Z, Kong D, et al. 2019a. Enhanced rice salinity tolerance via CRISPR/Cas9-targeted mutagenesis of the OsRR22 gene. *Molecular Breeding,* **39:** 47.

Zhang F, Yang J, Zhang N, Wu J & Si H. 2022. Roles of microRNAs in abiotic stress response and characteristics regulation of plant. *Front Plant Sci,* **13:** 919243.

Zhang L, Li X, Ma B, Gao Q, Du H, Han Y, et al. 2017. The tartary buckwheat genome provides insights into rutin biosynthesis and abiotic stress tolerance. *Mol Plant,* **10:** 1224–1237.

Zhang W, Xiao S & Ahn DU. 2013. Protein oxidation: Basic principles and implications for meat quality. *Crit Rev Food Sci Nutr,* **53:** 1191–1201.

Zhang Y, Li Z, Peng Y, Wang X, Peng D, Li Y, et al. 2015. Clones of FeSOD, MDHAR, DHAR genes from white clover and gene expression analysis of ROS-scavenging enzymes during abiotic stress and hormone treatments. *Molecules,* **20:** 22236–22240.

Zhang Y, Massel K, Godwin ID & Gao C. 2019b. Applications and potential of genome editing in crop improvement. *Genome Biol,* **20:** 13.

Zhao B, Liang R, Ge L, Li W, Xiao H, Lin H, et al. 2007. Identification of drought-induced microRNAs in rice. *Biochem Biophys Res Commun,* **354:** 585–590.

Zhao Y, Wang J, Huang W, Zhang D, Wu J, Li B, et al. 2023. Abscisic-acid-regulated responses to alleviate cadmium toxicity in plants. *Plants,* **12:** 1023.

Zhu X, Ma Z, Zhang X, Huang X, Liu J & Zhuang X. 2022. Effect of malondialdehyde-induced oxidation modification on physicochemical changes and gel characteristics of Duck Myofibrillar proteins. *Gels,* **8**.

Zhu Z, Chen J & Zheng H-L. 2012. Physiological and proteomic characterization of salt tolerance in a mangrove plant, Bruguiera gymnorrhiza (L.) Lam. *Tree Physiol,* **32:** 1378–1388.

Zitka O, Skalickova S, Gumulec J, Masarik M, Adam V, Hubalek J, et al. 2012. Redox status expressed as GSH:GSSG ratio as a marker for oxidative stress in paediatric tumour patients. *Oncol Lett,* **4:** 1247–1253.

Zulfiqar F, Ashraf M & Siddique KHM. 2022. Role of glycine betaine in the thermotolerance of plants. *Agronomy,* **12:** 276.

6 | Phytohormonal Regulation in Plants during Water Stress

Sujata Rathi[1*] and Deepak Kumar[2*]

[1] Department of Botany, Multanimal Modi College, Modinagar
[2] Graphic Era Deemed (to be) University, Dehradun, India

Plants respond to a complex array of abiotic stress such as water stress salinity and temperature by activating a signaling cascade involving metabolites, free radical scavengers, antioxidants, unsaturated fatty acids and phytohormones. Phytohormones are signaling molecules that are required in a very low concentration to regulate the growth and development of plants. These signaling molecules either work independently or synergistically. Secretion of these phytohormones fluctuates under abiotic stress to make plant stress tolerant. Phytohormones act in collaboration to make plants well-suited to all environmental stress. Sometimes they act at the genetic level by inducing the expression of the defense gene whose expression is needed for defense response against stress. They also regulate the level of other stress-responsive molecules. Phytohormones such as auxins (IAA), cytokinins (CKs), abscisic acid (ABA), ethylene (ET), gibberellins (GAs), salicylic acid (SA), brassinosteroids, and jasmonates play a pivotal role in preparing plants to sustain various kinds of abiotic stress such as water stress, salinity, cold, flooding and heat. Melatonin, a new addition to phytohormones, is known to alleviate water stress by regulating various physiological responses. In this chapter, we will critically assess the regulatory aspects of phytohormones in modulating growth and development to acclimate plants to water stress and their engineering to obtain stress-tolerant transgenic crops.

1. Introduction

Water stress is a major challenging issue in the current scenario as it reduces yield significantly by hampering photosynthesis due to stomatal closure. It limits the diffusion rate of stomata that results in less availability of carbon for production of sugar. Water stress also decreases chlorophyll by damaging the chloroplast

*Corresponding authors: srathi84@gmail.com, deepakphy@gmail.com

structure. It is also known to cause the formation of intracellular lipid droplets that has a negative impact on phloem loading and unloading (Kaiser et al., 1981; Burke, 2007; Hummel et al., 2010; Lemoine et al., 2013). Plants respond to moderate water stress by increasing the soluble carbohydrate concentration in the leaf and reducing the starch content for carbon availability due to the low rate of photosynthesis (Fahad et al., 2017; Liu et al., 2019). In severe water stress conditions the soluble fraction of carbohydrates also decreases (Rizwan et al., 2015; Fahad et al., 2017). Plant growth regulators or phytohormones are signaling molecules that are required in very low concentration within plants. These small lipophilic compounds possess an indispensable role in physiology and biochemistry in plant systems under optimal and stressful conditions. Phytohormones such as auxins (IAA), cytokinins (CKs), abscisic acid (ABA), ethylene (ET), gibberellins (GAs), salicylic acid (SA), brassinosteroids, and jasmonates play a pivotal role in preparing plants to sustain various kinds of environmental stress such as water stress, salinity, cold, flooding and heat. In recent years, our understanding of the regulatory function of these phytohormones at gene level has grown to a considerable extent. Phytohormones transduce baseline pathways to interact with a signaling cascade to make plants withstand abiotic stress such as water stress. This perception helps plants to change cellular dynamics to regulate growth and development. Thus a cross talk is established with the coordination of various phytohormones in the signaling network (Fig. 1). Water stress causes changes the redox potential in plants through the activation of various signaling pathways (Dat et al., 2000; Mittler, 2002; Ahanger et al., 2017; Ribba et al., 2020). Strigolactones (SL) and Brassinosteroids (BR) are gaining importance due to their effective role in the context of water stress (Brewer et al., 2013; Ha et al., 2014; Nolan et al., 2020).

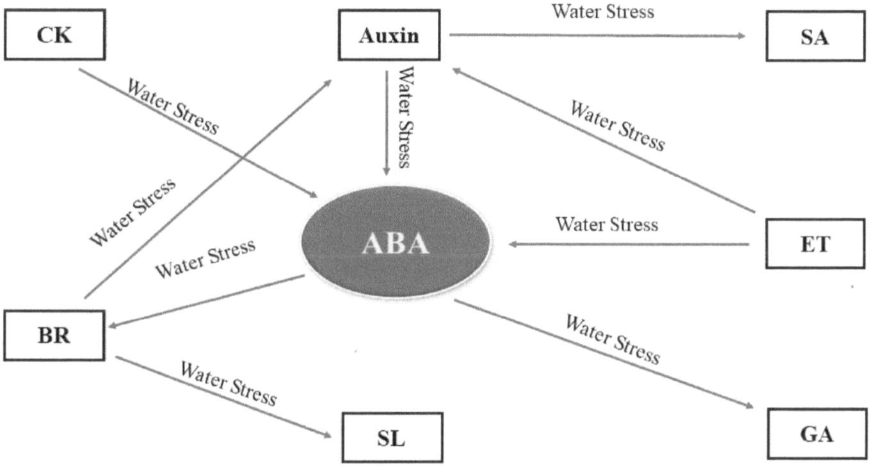

Figure 1: Phytohormones cross talking in water stress signaling cascade.

Melatonin, a new addition to the phytohormones, regulates various physiological responses such as root induction, growth, photosynthesis and regulation of water stress (Maheshwari et al., 2015). The plant system responds to water stress by activating a defense mechanism involving the cuticle as an external barrier and producing metabolites such as unsaturated fatty acids, free radical scavengers, and molecular chaperones for protein stability. For an effective defensive mechanism, these metabolites require signaling molecules to coordinate with each other in the form of phytohormones, polyamines, nitric oxide, hydrogen sulfide, and transcription factors to regulate gene expression (He et al., 2018). Phytohormones resist abiotic stress in combinations of more than one hormone in precise concentrations (Zhao et al., 2019). Phytohormones can be separated by using thin layers and high-performance liquid chromatography for understanding their physiological response more precisely (Floková et al., 2014; Stec et al., 2016). Phytohormone profiling is done by using GC-MS-based methods in *Citrus sinensis* L. under abiotic stress. SIM (specific ion monitoring) methods were used to extract various phytohormones in high percentages. With this technique, thirteen phytohormones were detected belonging to different groups. Among all the phytohormones studied, salicylic acid occurs predominantly (59-70%) in various tissues. Phytohormones play a crucial role in the normal functioning of the plant system by influencing various physiological functions such as growth, development and reproduction. Out of thirteen phytohormones investigated, five phytohormones, Auxin, ABA, CK, ET, and GA are referred to as key growth regulators while the rest have subsidiary functions (Su et al., 2017). The coping strategy of plants against water stress primarily involves phytohormones. The endogenous level of phytohormones can be increased by their exogenous application. The transport of water across the cellular membrane is facilitated by water channels called aquaporins. Entry of water inside the cell depends on the water potential difference between soil and the atmosphere (Scharwies and Dinneny, 2019).

Dry-state in plants refers to the unavailability of water in the soil for uptake and increased transpirational water loss leading to loss of turgidity (Scharwies and Dinneny, 2019). Thus water stress causes an increase in leaf osmotic pressure, a decrease in stomatal conductance, transpiration, turgidity and sap flow from the source to the sink (Mahdieh et al., 2008). Plants respond to water stress by utilizing a strategy that involves a change in osmotic and water potential, production of osmolytes and antioxidants and most essentially phytohormones (Fig. 2). Dry state in plants leads to the destruction of cellular membranes and biomolecules such as proteins DNA RNA photosynthetic pigments (Zandi and Schnug, 2022; Sofy et al., 2020). Water stress is the outcome of changing climate regimes posing a serious threat to increasing global food demand by influencing food productivity and security. Well drained soil is an essential requirement for healthy growth and development in various perennial fruit trees and crops. Significant reductions in growth and development are seen in these crops in poorly drained soils (Fahad et al., 2017; Fuller and Stevens, 2019). A marked fall in the carbon reservoir is

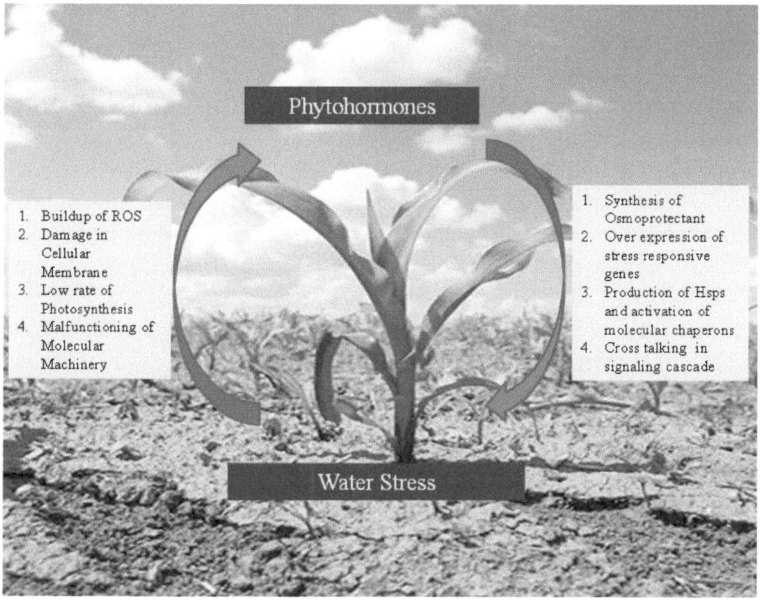

Figure 2: An overview of phytohormonal regulation of water stress.

noticed as a result of increased shoot respiration in response to water stress in citrus plants (Fahad et al., 2017).

Various conventional breeding approaches have been used to produce tolerant cultivars that are better suited to water stress. Several genes have been identified that are overexpressed in plants in drought stress. Plants transformed with these genes show better growth and yield. Phytohormones act in very low concentrations to change plants' reflex to stress and the signal perceived by them in one part is transmitted to the whole plant (Sah et al., 2016). Phytohormones are key mediators to tolerate and avoid the adverse impacts of water stress in plants by regulating the growth and development in the right out life span (Sah et al., 2016; Ullah et al., 2017). Phytohormones respond effectively to external stimuli and environmental strain in very low concentrations (Nowicka et al., 2018). This chapter aims to enhance the understanding of the role of various phytohormones in plant physiology under water stress and their metabolic engineering to confer tolerance in crop plants.

2. Phytohormones: Key Regulators in Plants Against Water Stress

2.1 Abscisic Acid

Abscisic acid, also known as abscin and dormin, plays a crucial role in breaking seed dormancy and inhibiting plant growth (Cornforth et al., 1965). However further investigation revealed its diverse role in cellular processes such as plant development, xylem fiber differentiation, anthocyanin synthesis in ripening fruit, seed maturation stomatal closing and opening, and abiotic stresses (Finkelstein, 2013; Campbell et al., 2018; Karppinen et al., 2018). Abscisic acid, a 15 carbon atom compound, is synthesized in a very low concentration in normal environmental conditions (Parveen et al., 2021). Abscisic acid synthesis occurs in plastids from carotenoid beta carotene that is converted into xanthophyll before the formation of abscisic acids through a series of reactions involving the formation of trans violaxanthin, neoxanthin, xanthoxin, and abscisic aldehyde. Roots are the sites of ABA synthesis under water stress and it is transported from them to the apical portion. Thus the concentration of ABA increases in leaves. The increased ABA concentration acts as a vital messenger to set a signaling cascade that affects the turgidity of guard cells (Danquah et al., 2013; Awan et al., 2017). ABA is also found to be associated with the changes in the expression of various transcription factors and stress-associated genes (Li et al., 2015). The concentration gradient of ABA in leaf epidermal tissue results in stomatal closure thereby reducing the transpirational loss of water in water stress (Peleg and Blumwald, 2011). ABA acts as a key mediator in plant-water economics and biochemistry. Various enzymes like phosphatases and kinases were also found to play a key role in the quick adjustment of a plant system in sub optimal water availability (Dar et al., 2017). One such ABA-responsive protein kinase, SnRK2 is found actively involved in stomatal dynamics (Yoshida et al., 2002). ABA causes stomatal closure as a result of the loss of turgidity of guard cells due to the efflux of K^+ ions (Kim et al. 2010; Sah et al., 2016). ABA also affects the expression profile of aquaporin genes involved in stomatal movements (Sharipova et al., 2016; Verma et al., 2016). Under water stress, ABA is rapidly synthesized by oxidative cleavage of beta carotene. ABA is also produced in plants exposed to water stress from an alternative pathway involving beta-glucosidase homologs (Xu et al., 2013; Llanes et al., 2014).

ABA also interacts with other growth regulators to confer tolerance in plants to moisture stress (He et al., 2018). ABA well equips plants against abiotic stress by stimulating the synthesis of various other defensive biochemicals such as proline, peroxidases, synthetases, antioxidants, hsps, and membrane strengthening with polyunsaturated fatty acids (Ashraf and Foolad, 2007; Chen et al., 2013; Lee and Suh, 2015; Hoque et al, 2016; Huang et al., 2016; Liu et al., 2018; Yin et al., 2018; Singhal et al., 2021). It is worth mentioning that ABA triggers the expression of various stress-responsive genes under abiotic stress situations in plants. ABA receptors that have been recognized as transducers of stress signals include PYRs and PYLs. Genes encoding these receptors have also been identified in many crops such as rice and Arabidopsis (Klingler et al., 2010). Transcription factors such as MYC, NAC and MYB are also found to be associated with the ABA signaling pathways during the stress response. Calcium-dependent protein kinases (CDPKs) have been identified to take part in ABA-associated signaling pathways in stress responses in various crops such as of Rice (29CDPKs),

Arabidopsis (34 CDPKs), Wheat (20CDPKs), and Maize 35 (CDPKs). There is a considerable increase (4 to 7 times higher) in the concentration of ABA in water deficit conditions in seaweed *Pyropia orbicularis* (Guajardo et al., 2016). In the drought-tolerant *Arabidopsis* gene, IbARF5 is responsible for the overexpression of the gene controlling ABA biosynthesis (Kang et al., 2018). Moreover, SAPK2, a key regulator of ABA-mediated response, enhances the expression of drought-responsive genes in rice such as OsLEA3, OsOREB1, OsRab16b, OsRab21, and OsbZIP23 (Lou et al., 2017). Several investigations with Arabidopsis indicate the direct involvement of phytohormone ABA in conferring drought tolerance by regulating stomatal movement to check transpirational water loss since genes involved in ABA-induced stress response have become a significant target for producing drought-tolerant cultivars in crop plants.

2.2 Auxin

Auxin is an important endogenous growth regulator. It regulates various morphogenetic actions such as root formation, apical dominance, tissue differentiation, embryogenesis, movement and leaf arrangements on the stem. Auxin also controls cell division and elongation in association with cytokinins (Tromas et al., 2009; Jurado et al., 2010). Indole Acetic Acid (IAA) is the most common form of auxin found in plants (Hamayun et al., 2021c). Auxin can be synthesized by two pathways, Tryptophan dependent pathway involving the formation of indole3-acetamide, indole-3-acetaldoxime, tryptamine, and indole-3-pyruvic acid and Tryptophan independent pathway (Woodward and Bartel, 2005; Taiz et al., 2015). In *Arabidopsis*, tryptophan aminotransferase converts tryptophan into indole 3-pyruvic acid that is finally converted into indole 3-acetic acid by the enzyme monooxygenase. IAA is metabolized into amino acid conjugates if present in excess in plants. In an alternative pathway, tryptophan is converted into indole 3-acetamide by the action of monooxygenase and amidase (Cheng et al., 2006; Zhao 2010; Mashiguchi et al., 2011; Du et al., 2013a, b). In response to water stress, auxin stimulates the expression of several genes called auxin-responsive genes. Auxin mediates cross-talks between other phytohormones in water deficit conditions. A number of genetic studies have revealed the vital role of auxin to sustain the normal development of plants in stressed conditions. In stress conditions reduction in IAA concentration causes a significant increase in ABA concentration. In drought stress, several auxin-coding genes are overexpressed in rice (Jung et al., 2015). A similar case was observed in the case of *Solanum tuberosum* and poplar with auxin coding gene YUC6 (Ke et al., 2015). In *Nicotiana tobaccum*, auxin induced root branching in seedlings under water deficit conditions (Verma et al., 2016; Wang et al., 2018). Drought-tolerant tomato plants showed overexpression of auxin-responsive genes due to the binding of ARFs (auxin-responsive transcription factors) to the promoter site (Bouzroud et al., 2018). In Kiwi drought stress induces temporal upregulation of various transcription factors such as those of AcARF1a, AcARF19a, AcARF26a, AcARF28a, and AcARF28b indicating the potential role that play in regulating drought stress (Su et al., 2021). These findings have evoked interest among researchers to use auxin-responsive genes as a biotechnological tool to raise drought-tolerant crops. Several reports indicate an alteration in the metabolic pathway of auxin synthesis action transport in plants exposed to water stress (Ljung, 2013). The evolutionary role of auxin to make the plants better adapted to the adverse environment has been well established with single-celled algae (De Smet et al., 2010). Transgenic *Arabidopsis* shows overexpression of ARF5 from sweet potatoes in drought tolerance (Kang et al., 2018). The action of ARFs is also influenced by other phytohormones as well (Wang et al.;, 2011). ABA is one of them to regulate ARFs in abiotic stress adaptation (Sah et al., 2016). Interestingly, genetic investigations proved the abundance of ABA-responsive elements present in the upstream region of ARF gene promoters (Xu et al., 2020). Many ARFs are unregulated or down regulated by ABA, indicating the plausible role of ARFs in ABA-mediated regulation water stress (Chen et al., 2021; Kang et al., 2018; Park et al., 2007; Wang et al., 2011; Xu et al., 2020). Auxin regulates a number of plant functions such as growth and differentiation, cell expansion, root initiation, floral architecture, and seed dormancy (Khosravi-nejad et al., 2022). Auxin also regulates the level of glucosinolates, one of the major secondary metabolites of the family Brassicaceae. The concentration of this compound affects the plant's suitability against abiotic stress. The biosynthesis of this compound is regulated genetically and environmentally and at various levels of development (Martínez Ballesta et al., 2013). It is well documented that glucosinolates are actively involved to confer tolerance in plants against non abiotic stresses (Champolivier and Merrien, 1996; Rask et al., 2000; Qasim et al., 2003; Radovich et al., 2005; Engelen Eigles et al., 2006; Velasco et al., 2007; Schonhof et al., 2007a; López Berenguer et al., 2009). The level of glucosinolates is regulated by a number of auxin-sensitive repressors such as IAA5, IAA6 and IAA19. These repressors maintain the optimal level of glucosinolates in plants under water deficit situations as many studies have proved that loss of the mutation function in any of these repressors significantly decreases the level of glucosinolates. This decreased level of glucosinolate in knock-out iaa 5 *Arabidopsis* can be recovered with an exogenous application of glucosinolates to fight against water stress (Salehin et al. 2019).

2.3 Salicylic Acid

Salicylic acid, a phenolic compound, was first isolated from the bark of willow trees in the form of salicyl alcohol glucoside called silicon. It is categorized under secondary metabolites that play an important role in several physiological processes such as carbon assimilation, stomatal movements, antioxidants production and ripening (Mir et al., 2022; Iqbal et al., 2022). Several reports also suggested its role in controlling the expression of a number of genes involved in water stress response. The endogenous level of SA affects the response of plants to water stress. Several investigations indicated that the higher the concentration of SA greater will be the ability of plants to tolerate abiotic stress. Water deficit directly affects the biosynthesis of SA. In the case of *Phillyrea angustifolia*, water stress significantly increases the endogenous concentration of SA (Munne-Bosch, 2003). Exogenous application of SA in *Hordeum vulgare* enhances water stress tolerance by stimulating stomatal diffusion of CO_2 that causes increased carbon assimilation (Bandurska and Stroinski, 2005). In the *Brassica rapa* level, SA also regulates the biosynthesis of several osmolytes such as Proline to prevent plants from oxidative damage. SA is synthesized from cinnamic acid (CA) in two ways. In one biosynthetic pathway, benzoic acid is produced by decarboxylation of the side chain of CA which further undergoes hydroxylation to produce salicylic acid while another pathway involves the production of coumaric acid which is further decarboxylated to produce salicylic acid (Yalpani et al., 1993; Silverman et al., 1995; Alibert and Ranjeva, 1971; Alibert et al., 1972). Water stress-induced oxidative stress can be tolerated in plants if they are pretreated with SA. This oxidative stress causes lipid peroxidation by generating reactive oxygen species (ROS) (Lee et al., 2019). In rice, exogenous treatment with SA substantially decreases the oxidative damage of seedlings grown in hydroponics by inducing the production of antioxidants (Ali et al., 2021b). In the case of *Portulaca oleracea* exogenous application of SA substantially increases the concentration of photosynthetic assimilate sunder water shortage by improving the pigment system, gaseous exchange, and synthesis of various osmolytes (Hamayun et al., 2021a). High SA levels reduce the stomatal opening in siz1 mutant to check the water loss under water stress (Miura et al., 2013). Similar results were also found with cpr5 and cpr6 mutants of *Arabidopsis*. SA also reduced the quantity of electrolyte leakage and malondialdehyde by improving antioxidant activities under drought stress. Overexpression of the transcription regulator of SA, CBP60g, causes plants to produce more SA and high ABA sensitivity thereby making the plant more drought resilient. Exogenous application of SA to the foliar region also confers resistance to water stress in plants. Several reports stated that *Triticum aestivum* exhibits increased catalase activity with exogenous application of SA under water stress (Maghsoudi et al., 2019). Studies on *Portulaca oleracea* L. reveals that SA helps in increasing biomass under water stress by enhancing photosynthetic activity with increased gas exchange, improved pigment system and production of osmolytes and antioxidants to maintain membrane integrity (Munsif et al., 2022, Saheri et al., 2020).

2.4 Cytokinin

Cytokinins (CKs), adenine derivatives having isoprene side chains at the N6 position, are key phytohormones found in plants in low concentration. Trans Zeatin was the first naturally occurring cytokinin obtained from maize (Miller, 1961). Folke Skoog and his coworkers isolated kinetin herring sperm DNA. They are known as ABA antagonists that work either individually or synergistically with other phytohormones to affect growth, development and tolerance under various environmental stresses (Pospisilova, 2003; Danilova et al., 2016). They are synthesized through an isoprenoid pathway in which the rate-limiting step is catalyzed by isopentenyl transferase (IPTs) that transfers the isopentenyl group from dimethylallyl diphosphate to ATP, ADP, or AMP (Kieber and Schaller, 2014). Transgenic tobacco plants show more tolerance to water stress by inducing leaf senescence carrying the IPT gene from *Agrobacterium tumefaciense* and promoter of the SARK gene from *Phaseolus vulgaris* (Rivero et al., 2010). Water stress causes substantial reduction in photosynthesis due to degradation in photosynthetic machinery. However, transgenic plants are able to protect photosynthetic machinery under water stress. The endogenous level of Cytokinin depends upon the duration and severity of water stress (Zwack and Rashotte, 2015). Transgenic drought-tolerant *Arabidopsis* showed delayed leaf senescence under water stress, independent of overexpression of a number of CK oxidases/dehydrogenases (CKX) (Werner et al., 2010; Hamayun et al., 2021b). CKs help plants tolerate osmotic stress by counteracting ABA-induced responses, preventing oxidative damage and lipid peroxidation (Gujjar et al, 2019). Increases in levels of CKs and their regulatory action on the biosynthesis of several metabolites under water stress provide potential clues to raise more plants tolerant to water stress. CKs play a diverse role in plants to affect growth and development by regulating photosynthesis, senescence, cell division and differentiation, floral development, seed and seedling growth, sink concentration and mineral uptake (Hwang et al., 2012). Numerous reports also indicated its role to confer tolerance in plants in stressed conditions (Pospisilova et al., 2000; Prerostova et al., 2018; Hai et al., 2020). Exogenous application of CKs increases the ability of plants to tolerate water stress to several folds. This application of aromatic CKs is more effective as compared to isoprenoid CKs as they are readily degraded by CKX (Rulcova and Pospisilova, 2001). Under water stress, roots absorb more water, leading to increased root growth. Simultaneously shoot growth in the aerial portion decreases to prevent water loss. These observations clearly hint at a remarkable decline in CKs in plants exposed to water stress (Hansen and Dörfing, 2003; Bano et al., 1994; Rivero et al., 2009). However, the role of CKs in water stress is not well understood but it increases the level of ABA causing stomatal closure to prevent excessive loss of water. CKs induce axillary growth by reducing apical dominance. In the case of sunflower root, the level of CKs reduces under water deficit conditions (Davies and Zhang, 1991; Mackova et al., 2013; Prerostova et al., 2018). CKs play a role in drought stress tolerance. Low CK levels in *Arabidopsis* mutants (such as ahk and ahp) contribute to adaptation under water stress conditions. Additionally CKs can negatively impact drought tolerance (Kang et al., 2012; Nguyen et al., 2016).

2.5 Gibberellins

Gibberellic acids play an important role in plant architecture by enhancing cell division and elongation. GAs are the class of tetracyclic diterpenoids carboxylic acids phytohormones. GA1 and GA4 are the most naturally occurring GAs in plants (Sedaghat et al., 2021). GA7 is the most abundant GA in plants derived from

GA12 (Farrow and Facchini, 2014). Studies on rice indicated the role of GAs in improving tolerance to water shortage by stimulating seed and seedling growth, internodal growth, and reproduction (Rasheed et al., 2020). Water stress affects the expression of dioxygenase genes that regulate the synthesis of GAs. Endogenous GAs act synergistically with other phytohormones to regulate the morphological and reproductive development of plants (Vishal & Kumar, 2018). Plants under water stress maintain water levels by restarting growth and development due to reduced biosynthesis of GAs. Thus phenotypic symptoms of the low endogenous level of GA are exactly similar to drought stress plants (Salazar-Cerezo et al., 2018). GA interacts with ethylene both positively and negatively depending on the type of signal received (Munteanu, 2014). Exogenous application of GAs causes stem elongation in the retarded growth exhibited in GA deficient mutants (Gilley and Fletcher, 1998; Vettakkorumakankav et al., 1999). GA synthesis requires the participation of plastids, endoplasmic reticulum and cytoplasm. Enzyme dioxygenase catalyzes the rate-limiting step of GA biosynthesis (Kasahara et al., 2002; Yamaguchi, 2008; Hedden and Thomas, 2012; Binenbaum et al., 2018). The GA regulates the gene expression of transcriptional regulators of the GRAS family. A few of these transcriptional regulators in *Arabidopsis* such as GAI and RGA (DELLA proteins) act as repressors to inhibit growth and development. Exogenous application of GA reverses these symptoms and facilitates growth and development by inhibiting growth repressors (Davière and Achard, 2013). Water stress causes the accumulation of DELLA proteins by lowering the GA content thereby inducing retarded growth in plants to tolerate stress (Achard et al., 2006; Magome et al., 2008; Wang E & Wang J, 2008). Another protein known as SPY acts as a negative regulator of the GA signaling cascade. Drought stress induces the accumulation of SPY protein by overexpression of the SPY protein gene that functions downstream of GA. SPY-1 and SPY-3 mutants in *Arabidopsis* are better suited to water stress by showing better growth. These findings clearly indicated the role of both DELA and SPY proteins in GA-mediated signaling (Qin et al., 2011). Exogenous application of GA3 enhances tolerance in Maize to moisture deficit conditions by maintaining water potential and membrane permeability (Kaya et al., 2006). Exogenous GA application overcomes oxidative stress by stimulating the production of the antioxidant H_2S (Kaya et al., 2020). Mutants having low GA content show reduced transpiration by producing smaller leaves and reducing stomatal dynamics (Nir et al., 2014). Tomatoes show reduced internodal elongation in low GA content under water stress (Li et al., 2012). In Tobacco plants overexpression of the PtGAox1 gene reduces the GA content in various parts to enable plants tolerate moisture deficit stress in various parts of plants. All these findings clearly indicate that in contrast to other phytohormones, GA is regarded as a negative regulator of water stress. In rice, overexpression of GA2ox6 increases yield under water stress (Lo et al., 2017). In Barley, GA maintains redox potential by controlling electron movement to acclimate plants to harsh environmental conditions (Mark et al., 2016).

2.6 Ethylene

Ethylene, a gaseous phytohormone, plays a crucial role in regulating various morphological and physiological aspects in the plant. It triggers a triple response in etiolated seedlings, causes fruit ripening and stimulates reproductive growth in plants. ET also plays a significant role in plants fighting abiotic stresses such as cold stress, salinity, and water stress (Awan et al., 2017). Quantification of ET has been done by considering various tissues in plants exposed to water stress (Kendrick and Chang, 2008). ET synthesis starts with S-Adenosyl methionine which is found in higher concentrations in fruit crops. The first step, the rate-limiting step in a reaction sequence converting SAM to ACC is catalyzed by 1-aminocyclopropane-1-carboxylic acid (ACC) synthase. ACC is further converted into intermediate methylthioadenosine (MTA) before forming L-methionine. The level of L-methionine remains unchanged irrespective of ET production. The concentration of ACC synthase fluctuates according to the requirement of ET (Maheshwari et al., 2015). Many ethylene-responsive genes such as AP2, EREBPs and APETALA2 have been identified under drought stress (Liu and Zhang, 2017). In the case of soybean and tobacco ethylene induces the overexpression of GmERF3 to provide drought tolerance by stimulating the production of proline, soluble sugars and reducing malondialdehyde content (Zhai et al., 2017). According to one investigation, some wheat genotypes showed an increase in ethylene and ABA content in tolerant plants with higher dry shoot weight under mild water stress response in comparison to the control (Valluru et al., 2016). Ethylene induces stomatal closure due to increased production of reactive oxygen species with increased NADPH oxidase activity tol mutant in *Arabidopsis* under water stress (Desikan et al., 2006; Tanaka et al., 2020). A similar case was observed in rice tol mutants due to overexpression of the OsETOL1 gene (Du et al., 2014). The gene controlling ethylene signaling under water stress could be a major target for recent investigation for a better understanding of ethylene action and to raise water stress-resistant transgenic plants.

2.7 Jasmonates

Jasmonates, derivatives of Jasmonic acid, are growth regulators in plants. It influences various morphological features such as root morphology, pollen morphology, tendrillar growth and fruit ripening (Kazan, 2015). Jasmonates also affect stomatal movement under water stress. The action of JA is regulated by the level of JAZ proteins that bind to several transcription factors in response to various stresses. A water deficit condition causes the destruction of JAZ proteins through the 26S proteasome pathway resulting in the overexpression of stress-responsive genes by the MYC2 transcription factor (Ghorbel M. et al., 2021). Phytohormones act synergistically rather than independently in controlling growth and development of a hostile environment in plants (Zaid &

Mohammad, 2018). Jasmonates, cyclopentanone phytohormones, are formed in the form of methyl jasmonates (MeJA) by the destruction of fatty acid content in the cell membrane in several plants (Campos-Rivero et al., 2017; Ali & Baek, 2020). In the case of tomato JAs inhibit seed germination under water deficit conditions (Zamani et al., 2020). Jasmonic acids (JAs) are best characterized among jasmonate phytohormones and are considered vital molecules involved in various defense mechanisms. Jasmonates include JA and its conjugates like MeJA, JA-Ile, and many more. They are synthesized in plastids, peroxisomes and cytosols. The first step of the reaction sequence in the formation of jasmonic acid starts with linolenic acid. Linolenic acid is converted into 12-OPDA through a series of catalyzed modifications involving various enzymes i.e lipoxygenase, synthases and cyclases. 12-OPDA is then lastly converted into 7-iso-JA (Hamberg and Gardner, 1992). The role of jasmontaes in regulating water stress was first proved in barley leaves that showed high JAs content with sorbitol treatment (Lehmann et al., 1995; Kramell et al., 2000). Exogenous application of JA helps to lower oxidative injury induced by water stress in many cereals and pulses (Li & Jäger, 1998; Bandurska et al., 2003; Anjum et al., 2011; Fahad et al., 2015). Jasmonates confer tolerance to plants against water stress by stimulating the secretion of osmolytes such as proline and increasing the activity of enzymes like superoxide dismutase and peroxidase in the ROS pathway to prevent damage to lipid membrane by reactive oxygen species (Anjum et al., 2011). In *Arabidopsis*, JAs induce the closing of stomata to prevent water loss under water stress (Munemasa et al., 2011; Riemann et al., 2015). Several other investigations conducted on *Arabidopsis*, papaya, tomato, and barley have confirmed the role of JAs in fighting environmental strain (Mahouachi et al., 2007; Arbona et al., 2010; Brossa et al., 2011).

2.8 Brassinosteroids

Brassinosteroids, derivatives of polyhydroxylated sterol, are ubiquitous in the plant kingdom. They were reported for the first time from pollen grains of *Brassica napus* (Saini et al., 2015). They are synthesized in every tissue where they are needed (Bishop and Yokota, 2001). They are analogous to animal steroid hormones. However, their level varies with age and plant species. They are produced in higher concentration in young tissues as compared to old (Clouse, 2011). Brassinosteroid follows the sterol pathway starting from 24-methylene diphenol (Du et al., 2017). Brassinosteroid phytohormone plays a significant role in regulating architectural modifications under optimal and sub optimal environmental conditions. It affects starch metabolism in guard cells to regulate stomatal opening (Li et al., 2020). It affects various physicochemical activities in plants such as cell elongation, division, root apical meristem size, root hair and lateral root formation, tropic movement, and vascular tissue differentiation (McGuiness et al., 2019; Wei and Li, 2016; Nolan et al., 2020). It induces the overexpression of several growth-promoting genes by the binding of BZR 1 transcription factors to the promoter site of PIF4 (Ibañez et al., 2018). They occur in both free form as well as in combination with other groups in tissue. More than 70 Brassinosteroids have been identified either in free or conjugated forms (Bajguz, 2011). Brassinolide is the most active brassinosteroid among all brassinosteroids. They are similar to Auxins in various working aspects. BRs are synthesized on the endoplasmic reticulum (ER) via the terpenoid pathway by cytochrome P450 (Ralston and Yu, 2006; Choe, 2007; Chung et al., 2011). Several reports from wheat, *Brassica* mentioned the role of BRs in relation to water stress (Divi et al., 2010). However, some investigations claimed the negative impact of BRs in growth and development under water stress by knocking out the BR1 gene which encodes a membrane-bound kinase (Feng et al., 2015). BRs interact with various transcription factors such as WRKY to regulate water stress in *Arabidopsis thaliana* (Clarke et al., 2004; Xue et al., 2013; Sánchez-Romera et al., 2014; Sánchez-Martín et al., 2015). BRs act in close proximity to their synthesis site but interact with other phytohormones such as auxins elsewhere at distant sites (Symons et al., 2008; Vriet et al., 2013). Several modifications are required to maintain bioactive levels of cellular BRs (Saini et al., 2015). BRs act in two ways: one involves kinases for phosphorylation of various intermediates of signaling cascade; the other involves the destruction of repressor proteins. A small amount of BRs is required to check growth and development under moisture deficit stress through signal amplification (Nolan et al., 2017). A number of investigations revealed the positive role of exogenous BRs in providing stress tolerance in plants (Kagale et al., 2007; Bajguz and Hayat, 2009; Yuan et al., 2010; Anjum et al., 2011; Divi et al., 2016). On the contrary, the application of exogenous BRs in higher concentrations exerts a negative impact on plant stature (Chaiwanon and Wang, 2015; Belda-Palazon et al., 2018). BRs regulate the level of cellular ROS in response to abiotic stress thus preventing oxidative damage to membranes and organelles (Kaur et al., 2018). In tomato, application of BRs enhances the antioxidant activity to overcome oxidative stress generated by sub-optimal environmental conditions (Ogweno et al., 2007). In *Arabidopsis*, a BR-responsive receptor BRL3 aids in drought tolerance (Fàbregas et al., 2018). The role of BRs depends on soil water content as confirmed by studies conducted with BR mutants (Northey et al., 2016). In *Arabidopsis*, mutants are less capable of mitigating water stress due to defective brassinosteroid biosynthesis. BES1 binds with the upstream region known as the BR response element of RD26 (NAC transcription factor) to inhibit the action of BRs in water stress conditions. These studies strongly support that the level of RD26 increases in response to water stress. The level of bioactive BRs in the cell is maintained by acylation, glycosylation, and various other modifications. BRs signaling starts with a membrane-bound receptor leucine rich kinase known as BRI1. BR action is concentration dependent as it negatively affects growth in higher concentrations while in low concentration it stimulates growth as confirmed with exogenous application of BRs.

2.9 Polyamines

Polyamines, a group of low molecular weight aliphatic nitrogenous compounds, are known due to their regulatory role in morphological development and working of plants as well as in providing resistance to plants to various physical stresses. They are polycationic with more than two amino groups. They regulate a number of developmental processes such as cell division and differentiation, gene expression and synthesis of biomolecules, organ development, and fruit ripening (Seiler and Raul, 2005; Alcázar et al., 2010; Igarashi and Kashiwagi, 2010; Childs et al., 2017). The most studied polyamines in plants include spermidine, spermine and putrescine. They are present either in free or conjugated form. Spermidine and spermine are derived from putrescine. Putrescine is derived from a different nitrogenous compound such as arginine, ornithine and citrulline (Chen et al., 2019). H_2O_2, produced due to polyamine oxidation, acts as a signaling molecule in response to water stress (Freitas et al., 2017; Mellidou et al., 2017; Sariyev et al., 2020). PA synthesis includes S-adenosyl methionine, the precursor of ethylene biosynthesis. NO, a key signaling molecule in regulating growth also generates during the metabolism of PA. Synthesis of PA is tissue-specific in plant systems. Putrescine is the most abundant PA found in leaves (Reginato et al., 2012; Takahashi et al., 2017). Transgenic plants showing the overexpression of gene encoding various enzymes such as spermidine synthase, and arginine decarboxylase of PA metabolism are better suited to water stress. Moreover, the exogenous application of PAs confers resistance to plants against several abiotic stresses (Alcázar et al., 2010; Qi et al., 2010; Minocha et al., 2014).

2.10 Strigolactones

Strigolactones are carotenoid derivatives and play a regulatory role in developmental and physiological processes in plants (Akiyama and Hayashi, 2006). The first strigolactone was isolated from the plant *Strigalutea* as an inducer of seed germination. Several other structural analogs of strigolactones such as GR5, GR7, and GR24 were also identified in plants. Chemical characterization of all these derivatives proved their controlling role in various developmental processes in plants such as organogenesis, reproduction, and senescence (Gomez-Roldan et al., 2008; Kapulnik et al., 2011; Kohlen et al., 2011; Ruyter-Spira et al., 2011). Strigolactones interact with other phytohormones such as auxin to regulate plant make up (Hayward et al., 2009). They are synthesized in plastids with the help of dioxygenase enzymes (Alder et al., 2012). Srigolactones exudation from roots stimulates hyphal growth of mycorrhizal fungi (Rochange et al., 2019). They also regulate stress response in plants in association with other phytohormones. Their cross-talking with other hormones establishes a regulatory network to control various physiological processes to adapt plants under abiotic stress (Marzec et al., 2013). Exogenous application of strigolactones reduces oxidative stress by activating the free radical scavenging machinery. They also enhance the rate of photosynthesis and stomatal diffusion (Ling et al., 2020).

2.11 Melatonin

Melatonin is a recently discovered plant hormone that is known to regulate various functions in plants (Wang et al., 2017). It serves as an antioxidant to act upon reactive oxygen species. Melatonin is also involved in controlling growth at several stages in plants by enhancing several physiological activities (Arnao and Hernández-Ruiz, 2015). The finding of the melatonin receptor gave new insight to establish the regulatory function of melatonin as a phytohormone. Melatonin signaling affects the level of intracellular Ca^{2+}, membrane permeability, stomatal dynamics and metabolism of biomolecules (Sharif et al., 2018). It maintains plant homeostasis against several physical stresses such as drought and salinity (Sun et al., 2015). Recently a melatonin-binding plasma membrane receptor CAND2/PMTR1 has been identified in *Arabidopsis*. It interacts with the G protein upon binding with melatonin to establish a signal cascade that results in stomatal closure due to enhanced K^+ influx and H_2O_2 production. Exogenous application of ABA-induced stomatal closure in CAND2 mutant suggests that the ABA binding receptor is different from the melatonin binding receptor (Guo et al., 2017; Arnao and Hernández-Ruiz, 2019).

3. Conclusion and Future Prospects

Abiotic stresses negatively affect developmental stages from seed germination to maturity. Phytohormones mitigate these negative influences of abiotic stresses by harnessing several coping strategies. They not only interact with each other but also interact with root microbiota to enhance growth in both optimal and sub-optimal environmental conditions. Their exogenous application also further improves plant stress-tolerating capacity. PHs are categorized into a number of categories. The cellular level of various phytohormones varies in response to stress. The signaling pathways of different hormones are well known. However, how they cross-talk in signaling pathways to alleviate the adverse effect of water stress is poorly understood and needs more investigation. Majority of phytohormones regulate the morphological and physiological changes of leaves and roots. PHs also affect stomatal dynamics to regulate water loss and carbon assimilation. They also activate various transcription factors to regulate the expression of stress-responsive genes participating in growth and development. All these findings give way to raising transgenic-resistant crop varieties that are better suited to the harsh environment. In addition to these various microbes (PGPR) also play a significant role to confer resistance to plants by changing the intracellular levels of phytohormones, osmoprotectants and other secondary metabolites. This environment-friendly approach gives a new way to crop improvement in a cost-effective manner. However, our understanding is still insufficient to know the intricate crosstalk of phytohormones to control drought stress. It needs more investigation by with an amalgamation of advanced biological tools such as genomics, proteomics, transcriptomics and bioinformatics.

References

Achard P, Cheng H, De Grauwe L, Decat J, Schoutteten H, Moritz T, et al. 2006. Integration of plant responses to environmentally activated phytohormonal signals. *Science,* **311:** 91–94.

Ahanger MA, Tomar NS, Tittal M, Argal S & Agarwal RM. 2017. Plant growth under water/salt stress: ROS production; antioxidants and significance of added potassium under such conditions. *Physiology and Molecular Biology of Plants,* **23(4):** 731–744.

Akiyama K & Hayashi H. 2006. Strigolactones: Chemical signals for fungal symbionts and parasitic weeds in plant roots. *Ann. Bot,* **97:** 925–931.

Alcázar R, Altabella T, Marco F, Bortolotti C, Reymond M, Koncz C, et al. 2010. Polyamines: Molecules with regulatory functions in plant abiotic stress tolerance. *Planta,* **231:** 1237–1249.

Alder A, Jamil M, Marzorati M, Bruno M, Vermathen M, Bigler P, et al. 2012. The path from β-carotene to carlactone, a strigolactonelike plant hormone. *Science,* **335:** 1348–1351.

Ali MS & Baek KH. 2020. Jasmonic acid signaling pathway in response to abiotic stresses in plants. *Int. J. Mol. Sci.,* **21:** 621.

Ali M, Kamran M, Abbasi GH, Saleem MH, Ahmad S, Parveen A, et al. 2021b. Melatonin-induced salinity tolerance by ameliorating osmotic and oxidative stress in the seedlings of two tomato (Solanum lycopersicum L.) cultivars. *J. Plant Growth Regulat,* **40:** 2236–2248.

Alibert G & Ranjȇva R. 1971. Recherches sur catalysant la biosynthese des acidesphenoliques Chezquercus pedunculata (Ehrh): I. Formation des premiers termes des series cinnamique et benzoique. *FEBS Lett,* **19:** 11–14.

Alibert G, Ranjeva R & Boudet A. 1972. II. Localisationintracellulaire de la phenylalanine ammoniaque-lyase, de la cinnamate 4-hydroxylase, et de la benzoate synthase. *Biochim. Biophys. Acta,* **279:** 282–289.

Anjum SA, Wang L, Farooq M, et al. 2011 Methyl jasmonate induced alteration in lipid peroxidation, antioxidative defence system and yield in soybean under drought. *J Agron Crop Sci,* **197:** 296–301.

Anjum SA, Wang LC, Farooq M, Hussain M, Xue LL & Zou CM. 2011. Brassinolide application improves the drought tolerance in maize through modulation of enzymatic antioxidants and leaf gas exchange. *J. Agron. Crop. Sci,* **197:** 177–185.

Arbona V, Argamasilla R & Gomez-Cadenas A. 2010. Common and divergent physiological, hormonal and metabolic responses of Arabidopsis thaliana and Thellungiella halophilato water and salt stress. *J. Plant Physiol,* **167:** 1342–1135. doi: 10.1016/j.jplph.2010.05.012

Arnao MB & Hernández-Ruiz J. 2019. Melatonin and reactive oxygen and nitrogen species: A model for the plant redox network. *Melatonin. Res.,* **2:** 152–168. doi: 10.32794/11250036

Ashraf M & Foolad MR. 2007. Roles of glycine betaine and proline in improving plant abiotic stress resistance. *Environ. Exp. Bot.,* **59:** 206–216. doi: 10.1016/j.envexpbot.2005.12.006

Awan FK', Khurshid MY & Mehmood AJIJIRB. 2017. Plant growth regulators and their role in abiotic stress management. *Int. J. Innov. Res. Biosci.,* **1:** 9–21.

Bajguz A & Hayat S. 2009. Effects of brassinosteroids on the plant responses to environmental stresses. *Plant Physiol. Biochem,* **47:** 1–8. doi: 10.1016/j.plaphy.2008.10.002

Bandurska H, Stroiński A & Kubiś J. 2003. The effect of jasmonic acid on the accumulation of ABA, proline and spermidine and its influence on membrane injury under water deficit in two barley genotypes. *Acta Physiol Plant,* **25:** 279–285.

Bano A, Hansen H, Dörfing K & Hahn H. 1994. Changes in the contents of free and conjugated abscisic acid, phaseic acid and cytokinins in xylem sap of drought stressed sunflower plants. *Phytochemistry,* **37:** 345–347.

Belda-Palazon B, Gonzalez-Garcia MP, Lozano-Juste J, Coego A, Antoni R, Julian J, et al. 2018. PYL8 mediates ABA perception in the root through noncell-autonomous and ligand-stabilization-based mechanisms. *Proc. Natl. Acad. Sci.,* U.S.A. **115:** E11857–E11863. doi: 10.1073/pnas.1815410115

Binenbaum J, Weinstain R & Shani E. 2018. Gibberellin localization and transport in plants. *Trends Plant Sci,* **23:** 410–421.

Bishop GJ & Yokota T. 2001. Plants steroid hormones, brassinosteroids: Current highlights of molecular aspects on their synthesis/metabolism, transport, perception and response. *Plant Cell Physiol.* Oxford Academic. doi: 10.1093/PCP/PCE018

Bouzroud S, Gouiaa S, Hu N, Bernadac A, Mila I, Bendaou N, et al. 2018. Auxin response factors (ARFs) are potential mediators of auxin action in tomato response to biotic and abiotic stress (Solanum lycopersicum). *PLoS One,* **13:** e0193517. doi: 10.1371/journal.pone.0193517

Brewer PB, Koltai H. & Beveridge CA. 2013. Diverse roles of strigolactones in plant development. *Mol. Plant,* **6:** 18–28.

Brossa R, López-Carbonell M, Jubany-Marí T & Alegre L. 2011. Interplay between abscisic acid and jasmonic acid and its role in water-oxidative stress in wild-type, ABA-deficient, JA-deficient, and ascorbate-deficient Arabidopsis plants. *J. Plant Growth Regul.,* **30:** 322–333.

Burke JJ. 2007. Evaluation of source leaf responses to water-deficit stresses in cotton using a novel stress bioassay. *Plant Physiology,* **143:** 108–121.

Campbell L, Etchells JP, Cooper M, Kumar M & Turner SR. 2018. An essential role for abscisic acid in the regulation of xylem fibre differentiation. *Dev.* **145**.

Campos-Rivero G, Osorio-Montalvo P, Sánchez-Borges R, Us-Camas R, Duarte-Aké F & De-la-Peña C. 2017. Plant hormone signaling in flowering: An epigenetic point of view. *J. Plant Physiol.,* **21:** 16–27.

Chaiwanon J & Wang Z-Y. 2015. Spatiotemporal brassinosteroid signaling and antagonism with auxin pattern stem cell dynamics in arabidopsis roots. *Curr. Biol.,* **25:** 1031–1042.

Champolivier L & Merrien A. 1996. Effects of water stress applied at different growth stages to Brassica napus L. var. oleifera on yield, yield components and seed quality. *Eur. J. Agron.,* **5:** 153160.

Chen D, Shao Q, Yin L, Younis A & Zheng B. 2019. Polyamine function in plants: Development, and roles in abiotic stress responses. *Front. Plant Sci,* **9:** 1945.

Chen H, Chen X & Zheng Y. 2013. The nuclear lamina regulates germline stem cell niche organization via modulation of EGFR signaling. *Cell Stem Cell,* **13:** 73–86. doi: 10.1016/j.stem.2013.05.003

Chen M, Zhu X, Liu X, Wu C, Yu C, Hu G, et al. 2021. Knockout of auxin response factor SlARF4 improves tomato resistance to water deficit. *International Journal of Molecular Sciences,* **22(7):** 3347.

Cheng Y, Dai X & Zhao Y. 2006. Auxin biosynthesis by the YUCCA favin monooxygenases controls the formation of foral organs and vascular tissues in Arabidopsis. *Genes Dev,* **20:** 1790–1799.

Childs C, Holdsworth RE, Christopher A-L, Jackson Manzocchi T, Walsh JJ & Yielding G. 2017. Introduction to the geometry and growth of normal faults. *Geol. Soc,* **439:** 1–9.

Choe S. 2007. Brassinosteroid biosynthesis and metabolism. pp. 156–178. *In:* Davies PJ (ed), Plant Hormones: Biosynthesis, Signal Transduction, Action, Dordrecht: Kluwer Academic Publishers. doi: 10.1007/978-1-4020-2686-7_8

Chung KY, Rasmussen SG, Liu T, Li S, DeVree BT, Chae PS, et al. 2011. Conformational changes in the G protein Gs induced by the β2 adrenergic receptor. *Nature,* **477:** 611–615. doi: 10.1038/nature10488

Clarke SM, Mur LA, Wood JE & Scott IM. 2004. Salicylic acid dependent signaling promotes basal thermotolerance but is not essential for acquired thermotolerance in Arabidopsis thaliana. *Plant J,* **38:** 432–447. doi: 10.1111/j.1365-313X.2004.02054.x

Clouse SD. 2011. Brassinosteroids. *Arab. B.,* **9:** e0151.

Cornforth JW, Milborow BV, Ryback G & Wareing PF. 1965. Identity of sycamore 'dormin' with abscisin II. *Nature,* **205:** 1269–1270.

Danilova MN, Kudryakova NV, Doroshenko AS, Zabrodin DA, Vinogradov NS & Kuznetsov VV. 2016. Molecular and physiological responses of Arabidopsis thaliana plants deficient in the genes responsible for ABA and cytokinin reception and metabolism to heat shock. *Russ. J. Plant Physiol.,* **63:** 308–318.

Danquah A, De Zelicourt A, Colcombet J & Hirt H. 2014. The role of ABA and MAPK signaling pathways in plant abiotic stress responses. *Biotechnol. Adv.,* **32:** 40–52.

Dar NA, Amin I, Wani W, Wani SA, Shikari AB, Wani SH & Masoodi KZ. 2017. Abscisic acid: A key regulator of abiotic stress tolerance in plants. *Plant Gene,* **11:** 106–111.

Dat J, Vandenabeele S, Vranová E, Van Montagu M, Inzé D & Van Breusegem F. 2000. Dual action of the active oxygen species during plant stress responses. *Cellular and Molecular Life Sciences,* **57:** 779–795.

Davière JM & Achard P. 2013. Gibberellin signaling in plants. *Dev,* **140:** 1147–1151.

Davies WJ & Zhang J. 1991. Root signals and the regulation of growth and development of plants in drying soil. *Annu Rev Plant Physiol Plant Mol Biol,* **42:** 55–76.

De Smet I, Voss U, Lau S, Wilson M, Shao N, Timme RE, et al. 2010. Unraveling the evolution of auxin signaling. *Plant Physiol,* **155:** 209–221.

Desikan R, Last K, Harrett-Williams R, et al. 2006. Ethylene-induced stomatal closure in Arabidopsis occurs via AtrbohF-mediated hydrogen peroxide synthesis. *Plant J,* **47:** 907–916. https://doi.org/ 10.1111/j.1365-313X.2006.02842.x

Divi UK, Rahman T & Krishna P. 2010. Brassinosteroid-mediated stress tolerance in Arabidopsis shows interactions with abscisic acid, ethylene and salicylic acid pathways. *BMC Plant Bio,* **110:** 151. doi: 10.1186/1471-2229-1 0-151

Divi UK, Rahman T. & Krishna P. 2016. Gene expression and functional analyses in brassinosteroid-mediated stress tolerance. *Plant Biotechnol. J.,* **14:** 419–432.

Du H, Liu H & Xiong L. 2013a. Endogenous auxin and jasmonic acid levels are differentially modulated by abiotic stresses in rice. *Front Plant Sci.,* https://doi.org/10.3389/fpls.2013.00397

Du H, Wu N, Chang Y, et al. 2013b. Carotenoid deficiency impairs ABA and IAA biosynthesis and differentially affects drought and cold tolerance in rice. *Plant Mol Biol,* **83:** 475–488.

Du H, Wu N, Cui F, et al. 2014. A homolog of ethylene overproducer 1, differentially modulates drought and submergence tolerance in rice. *Plant J,* **78:** 834–849. https://doi.org/10.1111/tpj.12508

Du J, Zhao B, Sun X, Sun M, Zhang D, Zhang S & Yang W. 2017. Identification and characterization of multiple intermediate alleles of the key genes regulating brassinosteroid biosynthesis pathways. *Front. Plant Sci,* **7:** 1893. doi: 10.3389/fpls.2016.01893

Engelen-Eigles G, Holden G, Cohen JD & Gardner G. 2006. The effect of temperature, photoperiod, and light quality on gluconasturtiin concentration in watercress (Nasturtium offcinale R. Br.). *J. Agric. Food Chem.,* **54:** 328334.

Fàbregas N, Lozano-Elena F, Blasco-Escámez D, Tohge T, Martínez Andújar C, Albacete A, et al. 2018. Overexpression of the vascular brassinosteroid receptor BRL3 confers drought resistance without penalizing plant growth. *Nat. Commun.,* **9:** 4680. doi: 10.1038/s41467-018-06 861-3

Fahad S, Bajwa AA, Nazir U, Anjum SA, Farooq A, Zohaib A, et al. 2017. Crop production under drought and heat stress: Plant responses and management options. *Front. Plant Sci.,* **8:** 1147.

Fahad S, Hussain S, Matloob A, et al 2015. Phytohormones and plant responses to salinity stress: A review. *Plant Growth Regul.,* **75:** 391–404. https://doi.org/10.1007/s10725-014-0013-y

Farrow SC & Facchini PJ. 2014. Functional diversity of 2-oxoglutarate/Fe(II)-dependent dioxygenases in plant metabolism. *Front. Plant Sci.,* **5:** 524.

Feng Y, Yin Y & Fei S. 2015. Down-regulation of BdBRI1, a putative brassinosteroid receptor gene produces a dwarf phenotype with enhanced drought tolerance in Brachypodium distachyon. *Plant Sci,* **234:** 163–173. doi: 10.1016/j.plantsci.2015.02.015

Finkelstein R. 2013. Abscisic acid synthesis and response. *Arab B,* **11:** e0166.

Floková K, Tarkowská D, Miersch O, Strnad M, Wasternack C & Novák O. 2014. UHPLC-MS/MS based target profiling of stress-induced phytohormones. *Phytochemistry,* **105:** 147–157.

Freitas AM, Nair VD, Harris WG, Mosquera-Losada MR & Nair PKR. 2017. Phosphorus release behavior of biosolids and corresponding biochars in abstract retrieved from ASA, CSSA and SSSA international annual meetings. Available online at: https://scisoc.confex.com/~ crops/2017am/webprogram/Paper106996.html (accessed August 20, 2021).

Fuller DQ & Stevens CJ. 2019. Between domestication and civilization: The role of agriculture and arboriculture in the emergence of the first urban societies. *Veg. Hist. Archaeobot.,* **28:** 263–282.

Ghorbel M, Brini F, Sharma A & Landi M. 2021. Role of jasmonic acid in plants: The molecular point of view. *Plant Cell Rep,* **40:** 1471–1494.

Gilley A & Fletcher RA. 1998. Gibberellin antagonizes paclobutrazol induced stress protection in wheat seedlings. *J Plant Physiol,* **153:** 200–207.

Gomez-Roldan V, Fermas S, Brewer PB, Puech-Pagès V, Dun EA. Pillot JP, et al. 2008. Strigolactone inhibition of shoot branching. *Nature,* **455:** 189–194. doi: 10.1038/nature07271

Guajardo E, Correa JA & Contreras-Porcia L. 2016. Role of abscisic acid (ABA) in activating antioxidant tolerance responses to desiccation stress in intertidal seaweed species. *Planta,* **24:** 767–781. doi: 10.1007/s00425-015-2438-6

Gujjar RS, Banyen P, Chuekong W, Worakan P, Roytrakul S & Supaibulwatana K. 2020. A synthetic cytokinin improves photosynthesis in rice under drought stress by modulating the abundance of proteins related to stomatal conductance, chlorophyll contents, and rubisco activity. *Plants,* **9:** 1106.

Guo D, Zhou Y, Li HL, Zhu JH, Wang Y, Chen XT, et al. 2017. Identification and characterization of the abscisic acid (ABA) receptor gene family and its expression in response to hormones in the rubber tree. *Sci Rep,* **7:** 45157. doi: 10.1038/srep45157

Ha CV, Leyva-Gonzalez MA, Osakabe Y & Phan Tran LS. 2014. Positive regulatory role of strigolactone in plant responses to drought and salt stress. *Proc Natl Acad Sci,* **111:** 851–856.

Hai NN, Chuong NN, Tu NHC, Kisiala A, Hoang XLT & Thao NP. 2020. Role and regulation of cytokinins in plant response to drought stress. *Plants (Basel),* **9(4):** 422. https://doi.org/10.3390/ plants9040422

Hamayun M, Hussain A, Iqbal A, Khan SA, Ahmad A, Gul S, et al. 2021a. Aspergillus foetidus regulated the biochemical characteristics of soybean and sunflower under heat stress condition: Role in sustainability. *Sustainability,* **13:** 7159.

Hamayun M, Hussain A, Iqbal A, Khan SA, Gul S, Khan H, et al. 2021. Penicillium glabrum acted as a heat stress relieving endophyte in soybean and sunflower. *Polish J. Env. Stud.,* **2021:** 30.

Hamayun M, Hussain A, Iqbal A, Khan SA, Khan MA & Lee I-J. 2021c. An endophytic fungus Gliocladium cibotii regulates metabolic and antioxidant system of glycine max and Helianthus annuus under heat stress. *Polish J. Env. Stud.,* **2021:** 30.

Hamberg M & Gardner HW. 1992. Oxylipin pathway to jasmonates: Biochemistry and biological significance. *Biochim Biophys Acta Lipids Lipid Metab,* **1165:** 1–18.

Hansen H & Dörfing K. 2003. Root-derived trans-zeatin riboside and abscisic acid in drought-stressed and rewatered sunflower plants: Interaction in the control of leaf diffusive resistance. *Funct Plant Biol,* **30:** 365–375.

Hayward A, Stirnberg P, Beveridge C & Leyser O. 2009. Interactions between auxin and strigolactone in shoot branching control. *Plant Physiol, 151:* 400–412.

He M, He C-Q & Ding NZ. 2018. Abiotic stresses: General defenses of land plants and chances for engineering multistress tolerance. *Front. Plant Sci,* **9:** 1771.

Hedden P & Thomas SG. 2012. Gibberellin biosynthesis and its regulation. *Biochem J,* **444:** 11–25.

Hoque T, Hossain MA, Mostofa MG, Burritt D, Fujita M & Tran L-S. 2016. Methylglyoxal: An emerging signaling molecule in plant abiotic stress responses and tolerance. *Front. Plant Sci.,* **7:** 1341. doi: 10.3389/fpls.2016.01341

Huang J, Mishra M, Palani S, Chew TG & Balasubramanian MK. 2016. Isolation of cytokinetic actomyosin rings from Saccharomyces cerevisiae and Schizosaccharomyces pombe. *Methods Mol. Biol.,* **1369:** 125–136. doi: 10.1007/978-1-4939-3145-3_10

Hummel I, Pantin F, Sulpice R, Piques M, Rolland G, Dauzat M, et al. 2010. Arabidopsis plants acclimate to water deficit at low cost through changes of carbon usage: An integrated perspective using growth, metabolite, enzyme, and gene expression analysis. *Plant Physiology,* **154:** 357–372.

Hwang I, Sheen J & Müller B. 2012. Cytokinin signaling networks. *Annu Rev Plant Biol,* **63:** 353–380.

Ibañez C, Delker C, Martinez C, Bürstenbinder K, Janitza P, Lippmann R, et al. 2018. Brassinosteroids dominate hormonal regulation of plant thermomorphogenesis via BZR1. *Curr. Biol,* **28:** 303-310. e3. doi: 10.1016/j.cub.2017.11.077

Igarashi K & Kashiwagi K. 2010. Modulation of cellular function by polyamines. *Int. J. Biochem. Cell Biol,* **42:** 39–51.

Iqbal N, Fatma M, Gautam H, Sehar Z, Rasheed F, Khan MIR, et al. 2022. Salicylic acid increases photosynthesis of drought grown mustard plants effectively with sufficient-N via regulation of ethylene, abscisic acid & nitrogen-use efficiency. *J. Plant Growth Regul,* 21–12.

Jung H, Lee DK, Do CY & Kim JK. 2015. OsIAA6, a member of the rice Aux/IAA gene family, is involved in drought tolerance and tiller outgrowth. *Plant Sci,* 236: 304–312.

Jurado S, Abraham Z, Manzano C, López-Torrejón G, Pacios LF & Del Pozo JC. 2010. The Arabidopsis cell cycle F-box protein SKP2A binds to auxin. *Plant Cell,* **22:** 3891–3904. doi: 10.1105/tpc.110.078972

Kagale S, Divi UK, Krochko JE, Keller WA & Krishna P. 2007. Brassinosteroid confers tolerance in Arabidopsis thaliana and Brassica napus to a range of abiotic stresses. *Planta,* **225:** 353–364.

Kaiser WM, Kaiser G, Schöner S & Neimanis S. 1981. Photosynthesis under osmotic stress. Differential recovery of photosynthetic activities of stroma enzymes, intact chloroplasts and leaf slices after exposure to high solute concentrations. *Planta,* **153:** 430–435.

Kang C, He S, Zhai H, Li R, Zhao N & Liu Q. 2018. A sweetpotato auxin response factor gene (IbARF5) is involved in carotenoid biosynthesis and salt and drought tolerance in transgenic Arabidopsis. *Front. Plant Sci,* **9:** 1307. doi: 10.3389/fpls.2018.01307

Kang G, Li G, Xu W, et al. 2012. Proteomics reveals the effects of salicylic acid on growth and tolerance to subsequent drought stress in wheat. *J Proteome Res,* **11:** 6066–6079. https://doi.org/ 10.1021/pr300728y

Kapulnik Y, Resnick N, Mayzlish-Gati E, Kaplan Y, Wininger S, Hershenhorn J, et al. 2011. Strigolactones interact with ethylene and auxin in regulating root-hair elongation in Arabidopsis. *J. Exp. Bot,* **62:** 2915–2924. doi: 10.1093/jxb/erq464

Karppinen K, Tegelberg P, Häggman H & Jaakola L. 2018. Abscisic acid regulates anthocyanin biosynthesis and gene expression associated with cell wall modification in ripening bilberry (Vaccinium myrtillus L.) fruits. *Front. Plant Sci,* **9:** 1259.

Kasahara H, Hanada A, Kuzuyama T, et al. 2002. Contribution of the mevalonate and methylerythritol phosphate pathways to the biosynthesis of Gibberellins in Arabidopsis. *J Biol Chem,* **277:** 45188–45194. https://doi.org/10.1074/jbc.M208659200

Kaur H, Sirhindi G, Bhardwaj R, Alyemeni MN, Siddique KHM & Ahmad P. 2018. 28-homobrassinolide regulates antioxidant enzyme activities and gene expression in response to salt and temperature-induced oxidative stress in Brassica juncea. *Sci. Rep.,* **28:** 8735.

Kaya C, Tuna AL & Alves AAC. 2006. Gibberellic acid improves water deficit tolerance in maize plants. *Acta Physiol Plant,* **28:** 331–337. https://doi.org/10.1007/s11738-006-0029-7

Kaya C, Sarıoğlu A, Ashraf M, Alyemeni MN & Ahmad P. 2020. Gibberellic acid-induced generation of hydrogen sulfide alleviates boron toxicity in tomato (Solanum lycopersicum L.) plants. *Plant Physiol. Biochem.,* doi: 10.1016/j.plaphy.2020.04.038

Kazan K. 2015. Diverse roles of jasmonates and ethylene in abiotic stress tolerance. *Trends Plant Sci,* **20:** 219–229. doi: 10.1016/j.tplants.2015.02.001

Ke Q, Wang Z, Ji C, et al. 2015. Transgenic poplar expressing Arabidopsis YUCCA6 exhibits auxin-overproduction phenotypes and increased tolerance to abiotic stress. *Plant Physiol Biochem,* **94:** 19–27. Elsevier. https://doi.org/10.1016/j.plaphy.2015

Kendrick MD & Chang C. 2008. Ethylene signaling: New levels of complexity and regulation. *Curr. Opin. Plant Biol.,* **11:** 479–485.

Khosravi-nejad F, Khavari-nejad RA, Moradi F & Najafi F. 2022. Cytokinin and abscisic acid alleviate drought stress through changing organic acids profile, ion immolation, and fatty acid profile to improve yield of wheat (Triticum aestivum L.) cultivars. *Physiol. Mol. Biol. Plants,* **28:** 1119–1129.

Kieber JJ & Schaller GE. 2014. Cytokinins. *Arab. B.*, **12:** e0168.

Kim TH, Böhmer M, Hu H, et al. 2010. Guard cell signal transduction network: Advances in understanding abscisic acid, CO_2, and Ca^{2+} signaling. *Annu Rev Plant Biol,* **61:** 561–591. https://doi. org/10.1146/annurev-arplant-042809-112226

Klingler JP, Batelli G & Zhu JK. 2010. ABA receptors: The START of a new paradigm in phytohormone signalling. *J. Exp. Bot.,* **61:** 3199–3210. doi: 10.1093/jxb/erq151

Kohlen W, Charnikhova T, Liu Q, Bours R, Domagalska MA, Beguerie S, et al. 2011. Strigolactones are transported through the xylem and play a key role in shoot architectural response to phosphate deficiency in nonarbuscular mycorrhizal host Arabidopsis. *Plant Physiol,* **155:** 974–987. doi: 10.1104/pp.110.164640

Kramell R, Miersch O, Atzorn R, et al. 2000. Octadecanoid-derived alteration of gene expression and the "oxylipin signature" in stressed barley leaves. Implications for different signaling pathways. *Plant Physiol,* **123:** 177–187.

Lee SB & Suh MC. 2015. Advances in the understanding of cuticular waxes in Arabidopsis thaliana and crop species. *Plant Cell Rep.*, **34:** 557–572. doi: 10.1007/s00299-015-1772-2

Lee YS, Lee DH, Jeong SY, Park SH, Oh SC, Park YS, et al. 2019. Ferroptosis-inducing agents enhance TRAIL-induced apoptosis through upregulation of death receptor 5. *J. Cell. Biochem.,* **120(1):** 928–939.

Lehmann J, Atzorn R, Brückner C, et al. 1995. Accumulation of jasmonate, abscisic acid, specific transcripts and proteins in osmotically stressed barley leaf segments. *Planta,* **197:** 156–162. https:// doi.org/10.1007/BF00239952

Lemoine R, La Camera S, Atanassova R, Dédaldéchamp F, Allario T, Pourtau N, et al. 2013. Source-to-sink transport of sugar and regulation by environmental factors. *Frontiers in Plant Science,* **4:** 272.

Li JG, Fan M, Hua W, Tian Y, Chen LG, Sun Y & Bai MY 2020. Brassinosteroid and hydrogen peroxide interdependently induce stomatal opening by promoting guard cell starch degradation. *Plant Cell,* **32:** 984–999. doi: 10.1105/tpc.19.00587

Li J, Sima W, Ouyang B, Wang T, Ziaf K, Luo Z, et al. 2012. Tomato SlDREB gene restricts leaf expansion and internode elongation by downregulating key genes for gibberellin biosynthesis. *J. Exp. Bot.,* **63:** 6407–6420. doi: 10.1093/jxb/ers295

Li L, Staden J & Jäger AK. 1998. Efects of plant growth regulators on the antioxidant system in seedlings of two maize cultivars subjected to water stress. *Plant Growth Regul.,* **25:** 81–87.

Li-Bei L, Ding-Wei Y, Feng-Li Z, et al. 2015. Genome-wide analysis of the calcium-dependent protein kinase gene family in Gossypium raimondii. *J Integr Agric,* **14:** 29–41.

Ling F, Su Q, Jiang H, et al. 2020. Effects of strigolactone on photosynthetic and physiological characteristics in salt-stressed rice seedlings. *Sci. Rep.,* **10:** 6183.doi: 10.1038/s41598-020-63352-6

Liu C & Zhang T. 2017. Expansion and stress responses of the AP2/EREBP superfamily in cotton. *BMC Genom,* **18:** 118.

Liu J, Hasanuzzaman M, Wen H, Zhang J, Peng T, Sun H & Zhao Q. 2019. High temperature and drought stress cause abscisic acid and reactive oxygen species accumulation and suppress seed germination growth in rice. *Protoplasma,* **256:** 1217–1227.

Liu Q, Yan S, Huang W, Yang J.-Y, Dong J, Zhang S, et al. 2018. NAC transcription factor ONAC066 positively regulates disease resistance by suppressing the ABA signaling pathway in rice. *Plant Mol. Biol.,* **98:** 289–302. doi: 10.1007/s11103-018-0768-z

Ljung K. 2013. Auxin metabolism and homeostasis during plant development. *Development,* **140:** 943–950.

Llanes A, Masciarelli O, Ordoñez R, Isla MI & Luna V. 2014. Differential growth responses to sodium salts involve different ABA catabolism and transport in the halophyte Prosopis strombulifera. *Biol. Plant,* **58:** 80–88. doi: 10.1007/s10535-013-0365-6

Lo S-F, Ho T-HD, Liu Y-L, Jiang M-J, Hsieh K-T, Chen K-T, et al. 2017. Ectopic expression of specific GA2 oxidase mutants promotes yield and stress tolerance in rice. *Plant Biotechnol. J.,* **15:** 850–864. doi: 10.1111/pbi.12681

López-Berenguer C, Martınez-Ballesta MC, Moreno DA, Carvajal M & Garcıa-Viguera C. 2009. Growing hardier crops for better health: Salinity tolerance and the nutritional value of broccoli. *J. Agric. Food Chem.,* **57:** 572578.

Lou D, Wang H, Liang G & Yu D. 2017. OsSAPK2 confers abscisic acid sensitivity and tolerance to drought stress in rice. *Front. Plant Sci.,* **8:** 993. doi: 10.3389/fpls.2017.00993

Macková H, Hronková M, Dobrá J, et al. 2013. Enhanced drought and heat stress tolerance of tobacco plants with ectopically enhanced cytokinin oxidase/dehydrogenase gene expression. *J Exp Bot,* **64:** 2805–2815.

Maghsoudi K, Emam Y, Ashraf M & Arvin MJ. 2019. Alleviation of field water stress in wheat cultivars by using silicon and salicylic acid applied separately or in combination. *Crop Pasture Sci,* **70:** 36–43.

Magome H, Yamaguchi S, Hanada A, et al. 2008. The DDF1 transcriptional activator upregulates expression of a gibberellin-deactivating gene, GA2ox7, under high-salinity stress in Arabidopsis. *Plant J,* **56:** 613–626. https://doi.org/10.1111/j.1365-313X.2008.03627.x

Mahdieh M, Mostajeran A, Horie T & Katsuhara M. 2008. Drought stress alters water relations and expression of PIP-Type aquaporin genes in nicotiana tabacum plants. *Plant Cell Physiol.,* **49:** 801–813.

Maheshwari DK, Dheeman S & Agarwal M. 2015. Phytohormone-producing PGPR for sustainable agriculture. pp. 159–182. *In:* Maheshwari DK (ed.), Bacterial Metabolites in Sustainable Agroecosystem, Cham: Springer. doi: 10.1007/978-3-319-24654-3_7

Mahouachi J, Arbona V. & Gómez-Cadenas A. 2007. Hormonal changes in papaya seedlings subjected to progressive water stress in this halophyte. *Plant Growth Regul.,* **53:** 43–51. doi: 10.1007/s10725-007-9202-2

Mark C, Zór K, Heiskanen A, Dufva M, Emnéus J & Finnie C. 2016. Monitoring intra- and extracellular redox capacity of intact barley aleurone layers responding to phytohormones. *Anal. Biochem.,* **515:** 1–8.

Martínez-Ballesta MC, Moreno DA & Carvaja M. 2013. The physiological importance of glucosinolates on plant response to abiotic stress in Brassica. *Int. J. Mol. Sci.,* **14:** 1160711625.

Marzec M, Muszynska A & Gruszka D. 2013. The role of strigolactones in nutrient-stress responses in plants. *Int. J. Mol. Sci,* **14:** 9286–9304.

Mashiguchi K, Tanaka K, Sakai T, et al. 2011. The main auxin biosynthesis pathway in

Arabidopsis. *Proc Natl Acad Sci* USA, **108:** 18512–18517. https://doi.org/10.1073/pnas.1108434108

McGuiness PN, Reid JB & Foo E. 2019. The role of gibberellins and brassinosteroids in nodulation and arbuscular mycorrhizal associations. *Front. Plant Sci.,* doi: 10.3389/fpls.2019.00269

Mellidou I, Karamanoli K, Beris D & Al E. 2017. Underexpression of apoplastic polyamine oxidase improves thermotolerance in Nicotiana tabacum. *J. Plant Physiol,* **218:** 171–174.

Miller CO. 1961. A kinetin-like compound in maize. *Proc. Natl. Acad. Sci.* USA, **47:** 170–174.

Minocha R, Majumdar R & Minocha SC. 2014. Polyamines and abiotic stress in plants: A complex relationship. *Front. Plant Sci,* **5:** 175.

Mir RA, Bhat BA, Yousuf H, Islam ST, Raza A, Rizvi MA, et al. 2022. Multidimensional role of silicon to activate resilient plant growth and to mitigate abiotic stress. *Front. Plant Sci.,* **13:** 819658.

Mittler R. 2002. Oxidative stress, antioxidants and stress tolerance. *Trends in Plant Science,* **7:** 405–410.

Miura K, Okamoto H, Okuma E, Shiba H, Kamada H, Hasegawa PM & Murata Y. 2013. SIZ1 deficiency causes reduced stomatal aperture and enhanced drought tolerance via controlling salicylic acid-induced accumulation of reactive oxygen species in Arabidopsis. *The Plant Journal,* **73:** 91–104.

Munemasa S, Mori IC & Murata Y. 2011 Methyl jasmonate signaling and signal crosstalk between methyl jasmonate and abscisic acid in guard cells. *Plant Signal Behav,* **6:** 939–941.

Munné-Bosch S & Peñuelas J. 2003. Photo- and antioxidative protection, and a role for salicylic acid during drought and recovery in field-grown Phillyrea angustifolia plants. *Planta,* **217:** 758–766.

Munsif F, Shah T, Arif M, Jehangir M, Afridi MZ, Ahmad I, et al. 2022. Combined effect of salicylic acid and potassium mitigates drought stress through the modulation of physio-biochemical attributes and key antioxidants in wheat. *Saudi J. Biol. Sci.,* **29:** 103294.

Munteanu V, Gordeev V, Martea R & Duca M. 2016. Effect of gibberellin cross talk with other phytohormones on cellular growth and mitosis to endoreduplication transition. *Int. J. Adv. Res. Biol. Sci.,* **1(6):** 136–153.

Nguyen KH, Van HC, Nishiyama R, et al. 2016. Arabidopsis type B cytokinin response regulators ARR1, ARR10, and ARR12 negatively regulate plant responses to drought. *Proc Natl Acad Sci* USA, **113:** 3090–3095. https://doi.org/10.1073/pnas.1600399113

Nir I, Moshelion M & Weiss D. 2014. The Arabidopsis gibberellin methyl transferase 1 suppresses gibberellin activity, reduces whole-plant transpiration and promotes drought tolerance in transgenic tomato. *Plant, Cell Environ.,* **37:** 113–123. doi: 10.1111/pce.12135

Nolan T, Chen J & Yin Y. 2017. Cross-talk of brassinosteroid signaling in controlling growth and stress responses. *Biochem. J,* **474:** 2641–2661.

Nolan T, Vukasinovi N, Liu D, Russinova E & Yina Y. 2020. Brassinosteroids: Multidimensional regulators of plant growth, development, and stress responses. *Plant Cell,* **32:** 295–318. doi: 10.1105/tpc.19.00335

Northey JG, Liang S, Jamshed M, Deb S, Foo E, Reid JB, et al. 2016. Farnesylation mediates brassinosteroid biosynthesis to regulate abscisic acid responses. *Nat. Plants,* **2:** 16114. doi: 10.1038/nplants.2016.114

Nowicka B, Ciura J, Szymańska R & Kruk J. 2018. Improving photosynthesis, plant productivity and abiotic stress tolerance – Current trends and future perspectives. *J. Plant Physiol.,* **231:** 415–433.

Ogweno JO, Song XS, Shi K, et al. 2007. Brassinosteroids alleviate heat-induced inhibition of photosynthesis by increasing carboxylation efficiency and enhancing antioxidant systems in lycopersicon esculentum. *J. Plant Growth Regul.,* **27:** 49–57. doi: 10.1007/s00344-007-9030-7

Park JE, Park JY, Kim YS, Staswick PE, Jeon J, Yun J, et al. 2007. GH3-mediated auxin homeostasis links growth regulation with stress adaptation response in Arabidopsis. *Journal of Biological Chemistry,* **282(13):** 10036–10046.

Parveen A, Ahmar S, Kamran M, Malik Z, Ali A, Riaz M, et al. 2021. Abscisic acid signaling reduced transpiration flow, regulated Na^+ ion homeostasis and antioxidant enzyme activities to induce salinity tolerance in wheat (Triticum aestivum L.) seedlings. *Env. Technol. Innov.,* **24:** 101808

Peleg Z & Blumwald E. 2011. Hormone balance and abiotic stress tolerance in crop plants. *Curr. Opin. Plant Biol.,* **14:** 290–295.

Pospíšilová J. 2003. Interaction of cytokinins and abscisic acid during regulation of stomatal opening in bean leaves. *Photosynthetica,* **41:** 49–56.

Pospisilova J, Synkova H & Rulcova J. 2000. Cytokinins and water stress. *Biol Plant,* **43:** 321–328.

Prerostova S, Dobrev PI, Gaudinova A, et al. 2018. Cytokinins: Their impact on molecular and growth responses to drought stress and recovery in Arabidopsis. *Front Plant Sci.* https://doi.org/10.3389/ fpls.2018.00655

Qasim M, Ashraf M, Ashraf MY, Rehman SU & Rha ES. 2003. Salt induced changes in two canola cultivars differing in salt tolerance. *Biol. Plantarum.,* **46:** 629632.

Qi YC, Wang FF, Zhang H & Liu WQ 2010. Overexpression of suadea salsa S-adenosylmethionine synthetase gene promotes salt tolerance in transgenic tobacco. *Acta Physiol. Plant,* **32:** 263–269.

Qin F, Kodaira KS, Maruyama K, Mizoi J, Tran LSP, Fujita Y, et al. 2011. SPINDLY, a negative regulator of gibberellic acid signaling, is involved in the plant abiotic stress response. *Plant Physiol.,* **157(4):** 1900–1913.

Radovich TJK, Kleinhenz MD & Streeter JG. 2005. Irrigation timing relative to head development influences yield components, sugar levels and glucosinolate concentrations in cabbage. *J. Am. Soc. Hortic. Sci.,* **130:** 943949.

Ralston L & Yu O. 2006. Metabolons involving plant cytochrome P450s. *Phytochem. Rev.,* **5:** 459. doi: 10.1007/s11101-006-9014-4

Rasheed A, Hassan MU, Aamer M, Batool M, Sheng F, Ziming WU, et al. 2020. A critical review on the improvement of drought stress tolerance in rice (Oryza sativa L.). *Not. Bot. Horti Agrobot. Cluj-Napoca,* **48:** 1756–1788.

Rask L, Andreasson E, Ekbom B, Eriksson S, Pontoppidan B & Meijer J. 2000. Myrosinase: Gene family evolution and herbivore defense in Brassicaceae. *Plant Mol. Biol.,* **42:** 93113.

Reginato M, Abdala GI, Miersch O, Ruiz OA, Moschetti E & Luna MV. 2011. Changes in the levels of jasmonates and free polyamines induced by Na_2SO_4 and NaCl in roots and leaves of the halophyte Prosopis strombulifera. *Biologia,* **67:** 689–697.

Ribba T, Garrido-Vargas F & Antonio O'Brien J. 2020. Auxin-mediated responses under salt stress: From developmental regulation to biotechnological applications. *Journal of Experimental Botany,* **71(13):** 3843–3853.

Riemann M, Dhakarey R, Hazman M, et al. 2015. Exploring jasmonates in the hormonal network of drought and salinity responses. *Front Plant Sci.* https://doi.org/10.3389/fpls.2015.01077

Rivero RM, Gimeno J, Van Deynze A, Walia H & Blumwald E. 2010. Enhanced cytokinin synthesis in tobacco plants expressing PSARK:IPT prevents the degradation of photosynthetic protein complexes during drought. *Plant Cell Physiol.,* **51:** 1929–1941.

Rivero RM., Shulaev V & Blumwald E. 2009. Cytokinin-dependent photorespiration and the protection of photosynthesis during water deficit. *Plant Physiol,* **150:** 1530–1540.

Rizwan M, Ali S, Ibrahim M, Farid M, Adrees M, Bharwana SA, et al. 2015. Mechanisms of silicon-mediated alleviation of drought and salt stress in plants: A review. *Environ. Sci. Pollut. Res. Int.,* **22:** 15416–15431.

Rochange S, Goormachtig S, Lopez-Raez JA & Gutjahr C. 2019. The role of strigolactones in plant-microbe interactions. pp. 121–142. *In:* Koltai H and Prandi C (eds.), Strigolactones – Biology and Applications, Cham: Springer.

Rulcova J & Pospisilova J. 2001.Effect of benzylaminopurine on rehydration of bean plants after water stress. *Biol Plant,* **44:** 75–81.

Ruyter-Spira C, Kohlen W, Charnikhova T, van Zeijl A, van Bezouwen L, de Ruijter N, et al. 2011. Physiological effects of the synthetic strigolactone analog GR24 on root system architecture in Arabidopsis: Another belowground role for strigolactones? *Plant Physiol,* **155:** 721–734. doi: 10.1104/pp.110.166645

Sah SK, Reddy KR & Li J. 2016. Abscisic acid and abiotic stress tolerance in crop plants. *Front. Plant Sci.,* **7:** 571.

Saheri F, Barzin G, Pishkar L, Boojar MMA & Babaeekhou L. 2020. Foliar spray of salicylic acid induces physiological and biochemical changes in purslane (Portulaca oleracea L.) under drought stress. *Biologia (Bratisl),* **75:** 2201. 10.2478/S11756-020-00571-2.

Saini S, Sharma I & Pati PK 2015. Versatile roles of brassinosteroid in plants in the context of its homoeostasis, signaling and crosstalks. *Front. Plant Sci,* **6:** 950. doi: 10.3389/fpls.2015.00950

Salazar-Cerezo S, Martínez-Montiel N, García-Sánchez J, Pérez-y-Terrón R, Martínez-Contreras RD. 2018. Gibberellin biosynthesis and metabolism: A convergent route for plants, fungi and bacteria. *Microbiol. Res,* **208:** 85–98.

Salehin M, Li B, Tang M, Katz E, Song L, Ecker JR, et al. 2019. Auxin-sensitive Aux/IAA proteins mediate drought tolerance in Arabidopsis by regulating glucosinolate levels. *Nat. Commun.,* **10**. doi: 10.1038/s41467-019-12002-1

Sánchez-Martín J, Heald J, Kingston-Smith A, Winters A, Rubiales D, Sanz M, et al. 2015. A metabolomic study in oats (Avena sativa) highlights a drought tolerance mechanism based upon salicylate signalling pathways and the modulation of carbon, antioxidant and photo-oxidative metabolism. *Plant Cell Environ,* **38:** 1434–1452. doi: 10.1111/pce.12501

Sánchez-Romera B, Ruiz-Lozano JM, Li G, Luu DT, Martínez-Ballestamdel C, Carvajal M, et al. 2014. Enhancement of root hydraulic conductivity by methyl jasmonate and the role of calcium and abscisic acid in this process. *Plant Cell Environ,* **37:** 995–1008. doi: 10.1111/pce.12214

Sariyev A, Barutcular C, Acar M, Hossain AEL & Sabagh A. 2020. Sub-surface drip irrigation in associated with H_2O_2 improved the productivity of maize under clay-rich soil of Adana, Turkey. *Phyton Int. J. Exp. Bot,* **89:** 519–528.

Scharwies JD & Dinneny JR. 2019. Water transport, perception, and response in plants. *J. Plant Res.,* **132:** 311–324.

Schonhof I, Klaring HP, Krumbein A & Schreiner M. 2007a. Interaction between atmospheric CO_2 and glucosinolates in broccoli. *J. Chem. Ecol.,* **33:** 105114.

Sedaghat M, Emam Y, Mokhtassi-Bidgoli A, Hazrati S, Lovisolo C, Visentin I, et al. 2021. The potential of the synthetic strigolactone analogue GR24 for the maintenance of photosynthesis and yield in winter wheat under drought: Investigations on the mechanisms of action and delivery modes. *Plants,* **10:** 1223.

Seiler N & Raul F. 2005. Polyamines and apoptosis. *J. Cell. Mol. Med.,* **9:** 623–642.

Sharif R, Xie C, Zhang H, Arnao MB, Ali M, Ali Q, et al. 2018. Melatonin and its effects on plant systems. *Molecules,* **23:** 2352. doi: 10.3390/molecules23092352

Sharipova G, Veselov D, Kudoyarova G, et al 2016 Exogenous application of abscisic acid (ABA) increases root and cell hydraulic conductivity and abundance of some aquaporin isoforms in the ABA-deficient barley mutant Az34. *Ann Bot,* **118:** 777–785.

Silverman P, Seskar M, Kanter D, Schweizer P, Metraux JP & Raskin I. 1995. Salicylic acid in rice. *Plant Physiol,* **108:** 633–639.

Singhal RK, Saha D, Skalicky M, Mishra UN, Chauhan J, Behera LP, et al. 2021. Crucial cell signaling compounds crosstalk and integrative multi-omics techniques for salinity stress tolerance in plants. *Front. Plant Sci.,* **12:** 670369. doi: 10.3389/fpls.2021.670369

Sofy MR, Aboseidah AA, Heneidak SA & Ahmed HR. 2021. ACC Deaminase containing endophytic bacteria ameliorate salt stress in Pisum Sativum through reduced oxidative damage and induction of antioxidative defense systems. *Environ. Sci. Pollut. Res.,* **28:** 40971–40991.

Stec N, Banasiak J & Jasiñski M. 2011. Abscisic acid – An overlooked player in plant-microbe symbioses formation? *Acta Biochim. Pol.,* **63:** 53–58.

Su Y, Xia S, Wang R & Xiao L. 2017. Phytohormonal quantification based on biological principles. pp. 431–470. *In:* Hormone Metabolism and Signaling in Plants. Elsevier Inc.

Sun W, Li Y, Zhao YX & Zhang H. 2015. The TsnsLTP4, a nonspecific lipid transfer protein involved in wax deposition and stress tolerance. *Plant Mol. Biol. Rep.,* **33:** 962–974.

Symons GM. Ross JJ, Jager CE & Reid JB. 2008. Brassinosteroid transport. *J. Exp. Bot.,* **59:** 17–24. doi: 10.1093/jxb/erm098

Taiz L, Zeiger E, Møller IM & Murphy A. 2015. Plant Physiology and Development, Sixth Edition. Massachusetts: Sinauer Associates, Inc.

Takahashi Y, Tahara M, Yamada, Y & Al E. 2017. Characterization of the polyamine biosynthetic pathways and salt stress response in Brachypodium distachyon. *J. Plant Growth Regul,* **37:** 625–634

Tanaka Y, Sano T, Tamaoki M, et al. 2020. Ethylene inhibits abscisic acid-induced stomatal closure in Arabidopsis 1. *Am Soc Plant Biol,* **138:** 2337–2343. https://doi.org/10.1104/pp.105.063503

Tromas A, Braun N, Muller P, et al. 2009. The auxin binding protein 1 is required for differential auxin responses mediating root growth. *PLoS ONE,* **4:** e6648. doi: 10.1371/journal.pone.0006648

Ullah A, Sun H, Yang X & Zhang X. 2017. Drought coping strategies in cotton: Increased crop per drop. *Plant Biotechnol. J.,* **15:** 271–284.

Valluru R, Davies WJ, Reynolds MP & Dodd IC. 2016. Foliar abscisic acid-to-ethylene accumulation and response regulate shoot growth sensitivity to mild drought in wheat. *Front Plant Sci,* **7:** 461.

Velasco P, Cartea ME, Gonzalez C, Vilar M & Ordas A. 2007. Factors affecting the glucosinolate content of kale (Brassica oleracea acephala group). *J. Agric. Food Chem.,* **55:** 955962.

Verma V, Ravindran P & Kumar PP. 2016. Plant hormone-mediated regulation of stress responses. *BMC Plant Biol,* **16:** 86.

Vettakkorumakankav NN, Falk D, Saxena P & Fletcher RA. 1999. A crucial role for gibberellins in stress protection of plants. *Plant Cell Physiol,* **40:** 542–548.

Vishal B & Kumar PP. 2018. Regulation of seed germination and abiotic stresses by gibberellins and abscisic acid. *Front. Plant Sci,* **9:** 838.

Vriet C, Russinova E & Reuzeau C. 2013. From squalene to brassinolide: The steroid metabolic and signaling pathways across the plant kingdom. *Mol. Plant,* **6:** 1738–1757. doi: 10.1093/mp/sst096

Wang C, Zhao Y, Gu P, Zou F, Meng L, Song W, et al. 2018. Auxin is involved in lateral root formation induced by drought stress in tobacco seedlings. *J. Plant Growth Regulat,* **37:** 539–549.

Wang E, Wang J, et al. 2008. Control of rice grain-filling and yield by a gene with a potential signature of domestication. *Nat Genet.* https://doi.org/10.1038/ng.220

Wang L, Hua D, He J, Duan Y, Chen Z, Hong X, et al. 2011. Auxin response factor2 (ARF2) and its regulated homeodomain gene HB33 mediate abscisic acid response in Arabidopsis. *PLoS Genetics,* **7:** e1002172.

Wang Y, Reiter RJ & Chan Z. 2017. Phytomelatonin: A universal abiotic stress regulator. *J. Exp. Bot,* **69:** 963–974. doi: 10.1093/jxb/erx473

Wang Y, Xu W, Yan X & Wang Y. 2011. Glucosinolate content and related gene expression in response to enhanced UV-B radiation in Arabidopsis. *African J. Biotechnol.,* **10:** 6481–6491.

Wei Z & Li J. 2016. Brassinosteroids regulate root growth, development, and symbiosis. *Mol. Plant,* doi: 10.1016/j.molp.2015.12.003

Werner T, Nehnevajova E, Köllmer I, Novák O, Strnad M, Krämer U, et al. 2010. Root-specific reduction of cytokinin causes enhanced root growth, drought tolerance, and leaf mineral enrichment in Arabidopsis and tobacco. *Plant Cell,* **22:** 3905–3920. doi: 10.1105/tpc.109.072694

Woodward AW & Bartel B. 2005. Auxin: Regulation and interaction. *Ann. Bot.,* **95:** 707–735.

Xu L, Wang D, Liu S, Fang Z, Su S, Guo C, et al. 2020. Comprehensive atlas of wheat (Triticum aestivum L.) auxin response factor expression during male reproductive development and abiotic stress. *Frontiers in Plant Science,* **11:** 586144.

Xu Z-Y, Kim SY, Hyeon DY, Kim DH, Dong T, Park Y, et al. 2013. The arabidopsis NAC transcription factor ANAC096 cooperates with bZIP-type transcription factors in dehydration and osmotic stress responses. *Plant Cell,* **25:** 4708–4724. doi: 10.1105/tpc.113.119099

Xue LJ, Guo W, Yuan Y, Anino EO, Nyamdari B, Wilson MC, et al. 2013. Constitutively elevated salicylic acid levels alter photosynthesis and oxidative state but not growth in transgenic populus. *Plant Cell,* **25:** 2714–2730. doi: 10.1105/tpc.113.112839

Yalpani N, Shulaev V & Raskin I. 1993. Endogenous salicylic acid levels correlate with accumulation of pathogenesis-related proteins and virus resistance in tobacco. *Phytopathology,* **83:** 702–708.

Yamaguchi S. 2008. Gibberellin metabolism and its regulation. *Annu Rev Plant Biol,* **59:** 225–251.

Yin Y, Jiang X, Ren M, Xue M, Nan D, Wang Z, et al. 2018. AmDREB2C, from Ammopiptanthus mongolicus, enhances abiotic stress tolerance and regulates fatty acid composition in transgenic Arabidopsis. *Plant Physiol. Biochem.,* **130:** 517–528. doi: 10.1016/j.plaphy.2018.08.002

Yoshida R, Hobo T, Ichimura K, Mizoguchi T, Takahashi F, Aronso J, et al. 2002. ABA-activated SnRK2 protein kinase is required for dehydration stress signaling in Arabidopsis. *Plant Cell Physiol,* **43:** 1473–1483.

Yuan GF, Jia CG, Li Z, Sun B, Zhang LP, Liu N, et al. 2010. Effect of brassinosteroids on drought resistance and abscisic acid concentration in tomato underater stress. *Sci. Hortic.,* **126:** 103–108.

Zaid A & Mohammad, F. 2018. Methyl jasmonate and nitrogen interact to alleviate cadmium stress in Mentha Arvensis. Regulating physio-biochemical damages and ROS detoxification. *J. Plant Growth Regul,* **37:** 1331–1348.

Zamani H, Arvin MJ, Jahromi AA, Abdossi V & Torkashvand AM. The effect of sodium silicate and methyl jasmonate on pigments and antioxidant activity of tomato (Solanum lycopersicum L.) under salinity stress. *Tarim Bilim. Derg.,* **26:** 479–487.

Zandi P & Schnug, E. 2022. Reactive oxygen species, antioxidant responses and implications from a microbial modulation perspective. *Biology (Basel),* **11:** 155.

Zhai Y, Shao S, Sha W, Zhao Y, Zhang J, Ren W, et al. 2017. Overexpression of soybean GmERF9 enhances the tolerance to drought and cold in transgenic tobacco. *Plant Cell Tiss Organ Cult,* **128:** 607–618.

Zhao J, Yu N, Ju M, Fan B, Zhang Y, Zhu E, et al. 2019. ABC transporter OsABCG18 controls the shootward transport of cytokinins and grain yield in rice. *J. Exp. Bot.,* **70:** 6277–6291.

Zhao Y. 2010. Auxin biosynthesis and its role in plant development. *Annu Rev Plant Biol,* **61:** 49–64. https://doi.org/10.1146/annurevarplant-042809-112308

Zwack PJ & Rashotte AM. 2015. Interactions between cytokinin signalling and abiotic stress responses. *J. Exp. Bot.,* **66:** 4863–4871.

7 | Genomic, Proteomic and Transcriptomic Approaches to Understanding Water Stress

Sunishtha Mishra and Girish Chandra Pandey*

Department. of Bioscience and Biotechnology, Banasthali Vidyapith, Rajasthan, India

The most common reasons for the loss of plants worldwide are abiotic factors including salinity, drought (water stress), temperature, and oxidative stress. Water stress poses a significant threat to global crop yields, impacting 80% of rainfed cropland and contributing to 40% of the world's food supply through irrigated agriculture. Abiotic factors such as salinity, drought, temperature, and oxidative stress are primary contributors to plant loss. Plants exhibit intrinsic stress responses at various levels to cope with water stress, affecting cell enlargement, growth, and reproduction.

Water stress disrupts plant water potential, turgor, and physiological processes like stomatal closure, gaseous exchange, transpiration, and carbon absorption. Genomic, proteomic, and transcriptomic approaches play a crucial role in understanding plant responses to abiotic stress and are vital for crop improvement.

Genomic techniques involve genome-wide analysis of multiple genes, while proteomic techniques identify and verify proteins for biomarker discovery. Transcriptomics, influenced by microarrays and RNA-Sequencing, capture sequences through high-throughput sequencing. Plants employ diverse tolerance mechanisms to combat abiotic stress, and a comprehensive understanding of biological and cellular pathways is facilitated by multi-omics technologies.

To create drought-resistant plant varieties for changing climates, integrating information across molecular pathways using big-omics data is imperative. Molecular insights into crop plants' water stress response pave the way for developing high-yielding, stress-tolerant cultivars. Omics technologies generate vast data, and integrated approaches are essential for genome-wide stress-regulation insights. This chapter emphasizes the significance of genomic, proteomic, and transcriptomic studies in plants under water stress, highlighting

*Corresponding author: girishchandrapandey@banasthali.in

novel tools and techniques for analyzing high-throughput data in intensive crop improvement studies.

1. Introduction

Plants experience several environmental challenges during their growth and development in both wild and cultivated environments (Lippmann et al., 2019). Abiotic stressors such as salt, drought, temperature, and oxidative stress are the primary causes of plant loss worldwide (Yadav et al., 2020; Chaudhry et al., 2022). Water is one of the most important factors influencing agricultural output, with rainfed farmland accounting for 80% of global cropland and irrigated cropland accounting for 40% of the global food supply (Thenkabail et al., 2010; Chen et al., 2019). One of the most important environmental issues affecting plant output is water stress. Water accounts for roughly 80-95% of the plant body's fresh biomass and plays a vital role in many physiological processes, including many aspects of plant growth, development, and metabolism (Lisar et al., 2012; Seleiman et al., 2021). Water stress is a prevalent issue for plants, particularly in locations with low water supplies or erratic rainfall patterns (Kar et al., 2011). Recent research has concentrated on determining the impacts of water stress on plant development, physiology, and metabolism. Water stress causes changes in root architecture, leaf morphology, and physiological processes including stomatal control and osmotic adjustment in plants (Ahluwalia et al., 2021; Seleiman et al., 2021; Wu et al., 2022). Reduced photosynthetic rates, lower leaf water potential, and changes in the expression of stress-responsive genes are all physiological and biochemical responses to water stress. Water stress causes morphological and anatomical changes such as smaller leaves, lower leaf thickness, and an increase in root-shoot ratio (Wahab et al., 2022; Bhandari et al., 2023). Water stress can also cause structural and functional changes in plant cells, such as creating root hairs or generating cuticular wax (Shepherd et al., 2006; Yang et al., 2021). Depending on the degree and length of the stress, water stress can significantly impact plant development and output. However, under moderate water stress circumstances, certain plants have been discovered to exhibit a degree of tolerance and may even demonstrate improved output. Plant resistance to water stress can be improved by genetic engineering, crop management practices, and the adoption of drought-tolerant crop types (Lisar et al., 2012; Parkash et al., 2021; Seleiman et al., 2021). More study is needed to comprehend the intricate connections between plants and water stress, as well as to create effective techniques for increasing plant resistance to water stress situations.

As a result, some feel that drought is the most significant environmental stressor for many plants, especially in drought-prone areas. A drought's impacts on agriculture are exacerbated by the loss of water resources and the increasing food demand due to an alarming worldwide population growth (FAO 2019; Seleiman et al., 2021). It is necessary to improve the plant's drought resilience under changing conditions. When the physical adaptation of roots and leaves is insufficient to cope with certain droughts, molecular signals such as the gene coding regularity protein, which expresses many other genes and signals via crosstalk via various regulatory mechanisms are used to improve water-use efficiency (Basu et al., 2016; Salehi-Lisar et al., 2016; Seleiman et al., 2021; Bandurska 2022). Further research into drought-tolerant plants and the application of cost-effective and beneficial agricultural practices will be crucial in meeting future food demand. Water stress is an abiotic stress that has an impact on plants at several organizational levels. It reduces the water potential and turgor of plant cells, causing cell hypertrophy, growth inhibition, and reproductive failure. It also influences plant water relations, stomatal closure, gaseous exchange, transpiration rates, and carbon absorption rates (Cramer et al., 2011; Lisar et al., 2012; Osakabe et al., 2014). Understanding plant sensitivity to abiotic stressors using genomic, proteomic, and transcriptome techniques is critical for crop development. On a genome-wide scale, genomic approaches include the simultaneous identification and study of multiple genes (Chantre Nongpiur et al., 2016; Hanko et al., 2020). Proteomic approaches involve the extraction, identification, and verification of proteins in Microarrays and RNA-sequences, which capture sequences using high-throughput sequencing and have had a significant impact on transcriptomics methodologies (Haider et al., 2013; Cui et al., 2022). To deal with abiotic stress, plants have evolved a variety of tolerance mechanisms. Currently, an umbrella of omics technologies, including genomic, proteomic, and transcriptomic technologies, allows for an accurate, complete, and systematic knowledge of the most essential biological and cellular pathways induced by stress in plants (Kaur et al., 2021; Rakkammal et al., 2022; Joshi et al., 2023). In comparison to a single omics method, this fusion of multi-omics techniques provides a more in-depth understanding of cellular dynamics under water or other stressful situations (Dai et al., 2022; Kwoji et al., 2023). There is a greater need to aggregate information (big-omics data) across various molecular pathways in order to generate drought-resistant plant types for cultivation in continually changing climates. Understanding the water stress response mechanism in agricultural plants at the molecular level opens the door to high-yielding, stress-tolerant cultivars. Omics technologies create vast amounts of data, and integrated omics approaches serve as essential for getting insights into genome-wide stress-regulation systems (Dai et al., 2022; Roychowdhury et al., 2023). This chapter discusses the significance of genomic, proteomic, and transcriptomic studies in plants under water stress, as well as novel tools and techniques for analyzing high-throughput genomic, proteomic, and transcriptomic data to reduce negative impacts and failures due to water stress in intensive crop improvement studies.

2. Cause of Water Stress in Plants

Water stress in plants can be caused by a variety of circumstances. The following are some of the most prevalent causes of water stress in plants:

Drought: A fundamental cause of water stress in plants is insufficient rainfall or a lack of available water. Plants struggle to acquire an adequate water source when the soil dries up (Seleiman et al., 2021; Lin et al., 2023).

High Temperatures: Hot weather and high temperatures increase evaporation rates, resulting in quicker water loss from plants and soil. Water stress can occur if plants are unable to recover lost water rapidly enough (Hatfield et al., 2015; Moore et al., 2021; Becker et al., 2023).

Poor Irrigation Techniques: Water stress can be caused by insufficient or inappropriate watering strategies. Overwatering or underwatering plants can upset the soil moisture balance and impair the plant's capacity to absorb water adequately (Rajanna et al., 2018).

Soil Conditions: Water stress can be exacerbated by certain soil conditions. Sandy soils, for example, quickly drain water, making it difficult for plants to retain moisture. Compacted soils can limit root development and water absorption (Barzegar et al., 2016; Abdul Rahman et al., 2021).

Salinity: Excessive salt in the soil might impair a plant's capacity to absorb water. This can be especially troublesome in locations with saline irrigation water or when salt accumulates in the soil (Vengosh et al., 2005; Shrivastava et al., 2015; Kamran et al., 2019; Otlewska et al., 2020).

Transpiration: The process through which plants lose water via their leaves is known as transpiration. High humidity, windy weather, or intense sunshine can all increase transpiration rates and cause water stress if plants are unable to replace the lost water.

Plant Size and Stage of Development: Larger plants with more leaves and a greater transpiration rate may be more vulnerable to water stress. Furthermore, some development phases, such as blooming or fruiting, may increase plant water requirements, making them more vulnerable to stress if water is limited (Seleiman et al., 2021; Yang et al., 2023).

It's vital to remember that various plant species tolerate water stress differently. Some plants have evolved to dry circumstances and are more drought tolerant than others. Water stress in plants may be reduced with proper water management, which includes effective irrigation systems and plant selection.

2.1 Effect of Water Stress on Plants

Water stress has a number of harmful consequences on plants. The following are some of the most prevalent impacts of water stress on plants:

Reduced Growth: Water stress affects normal plant growth and development. Water deficiency inhibits cell elongation and division, resulting in slowed development and reduced plant size. This can lead to decreased biomass production and poorer crop yields (Hussain et al., 2018; Zia et al., 2021).

Wilting: When plants are stressed by water, they may wilt, which is characterized by drooping and loss of turgidity in leaves, stems, and other plant components. Wilting happens as a result of the plant's effort to save water by shutting stomata (small holes in the leaves). This, however, restricts the plant's capacity to absorb carbon dioxide for photosynthesis (Seleiman et al., 2021).

Leaf Damage and Yellowing: Water stress can cause leaf damage and yellowing. Leaves may become dry and brittle, with burning or browning along the margins. In severe situations, the leaves may become yellow or brown and die (Seleiman et al., 2021).

Reduced Photosynthesis: Water stress hinders the process of photosynthesis, which is essential for plant development and energy generation. An inadequate water supply causes stomatal closure, which reduces carbon dioxide intake into leaves. As a result, the plant's capacity to manufacture carbohydrates and energy is reduced (Kapoor et al., 2020).

Imbalances in Nutrients: Water stress can interfere with the absorption and transport of critical nutrients in plants. Water scarcity disrupts the transport of nutrients from the soil to the roots, resulting in nutritional imbalances within the plant. This can lead to nutritional shortages or toxicities, risking plant health and production even more (Yadav et al., 2020; Kumari et al., 2022).

Decreased Yield and Quality: Water stress during important periods of plant growth, such as blooming or fruiting, can drastically impair crop yields and quality. Inadequate water supply impacts flower formation and size, fruit development, and seed output. It can also have an effect on the quality of harvested products, lowering nutritional content or risking marketability (Zia et al., 2021; Bijalwan et al., 2022).

Increased Disease and Pest Susceptibility: Water-stressed plants are frequently more susceptible to disease and pest infestations. The plant's debilitated state renders it more vulnerable to diseases and insect assaults. Water stress reduces the plant's capacity to protect itself and recover from such threats (Seleiman et al., 2021; Kansman et al., 2022).

It is crucial to note that the severity of the impacts might vary based on the plant species, the duration and intensity of the water stress, and the stage of growth of the plant. Proper water management and irrigation practices are required to mitigate these negative impacts and guarantee plant health and production.

3. Plant Responses to Water Stress

Plants have evolved responses to water stress that may be generally classified into two types: drought avoidance and drought tolerance.

3.1 Drought Avoidance

Reduced Transpiration: Plants may limit water loss through transpiration by shutting their stomata, which are tiny holes on the surface of their leaves. This lowers the amount of water vapor emitted by the plant (Van der Vyver et al., 2017; Meddya et al., 2023).

Leaf Rolling: Some plants may roll their leaves, decreasing the leaf surface's exposure to the atmosphere and therefore minimizing water loss (Ali et al., 2022).

Early blooming and Seed Formation: When exposed to water stress, certain plants accelerate their blooming and seed formation processes in order to finish their life cycle before a severe water shortage emerges (Wahab et al., 2022).

Root Growth and Extension: Plants can boost the development and expansion of their root systems in order to explore a wider soil volume in search of water. This aids in the extraction of water from deeper soil layers (Zhang et al., 2019; Kou et al., 2022).

Tolerance to Drought: Plants may collect osmolytes (compatible solutes) such as proline, carbohydrates, and amino acids, which assist sustain cellular osmotic potential even when exposed to water stress. This enables the plant to store water within its cells, preventing cellular dehydration (Singh et al., 2022).

Antioxidant Defense: Water stress can cause reactive oxygen species (ROS) to be produced in plants, causing oxidative damage. Plants respond by boosting the synthesis of antioxidants that neutralize ROS and safeguard cellular components of cells, such as superoxide dismutase (SOD), catalase (CAT), and peroxidase (POD) (Das et al., 2014; Hasanuzzaman et al., 2020).

Hormonal Regulation: Plant hormones, such as abscisic acid (ABA), play an important role in controlling water stress responses. Stomatal closure is triggered by ABA buildup, which reduces water loss. It also stimulates the production of defensive proteins and enzymes that aid in stress tolerance (Bharath et al., 2021).

Cell Membrane Stability: When plants are stressed by water, they create proteins and lipids that assist stabilization. It should be noted that various plant species may display variable degrees of responses and may utilize additional mechanisms unique to their adaptations and genetic composition. Furthermore, a plant's unique reaction to water stress is determined by the duration, severity, and timing of the stress, as well as the stage of development of the plant (Niu et al., 2018; Rogowska et al., 2020).

Crassulacean Acid Metabolism (CAM): Succulents and other plants employ CAM photosynthesis, which entails opening stomata at night to absorb carbon dioxide and storing it as an organic acid. This permits plants to undertake photosynthesis with closed stomata throughout the day, decreasing water loss (Heyduk et al., 2020).

It should be noted that various plant species may display variable degrees of such responses and may utilize additional mechanisms unique to their adaptations and genetic composition. Furthermore, a plant's unique reaction to water stress is determined by the duration, severity, and timing of the stress, as well as the stage of development of the plant.

4. Methods for Reducing the Negative Impact of Water Stress on Plants

Water stress can be detrimental to plants' growth, development, and general health. There are, however, numerous techniques to reduce the negative consequences of water stress in plants. Here are some effective approaches:

Irrigation Management: Using efficient irrigation practices can help plants cope with water stress. Drip irrigation and precision irrigation systems transfer water directly to the plant's root zone, decreasing water waste. Furthermore, utilizing sensors to monitor soil moisture levels can help with administering water when and where it is most required (Abioye et al., 2022; Obaideen et al., 2022).

Mulching: A layer of organic or inorganic mulch applied around plants helps maintain soil moisture by minimizing evaporation and discouraging weed development. Mulching also lowers soil temperature and increases water retention, creating a more favorable environment for plant development (Iqbal et al., 2020; El-Beltagi et al., 2022).

Crop Selection and Breeding: Selecting or creating crop types that are more resistant to water stress can considerably reduce its negative consequences. Scientists can improve features such as drought resistance, water-use efficiency, and root systems that can endure restricted water supply using traditional breeding or genetic engineering approaches (Fahad et al., 2017; Seleiman et al., 2021).

Soil Management: Improving soil quality and structure is critical for increasing water retention capacity. Organic matter (compost or manure), cover crops, and lowering soil compaction can all improve soil moisture retention and infiltration, allowing plants to access water more effectively (Koudahe et al., 2022).

Conservation Agriculture: Conservation agricultural practices, such as limited tillage or no-till farming, can help to decrease soil water loss due to evaporation and erosion. These practices improve soil structure, increase soil moisture retention, and increase water infiltration, all of which aid plant development and reduce water stress (Cárceles Rodríguez et al., 2022).

Water-saving Technology includes: By giving exact information on when and how much water to apply, modern technology like soil moisture sensors, automated irrigation systems, and weather-based irrigation controllers help optimize water consumption. These devices aid in preventing over-irrigation and ensuring that plants get enough water without wasting resources (Abioye et al., 2022; Obaideen et al., 2022).

Plant Water Management: Plant water management practices, such as altering watering schedules or using controlled deficit irrigation, can encourage plants to establish deeper root systems and become more resistant to water stress. Strategic water management approaches seek to maximize water efficiency while preserving plant health (Owino et al., 2022).

Plant Nutrition Management: Plants require proper nutritional management to deal with water stress. Nutrient balance, particularly nitrogen, and phosphorus, can assist plants in maintaining physiological functioning and improving their capacity to endure and recover from water stress (Kumari et al., 2022).

Biostimulants and Growth Regulators: Using biostimulants and growth regulators can help plants tolerate water stress. These compounds can improve root growth, nutrient absorption, and physiological processes, allowing plants to cope better with water deprivation (De Vasconcelos et al., 2019; Kisvarga et al., 2022).

Climate-resilient Farming Systems: Creating and implementing climate-resilient farming systems that incorporate numerous methods such as crop rotation, agroforestry, and water conservation practices can improve overall farm resistance to water stress. Diversifying crop selections and using integrated techniques can help to minimize susceptibility and boost adaptability to shifting water availability patterns (Jena et al., 2023).

It's vital to note that the efficacy of these methods varies based on factors including plant type, ambient circumstances, and the severity of water scarcity. As a result, combining various tactics that are suited to individual agricultural systems is frequently the most successful way.

5. Omics Approaches to Reducing the Negative Effects of Water Stress on Plants

Omics techniques, which entail studying numerous biological components on a broad scale, might give useful insights into the molecular processes behind water stress in plants (Zenda et al., 2021; Rakkammal et al., 2022; Joshi et al., 2023; Machuca-Sepúlveda et al., 2023). Researchers may use these tools to uncover critical genes, proteins, metabolites, and regulatory networks involved in plant responses to water stress. This knowledge may subsequently be applied to the development of solutions for mitigating the negative effects of water stress on plants. Here are some omics methodologies often utilized in water stress research, as well as their potential applications:

Genomics: Genomic studies entail examining an organism's full collection of genes. Researchers can find genes linked with water stress tolerance by comparing the genomes of water-stressed and well-watered plants. This knowledge can be utilized to breed or design drought-resistant plants (Khan et al., 2016; Wang et al., 2017; Wang et al., 2021).

Transcriptomics: Transcriptomics is the study of the transcriptome, which comprises all the RNA molecules in a cell. Researchers can discover genes that are differentially expressed under water stress situations utilizing techniques such as RNA sequencing. This can aid in the discovery of molecular pathways and regulatory systems involved in plant water stress responses (Lowe et al., 2017; Tyagi et al., 2022).

Proteomics: Proteomics is the study of the whole collection of proteins in a cell or organism. Researchers can discover proteins that are uniquely produced or changed under water-stress circumstances by comparing the proteomes of water-stressed and well-watered plants. This knowledge can help researchers understand the proteins involved in stress signaling, osmotic adjustment, and antioxidant defense (Al-Amrani et al., 2021; Marques et al., 2022).

Metabolomics: Metabolomics is the study of the whole collection of metabolites in a biological system. Researchers can uncover metabolites involved in stress tolerance processes by examining the metabolome of water-stressed plants. This can aid in comprehending the metabolic pathways and biochemical processes influenced by water stress (Kumar et al., 2021; Shen et al., 2023).

Epigenomics: Epigenomics is the study of changes in DNA and histone epigenetic modifications that can impact gene expression without changing the DNA sequence. Water stress can cause epigenetic alterations, which can give insights into the control of stress-responsive genes. Understanding the epigenetic processes behind water stress responses can aid in the development of techniques for modulating gene expression and increasing stress tolerance (Kumar 2018; Kumari et al., 2022).

Systems Biology: Systems biology is the study of complex interactions and networks within biological systems using data from several omics methods. Researchers may build complete models of plant responses to water stress by integrating genomes, transcriptomics, proteomics, and metabolomics data. These models can assist in identifying critical regulatory nodes and predicting the impact of genetic or environmental changes (Pinu et al., 2019; Pazhamala et al., 2021).

Overall, omics techniques provide valuable tools for investigating the molecular basis of plant water stress. These approaches can guide the development of strategies to alleviate the adverse effects of water stress by unraveling the complex networks and molecular mechanisms involved, such as breeding or engineering plants with enhanced drought tolerance or improving water management practices in agricultural systems, by unraveling the complex networks and molecular mechanisms involved.

6. Understanding Water Stress in Plants Using Genomic Approaches

Genomic techniques have transformed our understanding of plant responses to water stress. These techniques involve the investigation of plant genomes and their expression patterns under water stress settings. Genomic methods can shed light on the molecular processes underpinning plant responses to water stress.

This knowledge may be utilized to produce crops that are more resistant to water stress, which is critical for guaranteeing food security in a changing climate.

In this representation (Fig. 1) genomics methods are used to investigate the molecular features of water stress. These techniques include whole-genome sequencing, comparative genomics, and functional genomics.

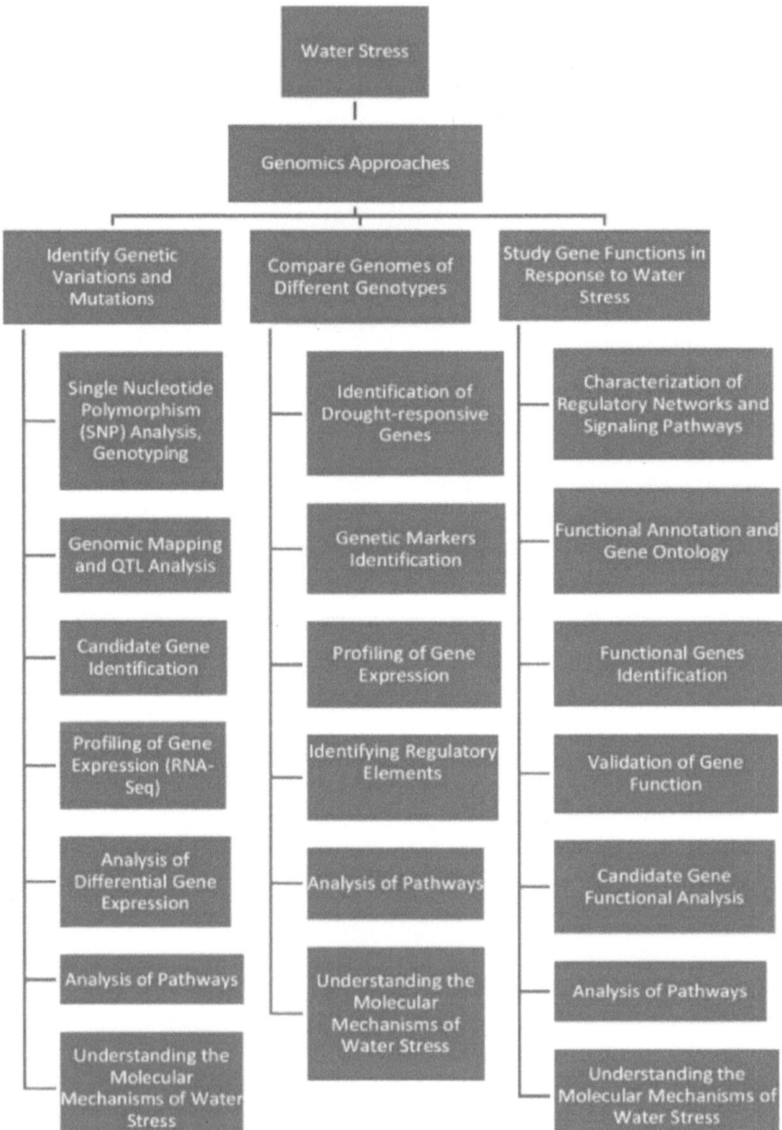

Figure 1: Genomics approaches to water stress.

Whole genome sequencing allows for the detection of genetic variants and mutations across the genome (Zhao et al., 2021). Single Nucleotide Polymorphisms (SNPs), which are differences in a single nucleotide within the DNA sequence, can be detected during this examination. To analyze and map these genetic differences, genotyping and genomic mapping techniques are utilized, as well as Quantitative Trait Locus (QTL) analysis to discover areas of the genome related to water stress tolerance (Grzegorczyk et al., 2021; Rahman et al., 2022).

Comparative genomics is the study of genomes from various genotypes in order to uncover common genetic components or changes linked with water stress tolerance. This study contributes to a better knowledge of the genetic underpinnings of water stress response and adaptability.

The study of gene activities in response to water stress is the focus of functional genomics. Identifying drought-responsive genes, characterizing regulatory networks and signaling pathways involved in water stress response, and undertaking functional annotation and gene ontology analysis to acquire insights into gene functions are all part of this effort.

Gene expression profiling, which is frequently performed using RNA-Seq technology, allows for the examination of gene expression patterns under water stress circumstances. Differential gene expression analysis aids in the identification of genes that are differently expressed in response to water stress, providing important information on the genes implicated in the stress response (Reis et al., 2020; Naldurtiker et al., 2023).

Pathway analysis is used to better understand how water stress affects biological processes. This study contributes to the understanding of the molecular processes underpinning water stress response and adaptability.

The ultimate objective of water stress genomics techniques is to discover the molecular pathways underpinning water stress tolerance and adaptation.

The ultimate objective of genomic methods to water stress is to discover the genetic pathways underpinning tolerance and adaptation to water stress. This includes identifying prospective genes and regulatory elements, validating gene functions, and analyzing these genes' activities in response to water stress.

7. Understanding Water Stress in Plants Using Proteomic Approaches

Proteomic methods can be utilized to better understand how plants respond to water stress. These procedures entail identifying and quantifying the proteins present in a plant under various water stress circumstances. Some proteomic techniques for studying plant water stress are as follows:

2D-GE (Two-dimensional Gel Electrophoresis): Proteins are separated using this approach based on their isoelectric point and molecular weight. 2D-GE may be used to discover proteins that are differently expressed in plants under water stress situations (Molina-Mora et al., 2020).

Mass Spectrometry (MS): This approach may be used to identify and quantify the proteins contained in a plant sample. Proteins are digested into tiny peptides before being analyzed using MS. This method may be used to discover proteins that are differently expressed in plants under water stress situations (Ahrens et al., 2022; Neagu et al., 2022).

Label-free Quantitative Proteomics: This approach quantifies proteins without using any labeling chemicals. It may be used to discover proteins that are differently expressed in plants under water stress situations (Abdallah et al., 2012; Malik et al., 2023).

iTRAQ (Isobaric Tags for Relative and Absolute Quantification): This is a labeling approach that enables the relative measurement of proteins in various samples. It may be used to discover proteins that are differently expressed in plants under water stress situations (Vélez-Bermúdez et al., 2016).

Overall, proteomic techniques may be utilized to discover important proteins and pathways involved in plant water stress responses, which can aid in the creation of water-stress-tolerant plant solutions. Proteomics methods are used in this representation to investigate protein-level alterations and responses to water stress. Proteomics is the extraction, separation, identification, and characterization of proteins that provide insights into the molecular aspects of water stress.

Protein extraction is the first phase, in which proteins are removed from plant tissues or cells that have been exposed to water stress. Protein separation techniques such as gel electrophoresis (2D-PAGE or SDS-PAGE) or liquid chromatography (LC) in conjunction with mass spectrometry (MS) (LC-MS/MS or HPLC-MS/MS) are then used. Proteins can be separated and identified using these approaches as shown in Fig. 2.

To discover differentially expressed proteins (DEPs), differential protein expression analysis examines the number of proteins under water stress situations. DEPs are quantified using quantitative proteomics methods such as label-free or labeling-based methodologies. Protein identification is accomplished by the use of mass spectrometry (MS) methods, which analyze the mass-to-charge ratio of peptides produced from proteins (Meng et al., 2020; Zhang et al., 2023). Protein databases are examined in order to identify proteins based on peptide sequences. The process of discovering the biological activities and pathways associated with the discovered proteins is known as the functional annotation of DEPs (Rao et al., 2014; Van den Broeck et al., 2023). This knowledge aids in comprehending the functional consequences of DEPs in water stress response. The biological pathways that are strongly influenced by water stress are identified via pathway analysis. It gives a systems-level knowledge of the protein interactions and stress response pathways. The identification of critical regulatory proteins aids in the identification of proteins that play important roles in the regulation of the response to water stress. Protein interaction networks are being studied to better understand the complex regulatory mechanisms that underpin water stress. Biological validation entails using experimental procedures to corroborate the results.

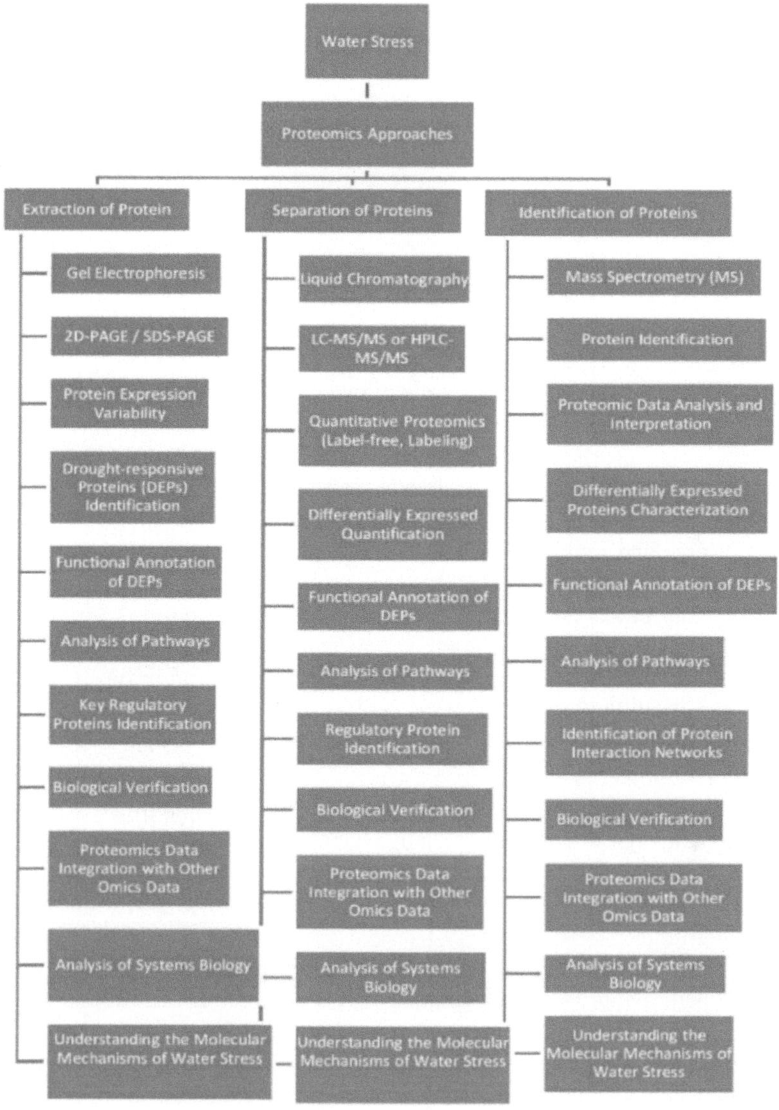

Figure 2: Proteomics approaches to water stress.

8. Understanding Water Stress in Plants Using Transcriptomic Approaches

Transcriptomic techniques are commonly employed to investigate plant responses to diverse environmental challenges, including water stress (Imadi et al., 2015; Tyagi et al., 2022). Water stress is a primary abiotic stress that affects plant growth and production. Thus, knowing the molecular processes behind the plant response to water stress is critical for designing drought-resistance techniques (Seleiman et al., 2021; Yang et al., 2021).

Transcriptomic techniques focus on the transcriptome, which is the entire collection of RNA molecules generated by a cell or tissue (Tyagi et al., 2022). RNA sequencing (RNA-seq) is a popular transcriptomic technique that includes sequencing RNA molecules to identify gene expression levels in a sample (Kukurba et al., 2015; Ko et al., 2023). RNA-seq may offer a complete picture of the transcriptome and be used to discover differentially expressed genes (DEGs) across water-stressed and controlled plants.

Transcriptomic techniques have been utilized in several research efforts to investigate the plant response to water stress (Li et al., 2021). Seki et al. (2002), for example, employed microarray analysis to detect DEGs in *Arabidopsis thaliana* under water stress (Seki et al., 2002; Rasheed et al., 2016). The researchers discovered many stress-responsive genes involved in a variety of functions, including ABA signaling, osmotic control, and antioxidant defense. Wang et al. (2016) utilized RNA-seq to investigate the response of soybean (Glycine max) to water stress. The involvement of several DEGs was discovered in a variety of biological activities, including ABA signaling, water transport, and stress response (Wang et al., 2016; Li & Gao, 2022). Transcriptomic techniques may be used to explore alternative splicing events and post-transcriptional alterations that contribute to the plant's response to water stress in addition to detecting DEGs. Palusa et al. (2007), for example, employed microarray analysis to uncover alternative splicing events in Arabidopsis under water stress (Palusa et al., 2007; Song et al., 2022).

Overall, transcriptomic techniques have offered important insights into the molecular pathways underpinning plant water stress response. These methods may be used to discover critical genes and processes involved in the response to water stress, as well as to design strategies to improve plant performance under drought conditions.

Transcriptomic methods are used in this representation (Fig. 3) to analyze gene expression patterns and the molecular response to water stress. Transcriptomics comprises RNA extraction, library preparation, high-throughput sequencing (RNA-Seq), and data processing.

The first step is to harvest RNA molecules from plant tissues or cells that have been exposed to water stress. After that, library preparation includes converting RNA into complementary DNA (cDNA) for sequencing.

To discover differentially expressed genes (DEGs), differential gene expression (DE) analysis examines the number of transcripts (mRNA) under water stress situations. Transcriptome analysis, also known as mRNA profiling, reveals gene expression patterns and alterations linked with water stress.

The discovery of drought-responsive DEGs aids in the identification of genes that are selectively regulated under water stress situations. These DEGs serve critical roles in water stress response and adaptability. Functional annotation of

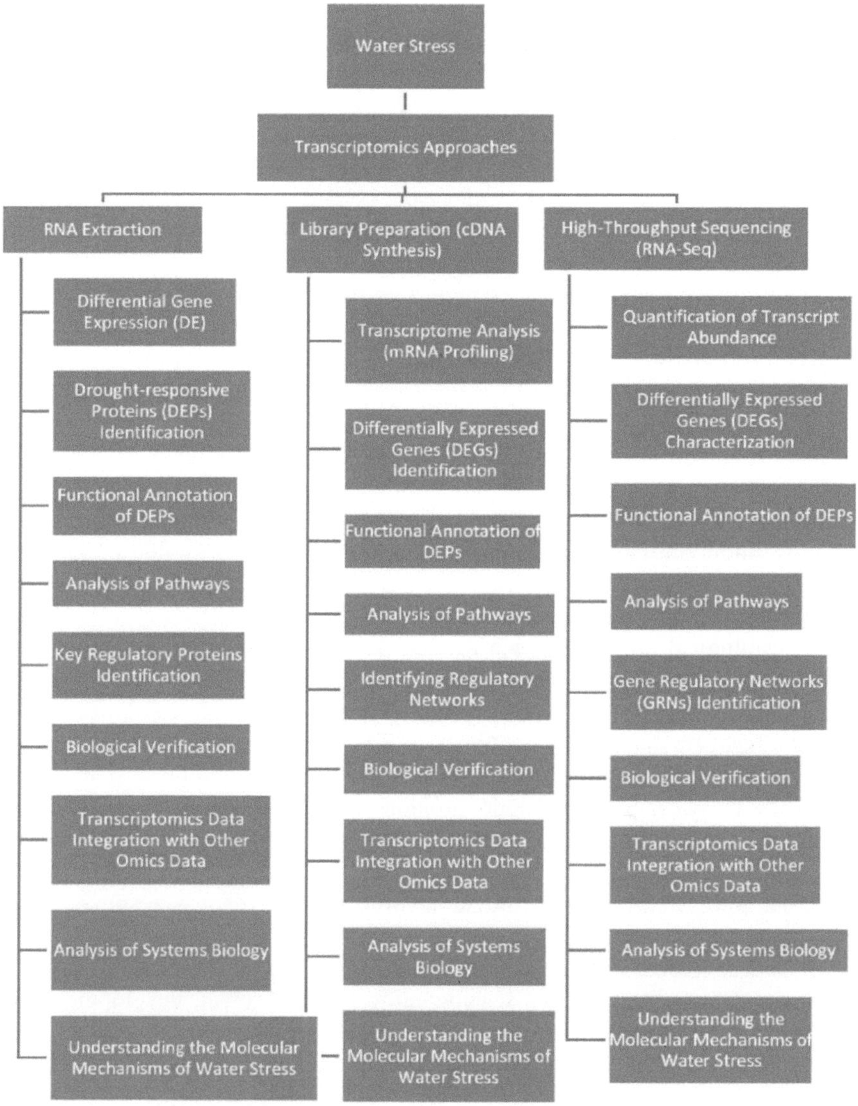

Figure 3: Transcriptomics approaches to water stress.

DEGs entails assigning biological activities to identified genes and categorizing them based on their functional annotations. This contributes to a better understanding of the functional significance of DEGs in water stress response.

The biological pathways that are strongly influenced by water stress are identified via pathway analysis. It elucidates the molecular pathways and networks involved in the stress response.

The identification of essential regulatory genes aids in the identification of genes that play critical roles in the regulation of the response to water stress. The complex regulatory processes behind water stress are studied using gene regulatory networks (GRNs).

Biological validation entails using experimental methodologies to check the outcomes of transcriptomics approaches, confirming the accuracy of the data.

The combination of transcriptomics and other omics data enables a thorough knowledge of the molecular response to water stress. This integration allows for the merging of transcriptome data with proteomics, metabolomics, and other omics layers to provide a comprehensive picture of the stress response.

Systems biology analysis entails integrating and interpreting complicated data from transcriptomics, proteomics, metabolomics, and other disciplines.

9. Conclusion

Plants experience a variety of environmental challenges, including abiotic stressors such as salt, drought, temperature, and oxidative stress. Water is one of the most important factors influencing agricultural output, with rainfed farmland accounting for 80% of global cropland and irrigated cropland accounting for 40%. Water stress can cause changes in root architecture, leaf morphology, and physiological processes, such as stomatal control and osmotic adjustment. Plant resistance to water stress can be improved by genetic engineering, crop management practices, and the adoption of drought-tolerant crop types. Omics technologies generate massive volumes of data, and integrated omics techniques are essential for learning about genome-wide stress-regulation systems. Plant responses to water stress include drought avoidance and drought tolerance, which involve reduced transpiration, leaf rolling, early blooming and seed formation, root growth and extension, and osmolytes. Plant hormones, such as ABA, play an important role in controlling water stress responses. Cell membrane stability and CAM photosynthesis are also important responses to water stress, which vary depending on the duration, severity, and timing of the stress. Water stress can be detrimental to plants' growth, development, and general health. To reduce its negative consequences, there are several effective approaches: Irrigation management, mulching, crop selection and breeding, soil management, and conservation agriculture. Water-saving technology, plant water management, plant nutrition management, biostimulants and growth regulators, and climate-resilient farming systems are all important strategies for reducing water stress. Omics techniques can be used to uncover critical genes, proteins, metabolites, and regulatory networks involved in plant responses to water stress. Genomic approaches involve the investigation of transcriptomics, proteomics, metabolomics, genomics, and epigenomics to identify genes that are differently expressed in response to water stress. Proteomic approaches can be used to better understand how plants respond to water stress. 2D-GE, mass spectrometry (MS), label-free quantitative proteomics, and iTRAQ are primary techniques used to identify and quantify proteins present in a plant under various water stress circumstances. Differentially expressed proteins (DEPs) are expressed proteins that are functionally annotated. Pathway analysis is used to analyze the pathways of these DEPs, while protein interaction networks are studied to better understand the complex regulatory mechanisms. Transcriptomic approaches are used to investigate plant responses to water stress, such as RNA sequencing (RNA-seq). Pathway analysis is used to identify regulatory genes, while biological validation is used to check the accuracy of transcriptomics approaches. Systems biology analysis integrates and interprets complicated data from various disciplines.

References

Abdallah C, Dumas-Gaudot E, Renaut J & Sergeant K. 2012. Gel-based and gel-free quantitative proteomics approaches at a glance. *International Journal of Plant Genomics,* **2012(1):** 494572.

Abdul Rahman NS, Abdul Hamid NW & Nadarajah K. 2021 Aug. Effects of abiotic stress on soil microbiome. *International Journal of Molecular Sciences,* **22(16):** 9036.

Abioye EA, Hensel O, Esau TJ, Elijah O, Abidin MS, Ayobami AS, et al. 2022 Feb. Precision irrigation management using machine learning and digital farming solutions. *AgriEngineering,* **4(1):** 70-103.

Ahluwalia O, Singh PC & Bhatia R. 2021 Sep. A review on drought stress in plants: Implications, mitigation and the role of plant growth promoting rhizobacteria. *Resources, Environment and Sustainability,* **5:** 100032.

Ahrens CH, Wade JT, Champion MM & Langer JD. 2022 Jan. A practical guide to small protein discovery and characterization using mass spectrometry. *Journal of Bacteriology,* **204(1):** e00353-21.

Al-Amrani S, Al-Jabri Z, Al-Zaabi A, Alshekaili J & Al-Khabori M. 2021 Sep. Proteomics: Concepts and applications in human medicine. *World Journal of Biological Chemistry,* **12(5):** 57.

Ali Z, Merrium S, Habib-ur-Rahman M, Hakeem S, Saddique MA & Sher MA. 2022 May. Wetting mechanism and morphological adaptation; leaf rolling enhancing atmospheric water acquisition in wheat crop—A review. *Environmental Science and Pollution Research,* **29(21):** 30967-30985.

Bandurska H. 2022 Mar. Drought stress responses: Coping strategy and resistance. *Plants,* **11(7):** 922.

Barzegar A, Yousefi A & Zargoosh N. 2016 Apr. Water stress and soil compaction impacts on clover growth and nutrient concentration. *Eurasian Journal of Soil Science,* **5(2):** 139-145.

Basu S, Ramegowda V, Kumar A & Pereira A. 2016. Plant adaptation to drought stress. *F1000Research,* **5**.

Becker R, Schüth C, Merz R, Khaliq T, Usman M, Aus der Beek T, et al. 2023 May. Increased heat stress reduces future yields of three major crops in Pakistan's Punjab region despite intensification of irrigation. *Agricultural Water Management,* **281:** 108243.

Bhandari U, Gajurel A, Khadka B, Thapa I, Chand I, Bhatta D, et al. 2023 Mar. Morpho-physiological and biochemical response of rice (Oryza sativa L.) to drought stress: A review. *Heliyon,* **9(3)**.

Bharath P, Gahir S & Raghavendra AS. 2021 Mar. Abscisic acid-induced stomatal closure: An important component of plant defense against abiotic and biotic stress. *Frontiers in Plant Science,* **12:** 615114.

Bijalwan P, Sharma M & Kaushik P. Review of the effects of drought stress on plants: A systematic approach. preprints.org

Cárceles Rodríguez B, Durán-Zuazo VH, Soriano Rodríguez M, García-Tejero IF, Gálvez Ruiz B & Cuadros Tavira S. 2022 Nov. Conservation agriculture as a sustainable system for soil health: A review. *Soil Systems,* **6(4):** 87.

Chantre Nongpiur R, Lata Singla-Pareek S & Pareek A. 2016 Aug. Genomics approaches for improving salinity stress tolerance in crop plants. *Current Genomics,* **17(4):** 343-357.

Chaudhry S & Sidhu GP. 2022 Jan. Climate change regulated abiotic stress mechanisms in plants: A comprehensive review. *Plant Cell Reports,* **41(1):** 1-31.

Chen Y, Feng X, Fu B, Shi W, Yin L & Lv Y. 2019 May. Recent global cropland water consumption constrained by observations. *Water Resources Research,* **55(5):** 3708-3738.

Cramer GR, Urano K, Delrot S, Pezzotti M & Shinozaki K. 2011 Dec. Effects of abiotic stress on plants: A systems biology perspective. *BMC Plant Biology,* **11:** 1-4.

Cui M, Cheng C & Zhang L. 2022 Nov. High-throughput proteomics: A methodological mini-review. *Laboratory Investigation,* **102(11):** 1170-1181.

Dai X & Shen L. 2022 Jul. Advances and trends in omics technology development. *Frontiers in Medicine,* **9:** 911861.

Das, K. & Roychoudhury A. 2014. Reactive oxygen species (ROS) and response of antioxidants as ROS-scavengers during environmental stress in plants. *Frontiers in Environmental Science,* **2:** 53.

De Vasconcelos AC & Chaves LH. 2019 Nov. Biostimulants and their role in improving plant growth under abiotic stresses. *Biostimulants in Plant Science,* **7:** 3-16.

El-Beltagi HS, Basit A, Mohamed HI, Ali I, Ullah S, Kamel EA, et al. 2022 Aug. Mulching as a sustainable water and soil saving practice in agriculture: A review. *Agronomy,* **12(8):** 1881.

Fahad S, Bajwa AA, Nazir U, Anjum SA, Sadia S, Nasim W, et al. 2017 June. Crop production under drought and heat stress: Plant responses and management options. *Frontiers in Plant Science,* **8:** 265598.

FAO. (2019). Water Scarcity – One of the Greatest Challenges of Our Time.

Grzegorczyk J, Gurgul A, Oczkowicz M, Szmatoła T, Fornal A & Bugno-Poniewierska M. 2021 Jul. Single nucleotide polymorphism discovery and genetic differentiation analysis of geese bred in Poland, using genotyping-by-sequencing (GBS). *Genes,* **12(7):** 1074.

Haider S & Pal R. 2013 Apr. Integrated analysis of transcriptomic and proteomic data. *Current Genomics,* **14(2):** 91-110.

Hanko EK, Paiva AC, Jonczyk M, Abbott M, Minton NP & Malys N. 2020 Mar. A genome-wide approach for identification and characterisation of metabolite-inducible systems. *Nature Communications,* **11(1):** 1213.

Hasanuzzaman M, Bhuyan MB, Zulfiqar F, Raza A, Mohsin SM, Mahmud JA, et al. 2020 Jul. Reactive oxygen species and antioxidant defense in plants under abiotic stress: Revisiting the crucial role of a universal defense regulator. *Antioxidants,* **9(8):** 681.

Hatfield JL & Prueger JH. 2015 Dec. Temperature extremes: Effect on plant growth and development. *Weather and Climate Extremes,* **10:** 4-10.

Heyduk K. 2022 Sep. Evolution of Crassulacean acid metabolism in response to the environment: Past, present, and future. *Plant Physiology,* **190(1):** 19-30.

Hussain HA, Hussain S, Khaliq A, Ashraf U, Anjum SA & Wang L. 2018 Apr. Chilling and drought stresses in crop plants: Implications, cross talk, and potential management opportunities. *Frontiers in Plant Science,* **9:** 348835.

Imadi SR, Kazi AG, Ahanger MA, Gucel S & Ahmad P. 2015 Sep. Plant transcriptomics and responses to environmental stress: An overview. *Journal of Genetics,* **94:** 525-537.

Iqbal R, Raza MA, Valipour M, Saleem MF, Zaheer MS, Ahmad S, et al. 2020 Dec. Potential agricultural and environmental benefits of mulches—A review. *Bulletin of the National Research Centre,* **44:** 1-6.

Jena PR, Tanti PC & Maharjan KL. 2023 Dec. Determinants of adoption of climate resilient practices and their impact on yield and household income. *Journal of Agriculture and Food Research,* **14:** 100659.

Joshi S, Patil S, Shaikh A, Jamla M & Kumar V. 2023 Nov. Modern omics toolbox for producing combined and multifactorial abiotic stress tolerant plants. *Plant Stress,* **21:** 100301.

Kamran M, Parveen A, Ahmar S, Malik Z, Hussain S, Chattha MS, et al. 2019 Dec. An overview of hazardous impacts of soil salinity in crops, tolerance mechanisms, and amelioration through selenium supplementation. *International Journal of Molecular Sciences,* **21(1):** 148.

Kansman JT, Basu S, Casteel CL, Crowder DW, Lee BW, Nihranz CT, et al. 2022 Apr. Plant water stress reduces aphid performance: Exploring mechanisms driven by water stress intensity. *Frontiers in Ecology and Evolution,* **10.**

Kapoor D, Bhardwaj S, Landi M, Sharma A, Ramakrishnan M & Sharma A. 2020 Aug. The impact of drought in plant metabolism: How to exploit tolerance mechanisms to increase crop production. *Applied Sciences,* **10(16):** 5692.

Kar RK. 2011 Nov. Plant responses to water stress: Role of reactive oxygen species. *Plant Signaling & Behaviour,* **6(11):** 1741-1745.

Kaur B, Sandhu KS, Kamal R, Kaur K, Singh J, Röder MS, et al. 2021 Sep. Omics for the improvement of abiotic, biotic, and agronomic traits in major cereal crops: Applications, challenges, and prospects. *Plants,* **10(10):** 1989.

Khan A, Sovero V & Gemenet D. 2016 Aug. Genome-assisted breeding for drought resistance. *Current Genomics,* **17(4):** 330-342.

Kisvarga S, Farkas D, Boronkay G, Neményi A & Orlóci L. 2022 Apr. Effects of biostimulants in horticulture, with emphasis on ornamental plant production. *Agronomy,* **12(5):** 1043.

Ko B & Van Raamsdonk JM. 2023 Jun. RNA sequencing of pooled samples effectively identifies differentially expressed genes. *Biology,* **12(6):** 812.

Kou X, Han W & Kang J. 2022 Dec. Responses of root system architecture to water stress at multiple levels: A meta-analysis of trials under controlled conditions. *Frontiers in Plant Science,* **13:** 1085409.

Koudahe K, Allen SC & Djaman K. 2022 Sep. Critical review of the impact of cover crops on soil properties. *International Soil and Water Conservation Research,* **10(3):** 343-354.

Kukurba KR & Montgomery SB. 2015 Nov. RNA sequencing and analysis. *Cold Spring Harbor Protocols,* **2015(11):** pdb-top084970.

Kumar M, Kumar Patel M, Kumar N, Bajpai AB & Siddique KH. 2021 Aug. Metabolomics and molecular approaches reveal drought stress tolerance in plants. *International Journal of Molecular Sciences,* **22(17):** 9108.

Kumar S. 2018 Mar. Epigenomics of plant responses to environmental stress. *Epigenomes,* **2(1):** 6.

Kumari P, Khan S, Wani IA, Gupta R & Verma S. 2022 May. Unravelling the role of epigenetic modifications in development and reproduction of angiosperms: A critical appraisal. *Frontiers in Genetics,* **13:** 819941.

Kumari VV, Banerjee P, Verma VC, Sukumaran S, Chandran MA, Gopinath KA, et al. 2022 Jul. Plant nutrition: An effective way to alleviate abiotic stress in agricultural crops. *International Journal of Molecular Sciences,* **23(15):** 8519.

Kwoji ID, Aiyegoro OA, Okpeku M & Adeleke MA. 2023 June. 'Multi-omics' data integration: Applications in probiotics studies. *npj Science of Food,* **7(1):** 25.

Li M & Gao X. 2022 Nov. Transcriptome analysis reveals key drought-stress-responsive genes in soybean. *Frontiers in Genetics,* **13:** 1060529.

Li Y, Shi LC, Pei NC, Cushman SA & Si YT. 2021 Jun, Transcriptomic responses to drought stress among natural populations provide insights into local adaptation of weeping forsythia. *BMC Plant Biology,* **21(1):** 273.

Lin Z, Zhang H & Liu Y. 2023 Jul. Drought adaptation of plants based on improved Lotka-Volterra model. *Highlights in Science, Engineering and Technology,* **60:** 70-77.

Lippmann R, Babben S, Menger A, Delker C & Quint M. 2019 Dec. Development of wild and cultivated plants under global warming conditions. *Current Biology,* **29(24):** R1326-R1338.

Lisar SY, Motafakkerazad R, Hossain MM & Rahman IM. 2012 Jan. Causes, effects and responses. *Water Stress,* **25(1):** 33.

Lowe R, Shirley N, Bleackley M, Dolan S & Shafee T. 2017 May. Transcriptomics technologies. *PLoS Computational Biology,* **13(5):** e1005457.

Machuca-Sepúlveda J, Miranda J, Lefin N, Pedroso A, Beltrán JF & Farias JG. 2023 Jun. Current status of omics in biological quality elements for freshwater biomonitoring. *Biology,* **12(7):** 923.

Malik A, Mor VS, Punia H, Duhan DS, Tokas J, Bhuker A, et al. 2023 Feb. Development and optimization of label-free quantitative proteomics under different crossing periods of bottle gourd. *Current Issues in Molecular Biology,* **45(2):** 1349-1372.

Marques I, Gouveia D, Gaillard JC, Martins S, Semedo MC, Lidon FC, et al. 2022 Jan. Next-generation proteomics reveals a greater antioxidative response to drought in Coffea arabica than in Coffea canephora. *Agronomy,* **12(1):** 148.

Meddya S, Meshram S, Sarkar D, Datta R, Singh S, Avinash G. et al. 2023 Sep. Plant stomata: An unrealized possibility in plant defense against invading pathogens and stress tolerance. *Plants,* **12(19):** 3380.

Meng JY, Ntambo MS, Rott PC, Fu HY, Huang MT, Zhang HL, et al. 2020 Jan. Identification of differentially expressed proteins in sugarcane in response to infection by Xanthomonas albilineans using iTRAQ quantitative proteomics. *Microorganisms,* **8(1):** 76.

Molina-Mora JA, Chinchilla-Montero D, Castro-Peña C & García F. 2020 Dec. Two-dimensional gel electrophoresis (2D-GE) image analysis based on CellProfiler: Pseudomonas aeruginosa AG1 as model. *Medicine,* **99(49):** e23373.

Moore CE, Meacham-Hensold K, Lemonnier P, Slattery RA, Benjamin C, Bernacchi CJ, et al. 2021 Apr. The effect of increasing temperature on crop photosynthesis: From enzymes to ecosystems. *Journal of Experimental Botany.* **72(8):** 2822-2844.

Naldurtiker A, Batchu P, Kouakou B, Terrill TH, McCommon GW & Kannan G. 2023 Feb. Differential gene expression analysis using RNA-seq in the blood of goats exposed to transportation stress. *Scientific Reports,* **13(1):** 1984.

Neagu AN, Jayathirtha M, Baxter E, Donnelly M, Petre BA & Darie CC. 2022 Apr. Applications of tandem mass spectrometry (MS/MS) in protein analysis for biomedical research. *Molecules,* **27(8):** 2411.

Niu Y & Xiang Y. 2018 Jul. An overview of biomembrane functions in plant responses to high-temperature stress. *Frontiers in Plant Science,* **9:** 366067.

Obaideen K, Yousef BA, AlMallahi MN, Tan YC, Mahmoud M, Jaber H, et al. 2022 Sep. An overview of smart irrigation systems using IoT. *Energy Nexus,* **7:** 100124.

Osakabe Y, Osakabe K, Shinozaki K & Tran LS. 2014 Mar. Response of plants to water stress. *Frontiers in Plant Science,* **5:** 76566.

Otlewska A, Migliore M, Dybka-Stępień K, Manfredini A, Struszczyk-Świta K, Napoli R, et al. 2020 Sep. When salt meddles between plant, soil, and microorganisms. *Frontiers in Plant Science,* **11:** 553087.

Owino L & Söffker D. 2022 Sep. How much is enough in watering plants? State-of-the-art in irrigation control: Advances, challenges, and opportunities with respect to precision irrigation. *Frontiers in Control Engineering,* **3:** 982463.

Palusa SG, Ali GS & Reddy AS. 2007 Mar. Alternative splicing of pre-mRNAs of Arabidopsis serine/arginine-rich proteins: Regulation by hormones and stresses. *The Plant Journal,* **49(6):** 1091-1107.

Parkash V, Singh S, Deb SK, Ritchie GL & Wallace RW. 2021 Jan. Effect of deficit irrigation on physiology, plant growth, and fruit yield of cucumber cultivars. *Plant Stress,* **1:** 100004.

Pazhamala LT, Kudapa H, Weckwerth W, Millar AH & Varshney RK. 2021 Jul. Systems biology for crop improvement. *The Plant Genome,* **14(2):** e20098.

Pinu FR, Beale DJ, Paten AM, Kouremenos K, Swarup S, Schirra HJ, et al. 2019 Apr. Systems biology and multi-omics integration: Viewpoints from the metabolomics research community. *Metabolites,* **9(4):** 76.

Rahman MM, Lim SJ & Park YC. 2022 Jan. Development of Single Nucleotide Polymorphism (SNP)-based triplex PCR marker for serotype-specific Escherichia coli detection. *Pathogens,* **11(2):** 115.

Rajanna GA, Dass A & Paramesha V. 2018. Excess water stress: Effects on crop and soil, and mitigation strategies. *Popular Kheti,* **6(3):** 48-53.

Rakkammal K, Priya A, Pandian S, Maharajan T, Rathinapriya P, Satish L, et al. 2022 Oct. Conventional and omics approaches for understanding the abiotic stress response in cereal crops—An updated overview. *Plants,* **11(21):** 2852.

Rao VS, Srinivas K, Sujini GN & Kumar GN. 2014. Protein-protein interaction detection: Methods and analysis. *International Journal of Proteomics,* **2014**.

Rasheed S, Bashir K, Matsui A, Tanaka M & Seki M. 2016 Feb. Transcriptomic analysis of soil-grown Arabidopsis thaliana roots and shoots in response to a drought stress. *Frontiers in Plant Science,* **7:** 183475.

Reis RR, Mertz-Henning LM, Marcolino-Gomes J, Rodrigues FA, Rockenbach-Marin S, Fuganti-Pagliarini R, et al. 2020 May. Differential gene expression in response to water deficit in leaf and root tissues of soybean genotypes with contrasting tolerance profiles. *Genetics and Molecular Biology,* **43:** e20180290.

Rogowska A & Szakiel A. 2020 Dec. The role of sterols in plant response to abiotic stress. *Phytochemistry Reviews,* **19(6):** 1525-1538.

Roychowdhury R, Das SP, Gupta A, Parihar P, Chandrasekhar K, Sarker U, et al. 2023 Jun. Multi-omics pipeline and omics-integration approach to decipher plant's abiotic stress tolerance responses. *Genes,* **14(6):** 1281.

Salehi-Lisar SY & Bakhshayeshan-Agdam H. 2016. Drought stress in plants: Causes, consequences, and tolerance. *In:* Drought Stress Tolerance in Plants, Vol 1: Physiology and Biochemistry, 1-6.

Seki M, Narusaka M, Ishida J, Nanjo T, Fujita M, Oono Y, et al. 2002 Aug. Monitoring the expression profiles of 7000 Arabidopsis genes under drought, cold and high-salinity stresses using a full-length cDNA microarray. *The Plant Journal,* **31(3):** 279-292.

Seleiman MF, Al-Suhaibani N, Ali N, Akmal M, Alotaibi M, Refay Y, et al. 2021 Jan. Drought stress impacts on plants and different approaches to alleviate its adverse effects. *Plants,* **10(2):** 259.

Shen S, Zhan C, Yang C, Fernie AR & Luo J. 2023 Jan. Metabolomics-centered mining of plant metabolic diversity and function: Past decade and future perspectives. *Molecular Plant,* **16(1):** 43-63.

Shepherd T & Wynne Griffiths D. 2006 Aug. The effects of stress on plant cuticular waxes. *New Phytologist,* **171(3):** 469-499.

Shrivastava P & Kumar R. 2015 Mar. Soil salinity: A serious environmental issue and plant growth promoting bacteria as one of the tools for its alleviation. *Saudi Journal of Biological Sciences,* **22(2):** 123-131.

Singh P & Choudhary KK. 2022 Sep. Salt stress resilience in plants mediated through osmolyte accumulation and its crosstalk mechanism with phytohormones. *Frontiers in Plant Science,* **13:** 1006617.

Song L, Pan Z, Chen L, Dai Y, Wan J, Ye H, et al. 2020 Dec. Analysis of whole transcriptome RNA-seq data reveals many alternative splicing events in soybean roots under drought stress conditions. *Genes,* **11(12):** 1520.

Thenkabail PS, Hanjra MA, Dheeravath V & Gumma M. 2010 Jan. A holistic view of global croplands and their water use for ensuring global food security in the 21st century through advanced remote sensing and non-remote sensing approaches. *Remote Sensing,* **2(1):** 211-261.

Tyagi P, Singh D, Mathur S, Singh A & Ranjan R. 2022 Dec. Upcoming progress of transcriptomics studies on plants: An overview. *Frontiers in Plant Science,* **13:** 1030890.

Van den Broeck L, Bhosale DK, Song K, Fonseca de Lima CF, Ashley M, Zhu T, et al. 2023 Aug. Functional annotation of proteins for signaling network inference in non-model species. *Nature Communications,* **14(1):** 4654.

Van der Vyver C & Peters S. 2017. How do plants deal with dry days. *Front. Young Minds,* **5(58):** 1-9.

Vélez-Bermúdez IC, Wen TN, Lan P & Schmidt W. 2016. Isobaric tag for relative and absolute quantitation (iTRAQ)-based protein profiling in plants. *Plant Proteostasis: Methods and Protocols,* 213-221.

Vengosh A. 2005. Environmental Geochemistry: Treatise on Geochemistry. Ben Gurion University of the Negev, Beer Sheva, Israel, **9:** 333.

Wahab A, Abdi G, Saleem MH, Ali B, Ullah S, Shah W, et al. 2022 June. Plants' physio-biochemical and phyto-hormonal responses to alleviate the adverse effects of drought stress: A comprehensive review. *Plants,* **11(13):** 1620.

Wang H & Qin F. 2017 June. Genome-wide association study reveals natural variations contributing to drought resistance in crops. *Frontiers in Plant Science,* **8:** 273925.

Wang H, Wang H, Shao H & Tang X. 2016 Feb. Recent advances in utilizing transcription factors to improve plant abiotic stress tolerance by transgenic technology. *Frontiers in Plant Science,* **7:** 178807.

Wang J, Li C, Li L, Reynolds M, Mao X & Jing R. 2021 Sep. Exploitation of drought tolerance-related genes for crop improvement. *International Journal of Molecular Sciences,* **22(19):** 10265.

Wu J, Wang J, Hui W, Zhao F, Wang P, Su C, et al. 2022 Feb. Physiology of plant responses to water stress and related genes: A review. *Forests,* **13(2):** 324.

Yadav S, Modi P, Dave A, Vijapura A, Patel D & Patel M. 2020 June. Effect of abiotic stress on crops. *Sustainable Crop Production,* **3(17):** 5-16.

Yang X, Lu M, Wang Y, Wang Y, Liu Z & Chen S. 2021 Mar. Response mechanism of plants to drought stress. *Horticulturae,* **7(3):** 50.

Yang Z, Qiu R, Agathokleous E & Hu W. 2023 Jan. Heat stress decreased transpiration but increased evapotranspiration in gerbera. *Frontiers in Plant Science,* **14:** 1119076.

Zenda T, Liu S, Li J, Wang Y, Wang N & Duan H. 2021 Dec. Omics-facilitated crop improvement for climate resilience and superior nutritive value. *Frontiers in Plant Science,* **12:** 774994.

Zhang J, Wang J & Liu J. 2019 May. Soil moisture determines horizontal and vertical root extension in the perennial grass Lolium perenne L. growing in Karst soil. *Frontiers in Plant Science,* **10:** 429960.

Zhang Y, Han Y, Shang Y, Wang X & Sun J. 2023 Apr. Proteomics identifies differentially expressed proteins in glioblastoma U87 cells treated with hederagenin. *Proteome Science,* **21(1):** 7.

Zhao Y, Fang LT, Shen TW, Choudhari S, Talsania K, Chen X, et al. 2021 Nov. Whole genome and exome sequencing reference datasets from a multi-center and cross-platform benchmark study. *Scientific Data,* **8(1):** 296.

Zia R, Nawaz MS, Siddique MJ, Hakim S & Imran A. 2021 Jan. Plant survival under drought stress: Implications, adaptive responses, and integrated rhizosphere management strategy for stress mitigation. *Microbiological Research,* **242:** 126626.

8 | Water Stress in Crop Plants and Its Management: A Metabolomic Insight

Rajesh Saini[1], Awadhesh Kumar Mishra[2], Chandrashekhar Singh[3], Pradeep Kumar[1], Praveen Kumar Shukla[1], Jyoti Vishwakarma[1] and Kavindra Nath Tiwari[1*]

[1] Department of Botany, MMV, Banaras Hindu University, Varanasi,India
[2] Department of Botany, Ramjas College, University of Delhi, Delhi, India
[3] Department of Pharmaceutical Engineering and Technology, IIT, Banaras Hindu University, Varanasi, India

A majority of crops are damaged by climate change and under abiotic stress plants lose their quality and yield. Water stress is the one devastating abiotic stress agent which covers 2.97 lakh hectares in the form of drought and 2.5 lakh hectares as water logged area. Plants alter their metabolites under the effect of water stress to face the challenges created by water stress. The primary metabolites retain the osmolytes of the plant to balance the osmoticum under the influence of drought and the water logged phase. On the other hand secondary metabolites help in providing the plant an ability to resist drought and submergence. Since genes control every reaction involving metabolites, understanding the precise mechanism is essential to obtain the highest crop yield and best quality. The signaling pathway provides knowledge about the particular transcription factor and promoter. It also sheds light on the overexpression of that part of the operon system that can enhance the resistance against water stress in crop plants.

1. Introduction

The plants adjust their growth and defense strategies based on the availability of environmental resources. When resources are abundant, plants allocate them toward growth. Conversely, during resource scarcity, they prioritize strengthening their defense mechanisms (Mundim and Pringle, 2018). They prioritize growth

*Corresponding author: kntiwaribhu@gmail.com

and defense based on the environmental resources. The increasing occurrence of unpredictable severe weather events that leads to climate change and significant reductions in crop yields. Globally, nearly 50-70% of crop yield reductions are caused by abiotic stress factors (Toulotte et al., 2022). According to Food and Agriculture Organization Statistics (FAOSTAT) and extreme weather disaster database (EM-DAT), in the past 5 decades (1961–2018), severe drought and heat waves in Europe have caused a threefold reduction in crop productivity (Brás et al., 2021). However plants acclimatize themselves to sustain the devastating environment. The various methods of defense include physiological and molecular mechanisms to cope with the stress. Plants acquire a series of changes under water stress. These include physiological, biochemical and molecular changes to sustain such adverse environments (Shao et al., 2009). Water stress affects almost all the biological processes of the plant. Under water scarcity stomata get closed and it leads to reduced transpiration and affects the carbon dioxide uptake. Ultimately the photosynthetic rate decreases (Lamaoui et al., 2018). Osmotic modifications are an essential strategy to resist the detrimental effects of water scarcity, which include the build-up of sugars, glycine-betaine, polyols plus additional organic substances (Gugliuzza et al., 2020). Plants under stress from drought are protected from oxidative damage and lipid peroxidation by secondary metabolites that scavenge the increased amount of reactive oxygen species (Yadav et al., 2021). While secondary metabolites serve different functions, primary metabolites are utilized by plants for growth and development. Flood creates water submergence which creates elevation of the ground water level and ultimately to a condition known as hypoxia in the rhizosphere. The low oxygen level around the root surface causes an anaerobic environment which leads to plant death. Primary and secondary metabolites support the plant in withstanding this negative impact while preserving its life and productivity (Pan et al., 2021).

2. Metabolites and Metabolome

Metabolites are the end products of various physiological and biochemical processes that occur within plant cells. Collectively, these metabolites for the plant's metabolome which plays a crucial role in enabling plants to survive and reproduce under stress conditions. Metabolite profiling approaches are an efficient tool to identify and classify the metabolites released when there is a drought (Kumar et al., 2021). Given that drought stress largely affects the plant's osmotic balance, a study estimated that a total of 103 metabolites synthesized in wheat cultivars to keep the crop plant's osmotic potential intact. Among the 103 metabolites, 26 were sugars, 45 were organic and fatty acids, and 17 were primary amino acids (Bowne et al., 2012). JingDong-17 genotype (drought toletant) and JingDong-8 genotype (drought sensitive) of wheat showed release of 58 metabolites after 15 days drought treatment as compared to control. In drought tolerant genotype the accumulation of metabolites which maintains osmolytes were rich as compared to drought sensitive genotype. The JungDong-17 contains the sugars fructose, mannose, tagatose, sucrose, oxalic acid, fumaric acid, and malic acid which are known to keep plants' osmotic balance in check (Guo et al., 2018). In chickpea (*C. arietinum*) a similar result was reported (Khan et al., 2019). A 25-day period of drought stress was applied to the crop, which included the drought-tolerant variety 93127 (G2) and the drought-vulnerable variety Noor Punjab 2009 (G1). A 25-day-old plant was subjected to drought stress, and subjected to metabolomic analysis using UPLC-HRMS. Approximately, 20 metabolites were recorded which to be expressed at higher levels compared to the control plant. These metabolites included L-isoleucine, Proline, L-histidine, L-arginine, allantoin, tryptophan, and tyrosine. When drought and flood stress imposed on Wufengyou 286 rice plant (*Oryza sativa* L.), it reduced photosynthetic rate and decreased its stomatal conductance with the help of secondary metabolites. Despite this reduction flavonoids continue to mitigate reactive oxygen species. Under flood stress the electron transport system (ETS) gets inhibited and causes photosynthesis inhibition. The drought treatement was imposed on the plant for 10 days, followed by 8 days of submergence. Subsequently, these plants were used for metabolite analysis. In drought treated rice plants a total of 107 metabolites exhibited changes compared to the control. Among these, 43 metabolites increased, while 64 decreased. Similarly, submerged plants showed 113 metabolite changes compared to the control, 55 metabolites increasing 58 metabolites decreasing. These results suggested that the submerged and drought plants responded differently to water stress (Xiong et al., 2019).

3. Role of Primary Metabolites in Regulating Water Stress

Water supply barriers for agricultural plants can significantly impact crop health, development and overall yield. Regarding primary metabolites, crop plants undergo alterations in the content of mannitol, sorbitol and other carbohydrate alcohols. Under water stress, soluble carbohydrates-especially glucose and sucrose-are affected, while amino acids like proline, glycine, adenine, also experience changes (Guedes et al., 2023). When exposed to drought stress along with other abiotic stresses, the barley cultivar Golden Promise exhibited changes in metabolite levels. In shoots three metabolites – putrescine, pipecolic acid and quinic acid – increased, while in roots seven metabolites proline, psicose, quinic acid, fructose, leucine, pipecolic acid and raffinose were expressed at higher levels. These results were common across all stresses, including drought stress. All metabolites expressed during stress belong to the primary (1°) metabolite category (Zhao et al., 2021). When comparing the metabolites of two different chili species – Capsicum annuum L. var. 'Chili-AS Rot' and Capsicum chinense Jacq. var. 'Naga Morich' – under water stress, variations were observed. To observe the primary metabolite changes in the chilies, two distinct watering systems were used; optimal irrigation and deficit irrigation (This is comparable to drought). Both varieties of *C. annuum* had higher quantities of glucose, sucrose and fructose under field capacity irrigation. However, in deficit irrigation, these sugars were found to be 50% lower than those under full capacity irrigation. Additionally, *C. chinense* 'Naga Morich' had a lower concentrations of fumaric acid and citric acid under field capacity irrigation, whereas *C. annuum 'Chili-AS Rot'* exhibited higher concentrations of both. Further research has shown that drought stress increases citric acid levels. Moreover, *C. chinense* ('Naga Morich') displayed increased levels of quinic, fumaric, shikimic and mallic acid under deficit irrigation (Zamljen et al., 2020). Stress in durum wheat associated with water imposed on genotype [(*Triticum turgidum* subsp. durum (Desf.)] cultivars (Mexa, Euroduro, Kiko Nick, Don Ricardo, and Haristide) was examined during the grain-filled stage. Additionally, the experiment was designed with the control group receiving an abundance of water when needed and the test group receiving it under water stress (rainfed - during scarcity of rain). Results revealed that sucrose and starch concentrations were higher in rainfed treated plants regardless of growth stage whereas malate was found reduced in them. According to Martínez-Peña et al. (2023), the peduncle of the plant had quantifiably higher levels of glucose and fructose compared to other organs.

4. Role of Secondary Metabolites in Regulating Water Stress

Plant secondary metabolites are considered as substances through which plants interact with the external environment. This helps them to mitigate stress conditions. This occurs despite their significance for defense mechanisms in plants. Potential sources of flavors, food additives, and pharmaceuticals are secondary metabolites (Akula and Ravishankar, 2011). *Citrus mandarin* (*Citrus reshni* Hort. Ex Tan.) and *Carrizo citrange* (*Poncirus trifoliata* L. Raf. x *Citrus sinensis* L. Osb.) were subjected to a 24-hour stress. Two groups of plants were exposed to two different temperatures: The control group at 25 °C received proper irrigation, while the second group at 40 °C was irrigated three times a week. The Carrizo citrange plant exhibited a 56% enhancement in metabolite biosynthesis under heat and water stress conditions, while in Cleopatra mandarin, a total 68 metabolites increased after exposure to heat and water stress. Among 68 metabolites which are increased after exposure of heat and water stress, majority of the compounds belong to the TCA (Tri-carboxylic acid cycle) and glycolysis cycle. In *Cleopatra mandarin*, glucose-6-phosphate found accumulated under heat and water stress. In *Carrizo citrange* plant dihydroxyacetone-phosphate and glycerate-3-phosphate found accumulated under heat and water stress (Zandalinas et al., 2017). Stress due to drought for three distinct periods of 5, 10 and 15 days was applied on 20 varieties of rice (*Oryza sativa* L.) and a control experiment was set up to see the effect on phenolic compounds and other important contents. Only three phenolic compounds (benzoic acid, ferulic acid, and p-coumaric acid) were found in higher concentrations in the Q2 and Q8 cultivars, compared to six phenolic compounds (ferulic acid, syringic acid, vanillin, p-coumaric acid, benzoic acid, and cinnamic acid) in the control plants. Furthermore, only the Q8 cultivar had p-hydroxybenzoic acid, indicating that it plays a role in controlling the effects of water stress (Quan et al., 2016). Similarly flood stress also induced several secondary metabolites which play an important role during mitigation against stress. Coutinho et al. (2018) performed an experiment on soybean cultivars BR4 (flooding stress sensitive) and E45 (flooding stress tolerant). Flood treatment was provided up to 3-4 cm above the soil for 15 days and results revealed that several metabolite quantities decreased whereas in roots they increased. Alanin was found accumulated in the roots of flood sensitive cultivars during flood stress while GABA (tricarboxylic acid) was found higher in the roots of flood stress tolerant (E45) cultivars. The BR4 cultivar's roots and leaves both have higher levels of sugar. Higher accumulation of citrate, succinate and acetate were reported in roots of BR4 but the amount of fumarate and malate found decreased. The abiotic stress becomes drastic when occurring in combination with the plant like alternate drought and water logged conditions (Xiong et al., 2019). Das et al. (2017) designed an experiment where they imposed heat (43/35 °C with 50% humidity) and drought (attained a volumetric water content (VWC) of 10% on day seven) stress separately on Soybean (Glycine max L. cultivar: Williams-82 for 7 days and metabolic profiling (UPLC-MS/MS) was performed. Metabolic profiling indicated the accumulation of sucrose, fructose, glucose and raffinose with respect to the control plants. The accumulation of glycine and proline indicated its role in osmotic balance. An increase of alanine and valine was also seen. Ullah et al. (2017) elucidated the drought tolerant ability of seven Triticeae species by using the secondary metabolite (Ullah et al., 2017). Three groups participated in the experiment namely, control which received 80% water field capacity, mild water stress plants which received 50% water field capacity and the drought stress plant which received 30% water field capacity. GC-MS analysis revealed that 45 metabolites accumulated within the foliage and stems of the stress-treated and control plants. The TR39477, IG132864 and Bolal genotype of triticeae under drought stress showed an increase in amino acids i.e, serine, valine, methionine, asparagines, glycine, proline and threonine and organic acids i.e. succinate, aspartate, citrate and gluconate in both roots and leaves.

5. Metabolites and their Mechanistic Understanding

Abiotic stress has a drastic impact on the yield, quality and quantity which is limiting agricultural productivity. It has been proved that secondary metabolites help the plants to mitigate the adverse conditions. This elucidates the significance of the mechanism of metabolite response against water stress (Ghasemi et al., 2023). A week's transcriptome analysis of a water deficit rice plant showed

1,563 upregulated genes using GeneChip. The experiment's findings showed that transcription factors from the ABA-dependent family (homeodomain, MYB, bHLH and bZIP) and ABA-independent family (bHLH, NAC, and AP2/ERF) were both upregulated (Ray et al., 2011) (Fig. 1). Several reports suggested that over-expression of OsNAC6, OsNAC045 and OsNAC10 genes conferred resistance to the transgenic rice stressed by a drought (Zheng et al., 2004). The AP2/ERF type transcription factors play key roles under abiotic stress. OsEREBP1 belongs to the AP2/ERF transcription family and provides plant tolerance to flooding and drought in transgenic rice where overexpression of this gene was tested (Jisha et al., 2015). Similar outcomes were observed in soybean following exposure to drought and flood by Chen et al. (2016). 289 transcription factors were upregulated; these belong to the family Ethylene Response Factors (ERFs), myeloblastosis (MYB), Basic Helix-loop Helix (bHLH), No apical meristem (NAC) and WRKY amino acid motif (WRKY). The response of plants to flood tolerance has been shown in Fig. 2. HaWRKY76, a transcription factor associated with the WRKY family, reduced the impact of floods and droughts in transgenic *Arabidopsis thaliana* plant (Raineri et al., 2015). Gahlaut et al. (2016) state that the wheat plant is immune to each and every one of the previously listed genes. TSRF1 gene belongs to the ERF transcription factor family and confers resistance to osmotic and drought induced imbalance in *Oryza sativa* subsp. *Japonica* cv. Nipponbare. The TSRF1 gene promotes

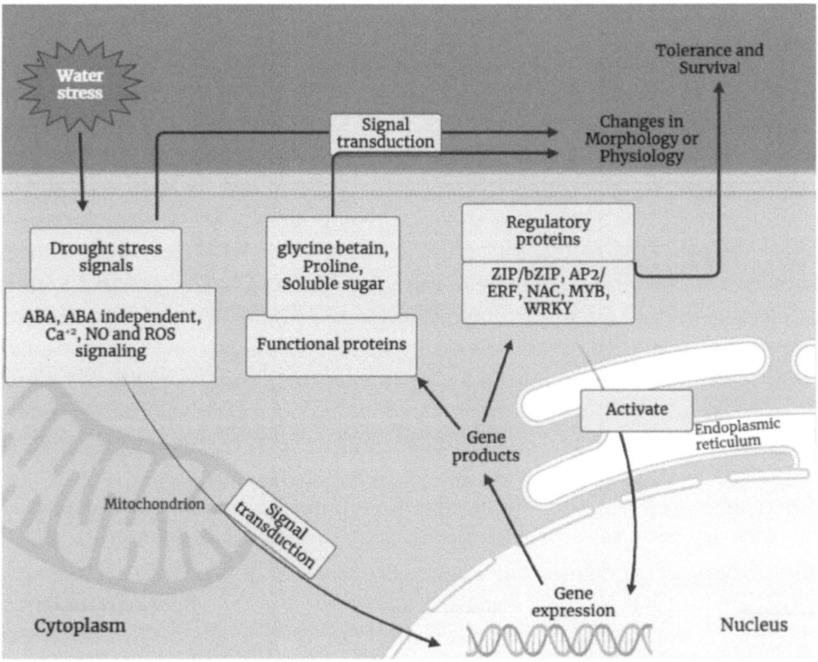

Figure 1: The pathway of drought tolerance.

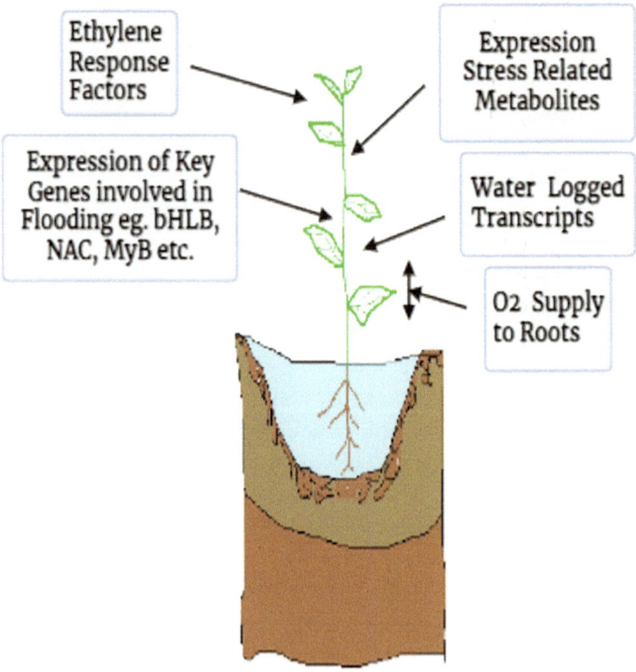

Figure 2: Molecular approach of flood tolerance in plants.

the growth of roots and shoots as well as seed germination, and renders the plant ABA sensitive (Klay et al., 2018). GmMYB84 a transcription factor discovered in soybean belongs to the R2R3-MYB family of transcription factors expressed in roots and flowers. GmMYB84 may be induced in ABA-mediated drought resistance mechanism because it is generated in response to ABA supplementation, salt stress and drought (Wang et al., 2017). As demonstrated recently around 75% of the transcription factors expressed during drought and flood stress are members of the bZIP family. Abscisic Acid-Responsive Element-Binding Protein 1 (AREB1) is one of the genes in the bZIP transcription family; overexpression of this gene progressively increases soybeans' resistance to drought and floods (Fuhrmann-Aoyagi et al., 2021). Additionally, nitric oxide (NO) is essential in drought stress and involves cGMP/PKG/cADPR cascade where NO and Ca^{2+} work cooperatively in an effort to lessen the impact of drought-related stress (Lau et al., 2021). Some species, including *O. sativa* ssp. indica cultivar FR13A, can withstand a completely submerged condition for up to two weeks, while the majority of rice cultivars perish within a week of total submersion. Genomic elucidation proved that it contained submergence 1 (Sub1) locus next to chromosome 9's centromere. The sub1 gene belongs to group VII in the ethylene response factor family. There are three ERF genes of sub1 locus, Sub1A, Sub1B and Sub1C among which sub1B and sub1C are present in all cultivars of indica and japonica (Fukao et al., 2009). The indica gene sub1A-1, when introduced into the submergence intolerant variety of *O. sativa* ssp. Japonica bearing sub1A-2 allele, confers tolerance to the intolerant japonica cultivar (Xu et al., 2006).

6. Conclusion

Water stress is the primary factor affecting the food and health of all living things, including humans. As per the UN report (UN-Water, 2021) ninety-three million people on earth live in countries that are severely water-stressed. Global warming is predicted to cause extreme weather events like severe drought, heavy rainfall, and flooding to occur more frequently, which could reduce the world's ability to produce food. Plants under stress need to coordinate their physiological and biochemical processes in a synergistic way to minimize the effects of ROS-induced lipid peroxidation, protein degradation, and DNA mutations as soon as possible. Plants can exhibit "drought avoidance" and "drought tolerance" mechanisms in response to drought stress, which enhances plant adaptation and increases yield. There were several genes playing a role as transcription factors that induced the resistance response against water stress in plants. All these metabolites (primary and secondary) help the plants to maintain the osmotic balance as well as provide assistance in getting resistant to stress. In order to deal with the water stress, plants employ an abundance of metabolites to regulate their stress response.

References

Akula R & Ravishankar GA. 2011. Influence of abiotic stress signals on secondary metabolites in plants. *Plant Signaling & Behavior,* **6(11):** 1720-1731.

Bowne JB, Erwin TA, Juttner J, Schnurbusch T, Langridge P, Bacic A, et al., 2012. Drought responses of leaf tissues from wheat cultivars of differing drought tolerance at the metabolite level. *Molecular Plant,* **5(2):** 418-429.

Brás TA, Seixas J, Carvalhais N & Jägermeyr J. 2021. Severity of drought and heatwave crop losses tripled over the last five decades in Europe. *Environmental Research Letters,* **16(6):** 1-13.

Chen W, Yao Q, Patil GB, Agarwal G, Deshmukh RK, Lin L, et al.. 2016. Identification and comparative analysis of differential gene expression in soybean leaf tissue under drought and flooding stress revealed by RNA-Seq. *Frontiers in Plant Science,* **7:** 1044-19.

Coutinho ID, Henning LM, Döpp SA, Nepomuceno A, Moraes LA, Marcolini-Gomes J, et al. 2018. Flooded soybean metabolomic analysis reveals important primary and secondary metabolites involved in the hypoxia stress response and tolerance. *Environmental and Experimental Botany,* **153:** 176-187.

Das A, Rushton PJ & Rohila JS 2017. Metabolomic profiling of soybeans (Glycine max L.) reveals the importance of sugar and nitrogen metabolism under drought and heat stress. *Plants,* **6(2):** 21.

Fuhrmann-Aoyagi MB, de Fátima Ruas C, Barbosa EG, Braga P, Moraes LA, de Oliviera AC, et al., 2021. Constitutive expression of Arabidopsis bZIP transcription factor AREB1 activates cross-signaling responses in soybean under drought and flooding stresses. *Journal of Plant Physiology,* **257:** 153338-10.

Fukao T, Harris T & Bailey-Serres J. 2009. Evolutionary analysis of the Sub1 gene cluster that confers submergence tolerance to domesticated rice. *Annals of Botany,* **103(2):** 143-150.

Gahlaut V, Jaiswal V, Kumar A & Gupta PK 2016. Transcription factors involved in drought tolerance and their possible role in developing drought tolerant cultivars with emphasis on wheat (*Triticum aestivum* L.). *Theoretical and Applied Genetics,* **129:** 2019-42.

Ghasemi S, Kumleh HH, Kordrostami M & Rezadoost MH. 2023. Drought stress-mediated alterations in secondary metabolites and biosynthetic gene expression in cumin plants: Insights from gene-specific and metabolite-level analyses. *Plant Stress,* **10:** 100241-12.

Guedes FA, Nascimento LB, Costa MP, Macrae A, Alves-Ferreira M, Caldana C, et al. 2023. Comparative primary metabolite profiling of Setaria viridis reveals potential markers to water limitation. *Agriculture,* **13(3):** 660-19.

Gugliuzza G, Talluto G, Martinelli F, Farina V & Lo Bianco R 2020. Water deficit affects the growth and leaf metabolite composition of young loquat plants. *Plants,* **9(2):** 274.

Guo R, Shi L, Jiao Y, Li M, Zhong X, Gu F, et al. 2018. Metabolic responses to drought stress in the tissues of drought-tolerant and drought-sensitive wheat genotype seedlings. *AoB Plants,* **10(2):** ply016. https://www.unwater.org/publications/summary-progress-update-2021-sdg-6-water-and-sanitation-all

Jisha V, Dampanaboina L, Vadassery J, Mithöfer A, Kappara S & Ramanan R. 2015. Overexpression of an AP2/ERF type transcription factor OsEREBP1 confers biotic and abiotic stress tolerance in rice. *PloS One,* **10(6):** 1-24.

Khan N, Bano A, Rahman MA, Rathinasabapathi B & Babar MA 2019. UPLC-HRMS-based untargeted metabolic profiling reveals changes in chickpea (Cicer arietinum) metabolome following long-term drought stress. *Plant, Cell & Environment,* **42(1):** 115-132.

Klay I, Gouia S, Liu M, Mila I, Khoudi H, Bernadac A, et al. 2018. Ethylene Response Factors (ERF) are differentially regulated by different abiotic stress types in tomato plants. *Plant Science,* **274:** 137-145.

Kumar M, Kumar Patel M, Kumar N, Bajpai AB & Siddique KH 2021. Metabolomics and molecular approaches reveal drought stress tolerance in plants. *International Journal of Molecular Sciences,* **22(17):** 9108-23.

Lamaoui M, Jemo M, Datla R & Bekkaoui F. 2018. Heat and drought stresses in crops and approaches for their mitigation. *Frontiers in Chemistry,* **6:** 1-14.

Lau SE, Hamdan MF, Pua TL, Saidi NB & Tan BC. 2021. Plant nitric oxide signaling under drought stress. *Plants,* **10(2):** 360.

Martínez-Peña R, Vergara-Díaz O, Schlereth A, Höhne M, Morcuende R, Nieto-Taladriz MT, et al. 2023. Analysis of durum wheat photosynthetic organs during grain filling reveals the ear as a water stress-tolerant organ and the peduncle as the largest pool of primary metabolites. *Planta,* **257(4):** 81-100.

Mundim FM & Pringle EG 2018. Whole-plant metabolic allocation under water stress. *Frontiers in Plant Science,* **9:** 852.

Pan J, Sharif R, Xu X & Chen X. 2021. Mechanisms of waterlogging tolerance in plants: Research progress and prospects. *Frontiers in Plant Science,* **11:** 1-16.

Quan NT, Anh LH, Khang DT, Tuyen PT, Toan NP, Minh TN, et al. 2016. Involvement of secondary metabolites in response to drought stress of rice (Oryza sativa L.). *Agriculture,* **6(2):** 1-14.

Raineri J, Ribichich KF & Chan RL 2015. The sunflower transcription factor HaWRKY76 confers drought and flood tolerance to Arabidopsis thaliana plants without yield penalty. *Plant Cell Reports,* **34:** 2065-80.

Ray S, Dansana PK, Giri J, Deveshwar P, Arora R, Agarwal P, et al. 2011. Modulation of transcription factor and metabolic pathway genes in response to water-deficit stress in rice. *Functional & Integrative Genomics,* **11:** 157-178.

Shao HB, Chu LY, Jaleel CA, Manivannan P, Panneerselvam R, Shao MA, et al. 2009. Understanding water deficit stress-induced changes in the basic metabolism of higher plants – Biotechnologically and sustainably improving agriculture and the ecoenvironment in arid regions of the globe. *Critical Reviews in Biotechnology,* **29(2):** 131-151.

Toulotte JM, Pantazopoulou CK, Sanclemente MA, Voesenek LA & Sasidharan R. 2022. Water stress resilient cereal crops: Lessons from wild relatives. *Journal of Integrative Plant Biology,* **64(2):** 412-430.

Ullah N, Yüce M, Neslihan Öztürk Gökçe Z & Budak H. 2017. Comparative metabolite profiling of drought stress in roots and leaves of seven Triticeae species. *BMC Genomics,* **18:** 1-2.

Wang N, Zhang W, Qin M, Li S, Qiao M, Liu Z, et al. 2017. Drought tolerance conferred in soybean (Glycine max. L) by GmMYB84, a novel R2R3-MYB transcription factor. *Plant and Cell Physiology,* **58(10):** 1764-1776.

Xiong Q, Cao C, Shen T, Zhong L, He H & Chen X. 2019. Comprehensive metabolomic and proteomic analysis in biochemical metabolic pathways of rice spikes under drought and submergence stress. *Biochimica et Biophysica Acta (BBA) - Proteins and Proteomics,* **1867(3):** 237-247.

Xiong QQ, Shen TH, Zhong L, Zhu CL, Peng XS, He XP, et al. 2019. Comprehensive metabolomic, proteomic and physiological analyses of grain yield reduction in rice under abrupt drought–flood alternation stress. *Physiologia Plantarum,* **167(4):** 564-584.

Xu K, Xu X, Fukao T, Canlas P, Maghirang-Rodriguez R, Heuer S, et al. 2006. Sub1A is an ethylene-response-factor-like gene that confers submergence tolerance to rice. *Nature,* **442(7103):** 705-708.

Yadav B, Jogawat A, Rahman MS & Narayan OP. 2021. Secondary metabolites in the drought stress tolerance of crop plants: A review. *Gene Reports,* **23:** 1-14.

Zamljen T, Zupanc V & Slatnar A. 2020. Influence of irrigation on yield and primary and secondary metabolites in two chilies species, Capsicum annuum L. and Capsicum chinense Jacq. *Agricultural Water Management,* 234: 106104-7.

Zandalinas SI, Sales C, Beltrán J, Gómez-Cadenas A & Arbona V. 2017. Activation of secondary metabolism in citrus plants is associated to sensitivity to combined drought and high temperatures. *Frontiers in Plant Science,* **7:** 1954-17.

Zhao H, Ni S, Cai S & Zhang G. 2021. Comprehensive dissection of primary metabolites in response to diverse abiotic stress in barley at seedling stage. *Plant Physiology and Biochemistry,* **161:** 54-64.

Zheng J, Zhao J, Tao Y, Wang J, Liu Y, Fu J, et al. 2004. Isolation and analysis of water stress induced genes in maize seedlings by subtractive PCR and cDNA macroarray. *Plant Molecular Biology,* **55:** 807-823.

9 | Plant Responses to Pathogens Under Water Stress

Javan Omondi Were[1*], **Becky Nancy Aloo**[2] **and Vishal Tripathi**[3]

[1] Department of Seed, Crop and Horticultural Sciences, University of Eldoret, Kenya
[2] Department of Biological Sciences, University of Eldoret, Kenya
[3] Department of Biotechnology, Grpahic Era (Deeemed to be University), Dehradun Uttarakhand, India

Climate change has led to a global increase in abiotic and biotic stresses like diseases and water stress, which simultaneously affect plant physiological and biochemical processes. Scientific documentation on exact responses by plants when subjected to water stress remains conflicting and highly variable depending on the interactions among plant species, prevailing environmental conditions, and the virulence status of the pathogenic organism involved. This chapter aims to expound on plants' intrinsic and extrinsic responses to diseases under water stress. Prior water stress exposure is a crucial determinant of whether the plant will exhibit susceptible or tolerant traits post-infection, depending on host-pathogen differentiation mechanisms and the signal transduction pathways activated by the plant-microbe interaction. Due to high specificity in the plant-microbe interaction sites, specific biochemicals such as stress hormones, phenolic compounds, and glucosinolates are synthesized in response by both plants to inhibit cell damage caused by the enzymes and toxins released by pathogens. The final phenotypic expression by the plant is strongly reliant on the compatibility and incompatibility resulting from the host plant – pathogen – degree of water stress conditions. This chapter concludes that water stress is vital in determining the plant's defense metabolism and can inhibit or promote the virulence and pathogenicity of pathogenic organisms.

1. Introduction

Water stress affects plants' physiological and biochemical processes, and this is significantly influenced by the growth stage at which the water deficit

*Corresponding author: javan.were@uoeld.ac.ke, javanwere2012@gmail.com

occurs (Huang et al., 2023). Particularly, under water deficiency within plant cells, the plant defense mechanisms, signaling pathways, and cross-talks by phytohormones remain negatively affected, and this impacts negatively on the known inherent characteristics like tolerance to biotic and abiotic stresses (Were et al., 2022a). However, to avoid further damage from water deficiency, most plants produce several biochemical substances like reactive oxygen species (ROS) and higher quantities of malondialdehyde (MDA). The produced ROS and MDA trigger the enhanced activity of Catalase (CAT), superoxide dismutase (SOD), and peroxidase to minimize the effects of cell damage due to water stress (Ye et al., 2020).

In the past decades, it has been documented that the interaction between a virulent pathogen and a vulnerable host plant alone does not necessarily cause disease development when environmental conditions are not favorable. In addition, for successful infection and microbe proliferation within the host plant, favorable environmental conditions must prevail during inoculation, and water plays a vital role at this stage post entry into a susceptible host (Aung et al., 2018). Before entering plants, pathogens need water, which plays key roles in rhizosphere microbes (Taketani et al., 2017), phyllosphere microbes, and stomatal defense (Melotto et al., 2017). After entering the plants, microbes require water for establishment, cell and tissue colonization, and reproduction within the host plant. In particular, bacterial pathogens require water within the host plant to form an aqueous habitat in the leaf apoplast. For fungal pathogens such as *Magnaporthe oryzae*, water is vital in the formation of water-soaked lesions during the initial stages of infection (Daigham et al., 2023; Xin et al., 2016).

The simultaneous occurrence of drought and bacterial pathogen stress reduces plant growth more than its occurrence (Pandey et al., 2017). However, for most fungal pathogens, including *Pyrenophora teres*, under the effect of exogenous phytohormones, early exposure to water shortage prior to infection resulted in a considerable reduction in disease severity (Were et al., 2022b). In contrast to *P. teres*, *Fusarium solani* and *Rhizoctonia bataticola,* causing black root rot and dry root rot, respectively in chickpeas showed increased severity under drought than under adequate water supply to the plants (Sinha et al., 2019). Based on these and many other conflicting findings on the relationship between water stress and plant disease severities, this chapter expounds on the knowledge surrounding global water stress and plant diseases. This knowledge can help in the sustainable management of plant diseases by understanding how water stress affects disease severity.

2. Effects of Drought and Water Stress on the Severity of Plant Infections

Drought and water stress can accelerate, reduce, or have no impact on the severity of plant infections and can be negative, positive, or neutral (Figure 1).

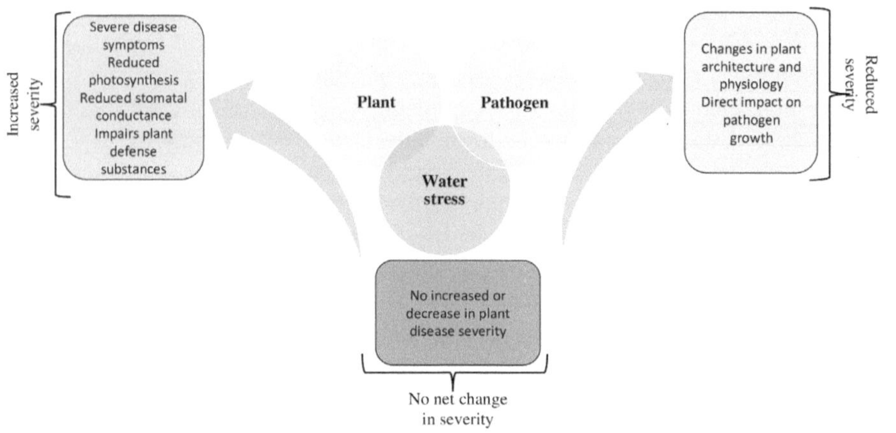

Figure 1: Effect of water stress on the severity of plant infections.

Nevertheless, these abiotic stresses are likely to have additional impacts on infected plants. In an earlier study, drought/water stress was shown to aggravate the harm brought about by xylem pathogens *Xyllela fastidiosa* in *Vitis vinifera*. This resulted in severe symptoms and a more significant reduction in physiological parameters such as total photosynthetic activity and stomatal conductance (Choi et al., 2013). Drought stress has also increased the incidence of dry root rot caused by *Macrophomina phaseolina* in chickpeas (Sharma & Pande, 2013). Sometimes, drought hinders the production of plant defense substances favoring the pathogen's development (Erbilgin et al., 2017).

Regarding reducing the severity of plant diseases, drought-induced alterations in physiology and plant architecture can decrease pathogen infection and plant growth (Choudhary & Senthil-Kumar, 2022). Thus, some diseases may be less severe under water stress. However, the reduction of plant disease susceptibility under water stress is largely reliant on the pathogen's nature and the presence and level of drought (El Aou-ouad et al., 2016; Gupta et al., 2016). Although numerous studies have focused on the influence of drought on plants' responses to pathogens, some have also pointed out the direct influence of drought on pathogen suppression within plants. In chickpeas, the decreased multiplication of *Ralstonia solanacearum* and disease severity were observed under water stress compared to those not under water stress (Sinha et al., 2016). In some instances, drought may not affect plant disease severity. For instance, combined drought and grapevine leafroll-associated virus infection in *V. vinifera* did not result in any additive damage to the plants (El Aou-ouad et al., 2016). However, individual droughts, viral infections, and combined stresses lower the total photosynthetic rate, stomatal conductance, and plant transpiration (El Aou-ouad et al., 2016). Examples of studies that have shown reduced, increased, or no effects on disease severity in plants under water stress are provided in Table 1.

Table 1: Effect of water stress on plant disease severity

Plant	Pathogen/Infection	Test conditions	References
Reduction of disease severity			
Arabidopsis thaliana	*P. syringiae*	Growth chamber	(Gupta et al., 2016)
Cicer arietinum	*R. solanacearum*	Potted	(Sinha et al., 2016)
Chickpea	*Macrophomina phaseolina, Fusarium oxysporum, Rhizoctonia solani*	Field	(Chilakala et al., 2023)
Tomato	*Oidium neolycopersici* causing powdery mildew	Greenhouse	(Sunarti et al., 2022)
Rice	Bacterial blight	Growth chamber	(Dossa et al., 2017)
Barley	*Pyrenophora teres*	Greenhouse	(Were et al., 2022b)
Increased disease severity			
Yellow passion	*Fusarium wilt*	Controlled	(Lima et al., 2019)
Common bean (*Phaseolus vulgaris*)	Fusarium root rot caused by *Fusarium solani*		(Pierre et al., 2023)
Grapevine (*Vitis vinifera*)	*Xylella fastidiosa*	Laboratory	(Choi et al., 2013)
Grapevine	Grapevine leafroll-associated virus	In vitro	(Buciumeanu et al., 2022)
Chickpea	*Macrophomina phaseolina* causing dry root rot	Controlled	(Sharma & Pande, 2013)
No change in disease severity			
V. vinifera	Grapevine leafroll-associated virus	Potted	(El Aou-ouad et al., 2016)

The consequences of drought stress on viral infections are even more complicated. According to Davis et al. (2015), the barley yellow dwarf pathosystem is quite complex, and chronic drought stress has no influence on plant development when applied to wheat plants infected with the barley yellow dwarf virus; however, acute water stress tends to make the plants more resistant to barley yellow dwarf disease infection (Davis et al., 2015).

3. Plant Resistance Mechanisms to Pathogen Infection Under Water Stress

3.1 Overview of Plant Resistance Mechanisms to Pathogens Under Water Stress

Plants can be simultaneously exposed to water stress and pathogens (Table 2).

Table 2: Combined exposure of plants to water stress and pathogenic attacks

Plant	Pathogen/Disease/Infection	Reference/Source
Barley (*Hordeum vulgare*)	*Pyrenophora teres*	(Were et al., 2022a)
Tomato (*Solanum lycopersicum*)	Tomato yellow leaf virus	(Mishra et al., 2022)
Rice (*Oryza sativa*)	Bacterial blight	(Vemanna et al., 2019)
Arabidopsis thaliana	*Pseudomonas syringae*	(Gupta et al., 2020)
Oak (*Quercus infectoria, Q. libani*)	*Obolarina persica, B. mediterranea*	(Ghanbary et al., 2021)
Mung bean (*Vigna radiata*)	*Macrophomina phaseolina*	(Kaur et al., 2023)
Chickpea	*Rhizoctonia bataticola,* black root rot (BRR) caused by *Fusarium solani*	(Sinha et al., 2019)
Chickpea	Foliar and xylem-invading bacterial pathogens	(Sinha et al., 2016)
Chickpea	Dry root rot caused by *Rhizoctonia bataticola*	(Sharma & Pande, 2013)
Chickpea	*Sclerotium rolfsii* Sacc. causing collar rot	(Tarafdar et al., 2018)
Chickpea	Dry root rot caused by *Rhizoctonia bataticola*	(Irulappan & Senthil-Kumar, 2018)
Grapevine	*Neofusicoccum parvum* causing Botryosphaeria dieback	(Galarneau et al., 2019)
Persina oak (*Quercus brantii*)	*Biscogniauxia mediterranea* or *Obolarina persica casuong chjarcoal* disease	(Ghanbary et al., 2018)

The intricate interplay between these stressors significantly challenges plant health and productivity. Water stress can profoundly impact various physiological processes within plants, including their interactions with pathogens (Li et al., 2021). In response, plants have devised various defense mechanisms to combat pathogenic threats even under water-stressed conditions. This section overviews plant resistance mechanisms or responses to pathogens under water stress.

When subjected to water stress, plants undergo complex physiological changes to conserve water and energy (Li et al., 2021; Sinha et al., 2016). These changes often trigger immune signaling pathways that overlap with pathogen attack responses. For instance, abscisic acid (ABA), a critical regulator of water stress responses, also modulates the immune responses of plants to pathogens through the modulation of stomata enclosure (Bharath et al., 2021). According to Choudhary and Senthil-Kumar (2022), the ABA pathway promotes decreased plant tolerance to pathogens under combined stress. The intersection between water stress and immune signaling highlights the intricate crosstalk within stressed plants. Details of plants' physiological and metabolic responses to pathogens under water stress are discussed in subsection 3.3.

Hormonal regulation is pivotal in water stress and defense against pathogens. For instance, jasmonic acid (JA) and salicylic acid (SA), which are the two main defense-related hormones of plants owing to their role in activating the plant's immune system, can shift under water stress, influencing the plant's overall resistance to pathogens (Yang et al., 2019). This chapter also provides more details concerning the synthesis of plant hormones and metabolites in response to pathogens under water stress in subsection 3.3.

Apart from the complex physiological changes and hormonal regulation that can ensue in plants in the wake of water stress and pathogenic attacks, plants can also enhance physical barriers to pathogens under water stress. According to Bharath et al. (2021), water stress can lead to stomatal closure to reduce transpiration but also affect the penetration of pathogens in the process. Stomatal closure restricts transpiration, photosynthetic CO_2 exchange, and pathogen invasion (Qi et al., 2018). Other studies have also elucidated the molecular responses of plants under water stress and pathogen attack, as elaborated in Section 3.4 of this Chapter. The general responses of plants to pathogenic attacks under water stress, like osmotic stress responses, production of reactive oxygen species (ROS) and antioxidants, and priming and induced resistance are highlighted in Figure 2.

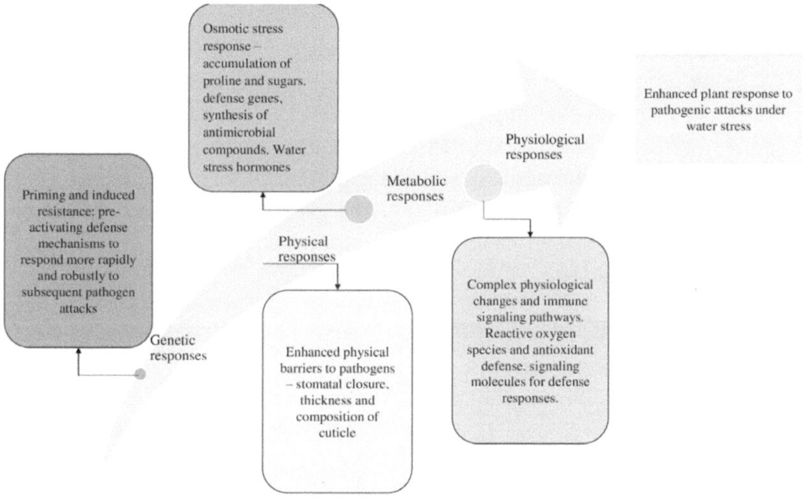

Figure 2: Schematic illustration of the response of plants to pathogens under water stress.

According to recent studies, plants subjected to mild or moderate water stress can exhibit a phenomenon known as priming, which involves the pre-activation of defense mechanisms to respond more rapidly to pathogen attacks. For instance, water stress-induced priming can enhance a plant's ability to mount effective defense responses under pathogenic challenges (Martínez-Ferri et al., 2019; Ramegowda et al., 2013). The interplay between water stress and plant-pathogen interactions is a complex phenomenon that continues to be an active research area. Cognizant of this, understanding the intricate molecular, physiological, and hormonal mechanisms underlying plant resistance to pathogens under water stress is pivotal for developing strategies for improving crop resilience and crop production in the face of changing climate patterns.

3.2 Phenolic Compounds and Glucosinolates in Plants Under Pathogen Infection

The primary characteristic of phenolic compounds is their antimicrobial properties, which are believed to act as agents against bacterial, fungal, and viral plant diseases. Currently, more than two hundred thousand phenolic compounds are produced naturally by plants at different stages of growth and development in varying environmental and growth conditions, especially under water stress (Gani & Shama, 2021). When produced through shikimate and malonate acetate pathways (Kołton et al., 2022), these phenolic compounds and glucosinolates act as the second line of defense against further damage to plants by several pathogenic organisms hence lowering disease severities by inducing inherent and acquired resistance and tolerance.

However, the type of phenolic compound and when it should be synthesized is highly reliant on factors both internal (genetic) and external (environmental) to the plant (Kołton et al., 2022). There are five major groups of phenolic chemicals found in most plants. These include phenolic acids, stilbenes, flavonoids, lignans, and tannins or proanthocyanidins (Pratyusha, 2022; Zhang et al., 2022). Further, in response to abiotic and biotic stressors, plants produce and accumulate these phenolics in the affected tissues, and the concentration of a specific phenolic compound is mainly dependent on season, growth stage, and development which has been observed in cases of *Agrobacterium* and *Rhizobium* infections (Bhattacharya et al., 2010).

3.3 Physiological, Morphological, and Metabolic Responses of Plants to Pathogens Under Water Stress

Under water stress, plants display a variety of physiological and metabolic responses to infections, including stomatal closure, reduced photosynthesis, suppression of leaf development, changes in root/shoot ratio, and the creation of plant hormones and metabolites (Figure 3). The synthesis of plant hormones is one of the physiological reactions of plants to pathogens, even under water stress. The effect of water stress on plant-pathogen interaction has often been

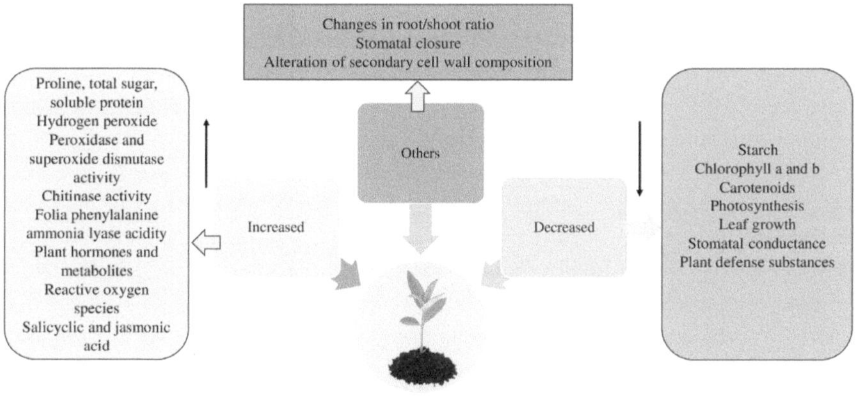

Figure 3: Physiological and metabolic responses of plants to pathogens under water stress.

linked to the cross-talk between the drought stress-induced ABA and pathogen-induced SA and JA pathways (Lievens et al., 2017). Interactions between phytohormone signaling pathways determine disease outcomes. While the SA route is primarily involved in plant defense responses to pathogens, the ABA pathway is also involved in plant adaptation to drought stress. Also, reciprocal antagonism between SA and ABA signaling is one of the methods employed by plants to respond to abiotic or biotic stress (Park et al., 2021).

The crosstalk between these hormones and their overall influence on plant response to pathogens under water stress is quite intricate. The profiling of the three hormones in *A. thaliana* to water stress and *P. syringae* infection established that ABA was suppressed when drought was the first stress. At the same time, SA and JA were elevated to reduce bacterial growth. However, when the plants were first exposed to accumulated pathogens, ABA, SA, and JA they did not alter bacterial growth (Gupta et al., 2017). Thus, suppression of ABA production seemed to cause resistance to pathogens under water stress. Furthermore, prolonged water stress can suppress plant defense hormones, leading to the vulnerability of plants to pathogens (Gupta et al., 2017). In a recent study by Kaur et al. (2023) that studied the morpho-physiological and biochemical effects of simultaneously exposing *Vigna radiata* to *Macrophomina phaseolina* and drought stress, differences in levels of SA were observed upon fungal infection under well-watered and drought conditions (Kaur et al., 2023).

Studies show that ROS, antioxidants, and defense-related hormones can accelerate plant responses to pathogen attacks under water stress. For example, drought stress-induced production of ROS in chickpea plants can contribute to decreased multiplication of *Pseudomonas syringae* and *R. solanacearum* (Sinha et al., 2016). According to Qi et al. (2018), water stress causes the production of ROS, which works as a secondary messenger in signal transduction and stimulates a defense reaction against pathogens at lower concentrations. The

ROS demodulates the host microbiome that participates in infection mitigation (Asghari et al., 2020). Nevertheless, the buildup of ROS, which is a crucial part of the plant's hypersensitive response to pathogens, might be hindered or imbalanced under water stress. A recent study has also shown that plants exhibit higher levels and activities of superoxide dismutase and peroxidase than non-inoculated seedlings under water stress (Ghanbary et al., 2018).

Drought and pathogen stress have been shown to change physio-morphological parameters like photosynthesis, transpiration rate, stomatal conductance, plant development, and root shape (Irulappan & Senthil-Kumar, 2018). In the study by Kaur et al. (2023) that assessed morpho-physiological and biochemical implications of simultaneously exposing *Vigna radiata* cultivars to *Macrophomina phaseolina* and water stress, resistance, and disease reduction were attributed to a stronger defense response through antioxidants in dual stress-treated plants. Soluble sugar-mediated control was dominant in combined stress, showing the intricate physiological and metabolic responses that plants can exhibit to pathogen attacks under water stress (Kaur et al., 2023).

Interestingly, secondary cell wall changes in *R. Solanacearum*-infected plants are also impacted differently by the severity of drought. Changes in the thickness and composition of the plant cuticle and cell wall may contribute to enhanced defense against pathogen invasion under water stress. Water stress can increase lignin production and lengthen drought-induced cellulose biosynthesis, which is linked to increased vulnerability (Sinha et al., 2017). The primary cell wall is mainly formed of polysaccharides, whereas lignin is a more rigid hydrophobic polymer composed of phenolic monomers (Novaković et al., 2018). Resistance and disease reduction have also been attributed to a more robust defense through lignin deposition in plants subjected to dual stress.

3.4 Genetic and Molecular Responses of Plants to Pathogens under Water Stress

Plants exhibit elaborate defense-related genes and proteins to counteract pathogen attacks during water stress. According to Gupta et al. (2020), plants exposed to water stress and pathogen attack exhibit more prominent molecular responses than those exposed to individual stressors (Gupta et al., 2020). Many genes are upregulated under water stress and pathogen attack to improve stress adaptation. According to Sinha et al. (2016), pathogenesis-related protein-5 and plant defense in genes, as well as serine-threonine protein kinases involved in pathogen defense, were shown to be strongly expressed under drought stress. The enhanced expression of CaPR4 in drought stress alone clearly prepped the plant for upcoming pathogen stress. Similarly, water stress has been shown to reduce *Arabidopsis thaliana* defense against *Pseudomonas syringae* by reducing the expression of CBP60g/SARD1 during combined stress (Choudhary & Senthil-Kumar, 2022). According to the transcriptome meta-analysis, drought adversely alters critical plant defense mechanisms.

Among the other molecular mechanisms of plants to pathogen resistance under water stress is the production of pathogenesis-related genes. According to Akbudak et al. (2020), these genes regulate the secretion of several peptides, proteins, or compounds that are toxic to pathogens or prevent their infections (Akbudak et al., 2020). Under combined stress, the expression of pathogenesis-related genes generally increases in diverse plants (De Pascali et al., 2019). Tomato plants exposed to water stress and *Fusarium oxysporum* infection yielded the same results (Akbudak et al., 2020). The hypersensitive response is a critical mechanism in plants' defense against pathogens that involves localized fast cell death at the site of infection. This response is often associated with recognizing specific pathogen molecules by plant resistance (R) genes. However, the effectiveness of this defense mechanism can be influenced by various environmental factors, including water stress.

4. Preventing and Managing Plant Pathogens Under Water Stress

Improving crop output in stressful conditions and sub-optimal locations will be critical to feed the world's rapidly rising population. One way to prevent and manage plant pathogens under water stress is to practice optimum irrigation management to create an environment less favorable for pathogen development. As long as plant-available water is not significantly decreased, methods that lower soil water content can restrict the spread of soilborne diseases.

Maintaining healthy soil conditions through cover cropping, crop rotation, and organic matter incorporation enhances soil structure and microbial activity, contributing to plant health and water retention. In this regard, rhizospheric bacteria have received considerable attention due to their potential to contribute to the reduction of abiotic stressors of host plants through various mechanisms, including exopolysaccharide production, lowering ethylene levels using 1-aminocyclopropane-1-carboxylate (ACC) deaminase, which is essential in plant growth under water stress (Robison et al., 2001). Other metabolites and PGP bacteria produce, phytohormones, siderophores, volatile compounds, and organic acids that enhance plant growth and control pathogens under stress conditions (Jha et al., 2023). Research has been conducted on the beneficial role of endophytic bacteria in improving drought stress tolerance in a range of crops, including soybean (Dubey et al., 2021), sorghum (Carlson et al., 2020), wheat (Samain et al., 2022), common beans (Martins et al., 2018), and others.

Apart from practicing optimal irrigation and the application of rhizobacteria to attenuate plant stress and fight pathogens, breeding of plants resistant to abiotic and biotic stressors can also be explored to minimize the impact of water stress and pathogen attacks (Pérez-Méndez et al., 2021).

5. Emerging Perspectives and Way Forward

Even without water stress, plants continually face various pathogens and diseases. The interplay between such stresses can be very complicated in an environment where plants are exposed to multiple abiotic and biotic stresses (Rivero et al., 2022). However, most stress research remains unidimensional, concentrating on the impact of specific stressors on plants. This technique does not represent actual natural ecosystems in which one stress element substantially impacts plant responses to another.

In this chapter, water stress has often been shown to exacerbate plant infections. Various perspectives can be studied to enhance plant reactions to pathogens under water stress. A study of the plant-pathogen interactome at the infection site under water stress may disclose the mechanism behind the altered defense under combined stress. The rhizosphere is another significant step of plant-pathogen interaction sensitive to water stress because plants generate antimicrobial chemicals to defend against pathogens. Thus, examining metabolites at the root interface under combined water and pathogen stress can give information on the control of small RNA under combined pressures. Furthermore, examining the stress-modulated metabolomes of plants and pathogens might give information on newly acquired plant defense methods against infections (Balmer et al., 2013; Castro-Moretti et al., 2020).

Water stress can have a detrimental effect on the R gene-mediated hypersensitive response in plants, which can disrupt defense signaling pathways, hormonal balance, resource availability, and the overall physiological condition of the plant (Akbudak et al., 2020). Understanding these interactions is crucial for developing strategies to enhance plant resistance to pathogens under water stress conditions, potentially through selecting drought-resistant plant varieties or applying irrigation and water management techniques.

The effect of water stress on pathogens after infection is mainly understood, and it will be fascinating to see if water stress affects viruses' capacity to overcome plant defenses and infect plants successfully. Because each plant-pathogen relationship is distinct and the effect of water stressors on plant-pathogen interactions is distinctive, investigating the effect of combined stress on specific plants will be critical. Although combined stress research is still in its early stages, it is rapidly unraveling the complicated systems of plant defense and opening options for further crop development under present climatic circumstances. Plant-drought-pathogen interaction can be either beneficial or negative. Furthermore, such interactions are more complicated in the field and depend on the soil microbiome-plant combination and meteorological elements in the production region. These various stress interaction possibilities and outcomes have yet to be thoroughly investigated using well-designed trials (Pandey et al., 2017).

6. Conclusion

Recent research has shown that water stress, combined with other abiotic stresses, induces unique but specific physiological, biochemical, and molecular reactions, which involve the accumulation of various metabolites, rewiring of the hormonal pathways, and signal induction for resistance/tolerance or suppression of plant immunity genes.

References

Akbudak MA, Yildiz S & Filiz E. 2020. Pathogenesis related protein-1 (PR-1) genes in tomato (*Solanum lycopersicum* L.): Bioinformatics analyses and expression profiles in response to drought stress. *Genomics,* **112(6):** 4089–4099, https://doi.org/10.1016/j.ygeno.2020.07.004.

Asghari, B, Khademian, R & Sedaghati, B. 2020. Plant growth promoting rhizobacteria (PGPR) confer drought resistance and stimulate biosynthesis of secondary metabolites in pennyroyal (*Mentha pulegium* L.) under water shortage condition *Scientia Horticulturae,* **263:** 109132, https://doi.org/10.1016/j.scienta.2019.109132.

Aung K, Jiang Y & He SY. 2018. The role of water in plant–microbe interactions. *The Plant Journal,* **93:** 771–780, doi: 10.1111/tpj.13795.

Balmer D, Flors V, Glauser G & Mauch-Mani B. 2013. Metabolomics of cereals under biotic stress: Current knowledge and techniques. *Frontiers in Plant Science,* **4:** 82, https://doi.org/10.3389/fpls.2013.00082.

Bharath P, Gahir S & Raghavendra AS. 2021. Abscisic acid-induced stomatal closure: An important component of plant defense against abiotic and biotic stress. *Frontiers in Plant Science,* **12:** 615114, https://doi.org/10.3389/fpls.2021.615114.

Bhattacharya A, Sood P & Citovsky V. 2010. The roles of plant phenolics in defence and communication during *Agrobacterium* and *Rhizobium* infection. *Molecular Plant Pathology,* **11(5):** 705–719, doi: 10.1111/J.1364-3703.2010.00625.X.

Buciumeanu E-C, Guţă IC, Vizitiu DE & Ciripan LM. 2022. The response of grapevine to water stress in the presence of virus infection. *Studii Si Cercetari Stiintifice. Seria Biologie* (Scientific Studies & Research. Series Biology), **31(1):** 70–73, https://pubs.ub.ro/dwnl.php?id=SCSB202201V31S01A0007.

Carlson R, Tugizimana F, Steenkamp PA, Dubery IA, Hassen AI & Labuschagne N. 2020. Rhizobacteria-induced systemic tolerance against drought stress in *Sorghum bicolor* (L.) Moench. *Microbiological Research,* **232:** 126388, https://doi.org/10.1016/j.micres.2019.126388.

Castro-Moretti FR, Gentzel IN, Mackey D & Alonso AP. 2020. Metabolomics as an emerging tool for the study of plant–pathogen interactions. *Metabolites,* **10(2):** 52, doi: 10.3390/metabo10020052.

Chilakala AR, Pandey P, Durgadevi A, Kandpal M, Patil BS, Rangappa K, et al. 2023. Drought attenuates plant responses to multiple rhizospheric pathogens: A study on a dry root rot-associated disease complex in chickpea fields. *Field Crops Research,* **298:** 108965, https://doi.org/10.1016/j.fcr.2023.108965.

Choi H-K, Iandolino A, daSilva FG & Cook DR. 2013. Water deficit modulates the response of *Vitis vinifera* to the Pierce's disease pathogen *Xylella fastidiosa*. *Molecular Plant-Microbe Interactions,* **26(6):** 643–657, http://dx.doi.org/10.1094/MPMI-09-12-0217-R.

Choudhary A & Senthil-Kumar M. 2022. Drought attenuates plant defence against bacterial pathogens by suppressing the expression of CBP60g/SARD1 during combined stress. *Plant, Cell & Environment,* **45(4):** 1127–1145, doi: 10.1111/pce.14275.

Daigham GE, Mahfouz MY, Abdelaziz AM, Nofel MM & Attia MS. 2023. Protective role of plant growth-promoting fungi *Aspergillus chevalieri* OP593083 and *Aspergillus egyptiacus* OP593080 as biocontrol approach against *Alternaria* leaf spot disease of *Vicia faba* plant. *Biomass Conversion and Biorefnery,* 1-17, https://doi.org/10.1007/s13399-023-04510-4.

Davis TS, Bosque-Pérez NA, Foote NE, Magney T & Eigenbrode SD. 2015. Environmentally dependent host–pathogen and vector–pathogen interactions in the Barley yellow dwarf virus pathosystem. *Journal of Applied Ecology,* **52(5):** 1392–1401, https://doi.org/10.1111/1365-2664.12484.

De Pascali M, Vergine M, Sabella E, Aprile A, Nutricati E, Nicolì F, et al. 2019. Molecular effects of *Xylella fastidiosa* and drought combined stress in olive trees. *Plants,* **8(11):** 437, https://doi.org/10.3390/plants8110437.

Dossa GS, Torres R, Henry A, Oliva R, Maiss E, Cruz CV, et al. 2017. Rice response to simultaneous bacterial blight and drought stress during compatible and incompatible interactions. *European Journal of Plant Pathology,* **147(1):** 115–127, https://doi.org/10.1007/s10658-016-0985-8.

Dubey A, Saiyam D, Kumar A, Hashem A, Abd-Allah EF & Khan ML. 2021. Bacterial root endophytes: Characterization of their competence and plant growth promotion in soybean (*Glycine max* (L.) Merr.) under drought stress. *International Journal of Environmental Research and Public Health,* **18(3):** 931, https://doi.org/10.3390/ijerph18030931.

El Aou-ouad H, Montero R, Medrano H & Bota J. 2016. Interactive effects of grapevine leafroll-associated virus 3 (GLRaV-3) and water stress on the physiology of *Vitis vinifera* L. cv. Malvasia de Banyalbufar and Giro-Ros. *Journal of Plant Physiology,* **196:** 106–115, doi:10.1016/j.jplph.2016.04.003.

Erbilgin N, Cale JA, Lusebrink I, Najar A, Klutsch JG, Sherwood P, et al. 2017. Water-deficit and fungal infection can differentially affect the production of different classes of defense compounds in two host pines of mountain pine beetle. *Tree Physiology,* **37(3):** 338–350, https://doi.org/10.1093/treephys/tpw105.

Galarneau ER, Lawrence DP, Travadon R & Baumgartner K. 2019. Drought exacerbates botryosphaeria dieback symptoms in grapevines and confounds host-based molecular markers of infection by *Neofusicoccum parvum*. *Plant Disease,* **103(7):** 1738–1745, https://doi.org/10.1094/PDIS-09-18-1549-RE.

Gani MA & Shama M. 2021. Phenolic compounds. pp. 223–240. *In:* Zepka LQ, Nascimento TC, Jacob-Lopes E (eds), Bioactive Compounds: Biosynthesis, Characterization and Applications, 1st ed. IntechOpen: London, UK.

Ghanbary E, Fathizadeh O, Pazhouhan I, Zarafshar M, Tabari M, Jafarnia S, et al. 2021. Drought and pathogen effects on survival, leaf physiology, oxidative damage, and defense in two middle eastern oak species. *Forests,* **12(2):** 247, https://doi.org/10.3390/f12020247.

Ghanbary E, Kouchaksaraei MT, Guidi L, Mirabolfathy M, Etemad V, Sanavi SAMM, et al. 2018. Change in biochemical parameters of Persian oak (*Quercus brantii* Lindl.)

seedlings inoculated by pathogens of charcoal disease under water deficit conditions. *Trees,* **32(6):** 1595–1608, https://doi.org/10.1007/s00468-018-1736-6.

Gupta A, Patil M, Qamar A & Senthil-Kumar M. 2020. Ath-miR164c influences plant responses to the combined stress of drought and bacterial infection by regulating proline metabolism. *Environmental and Experimental Botany,* **172:** 103998, https://doi.org/10.1016/j.envexpbot.2020.103998.

Gupta A, Hisano H, Hojo Y, Matsuura T, Ikeda Y, Mori IC, et al. 2017. Global profiling of phytohormone dynamics during combined drought and pathogen stress in *Arabidopsis thaliana* reveals ABA and JA as major regulators. *Scientific Reports,* **7(1):** 4017, doi:10.1038/s41598-017-03907-2.

Gupta A, Sarkar AK & Senthil-Kumar M. 2016. Global transcriptional analysis reveals unique and shared responses in *Arabidopsis thaliana* exposed to combined drought and pathogen stress. *Frontiers in Plant Science,* **7:** 686, https://doi.org/10.3389/fpls.2016.00686.

Huang C, Qin A, Gao Y, Ma S, Liu Z, Zhao B, et al. 2023. Effects of water deficit at different stages on growth and ear quality of waxy maize. *Frontiers in Plant Science,* **14:** 1–14, https://doi.org/10.3389/fpls.2023.1069551.

Irulappan V & Senthil-Kumar M. 2018. Morpho-physiological traits and molecular intricacies associated with tolerance to combined drought and pathogen stress in plants. *Biotechnologies of Crop Improvement,* **3:** 59–74, doi:10.1007/978-3-319-94746-4_4.

Jha Y, Yadav KA & Mohamed HI. 2023. Plant growth-promoting bacteria and exogenous phytohormones alleviate the adverse effects of drought stress in pigeon pea plants. *Plant and Soil,* 1–21, doi:10.1007/s11104-023-06155-8.

Kaur S, Kumari N, Parmar P & Sharma V. 2023. Simultaneous exposure of contrasting *Vigna radiata* cultivars to Macrophomina phaseolina infection and drought stress: Morpho-physiological and biochemical implications. *Plant Biosystems – An International Journal Dealing with All Aspects of Plant Biology*, **157(5):** 1100–1113.

Kołton A, Długosz-Grochowska O, Wojciechowska R & Czaja M. 2022. Biosynthesis regulation of folates and phenols in plants. *Scientia Horticulturae,* **291:** 110561, https://doi.org/10.1016/j.scienta.2021.110561.

Li C, Wang Z, Nong Q, Lin L, Xie J, Mo Z, et al. 2021. Physiological changes and transcriptome profiling in *Saccharum spontaneum* L. leaf under water stress and re-watering conditions. *Scientific Reports,* **11(1):** 5525, https://doi.org/10.1038/s41598-021-85072-1.

Lievens L, Pollier J, Goossens A, Beyaert R & Staal J. 2017. Abscisic acid as pathogen effector and immune regulator. *Frontiers in Plant Science,* **8:** 587, https://doi.org/10.3389/fpls.2017.00587.

Lima LKS, de Jesus ON, Soares TL, de Oliveira SAS, Haddad F & Girardi EA. 2019. Water deficit increases the susceptibility of yellow passion fruit seedlings to *Fusarium* wilt in controlled conditions. *Scientia Horticulturae,* **243:** 609–621, doi:10.1016/j.scienta.2018.09.017.

Martínez-Ferri E, Moreno-Ortega G, Van den Berg N & Pliego C. 2019. Mild water stress-induced priming enhances tolerance to *Rosellinia necatrix* in susceptible avocado rootstocks. *BMC Plant Biology,* **19(1):** 1–15, https://doi.org/10.1186/s12870-019-2016-3.

Martins SA, Schurt DA, Seabra SS, Martins SJ, Ramalho MAP, de Souza Moreira FM, et al. 2018. Common bean (*Phaseolus vulgaris* L.) growth promotion and biocontrol by rhizobacteria under *Rhizoctonia solani* suppressive and conducive soils. *Applied Soil Ecology,* **127:** 129–135, doi:10.1016/j.apsoil.2018.03.007.

Melotto M, Zhang L, Oblessuc PR & He SY. 2017. Stomatal defense a decade later. *Plant Physiology,* **174:** 561–571.

Mishra R, Shteinberg M, Shkolnik D, Anfoka G, Czosnek H & Gorovits, R. 2022. Interplay between abiotic (drought) and biotic (virus) stresses in tomato plants. *Molecular Plant Pathology,* **23(4):** 475–488, https://doi.org/10.1111/mpp.13172.

Novaković L, Guo T, Bacic A, Sampathkumar A & Johnson KL. 2018. Hitting the wall—Sensing and signaling pathways involved in plant cell wall remodeling in response to abiotic stress. *Plants,* **7(4):** 89, doi: 10.3390/plants7040089.

Pandey P, Muthukumar VI, Bagavathiannan V & Senthil-Kumar M. 2017. Impact of combined abiotic and biotic stresses on plant growth and avenues for crop improvement by exploiting physio-morphological traits. *Frontiers in Plant Science,* **8:** 537, https://doi.org/10.3389/fpls.2017.00537.

Park S-H, Lee B-R, La VH, Mamun MA, Bae D-W & Kim T-H. 2021. Drought intensity-responsive salicylic acid and abscisic acid crosstalk with the sugar signaling and metabolic pathway in *Brassica napus*. *Plants,* **10(3):** 610, doi: 10.3390/plants10030610

Pérez-Méndez N, Miguel-Rojas C, Jimenez-Berni JA, Gomez-Candon D, Pérez-de-Luque A, Fereres E, et al. 2021. Plant breeding and management strategies to minimize the impact of water scarcity and biotic stress in cereal crops under Mediterranean conditions. *Agronomy,* **12(1):** 75, https://doi.org/10.3390/agronomy12010075.

Pierre E, Fabiola YN, Vanessa ND, Tobias EB, Marie-Claire T, Diane YY, et al. 2023. The co-occurrence of drought and *Fusarium solani* f. Sp. *Phaseoli* Fs4 infection exacerbates the *Fusarium* root rot symptoms in common bean (*Phaseolus vulgaris* L.). *Physiological and Molecular Plant Pathology,* **127:** 102108.

Pratyusha S. 2022. Phenolic compounds in the plant development and defense: An overview. *In:* Hasanuzzaman M and Nahar K (eds), Plant Stress Physiology, doi: 10.5772/intechopen.102873.

Qi J, Song C, Wang B, Zhou J, Kangasjärvi J, Zhu J, et al. 2018. Reactive oxygen species signaling and stomatal movement in plant responses to drought stress and pathogen attack. *Journal of Integrative Plant Biology,* **60(9):** 805–826, doi: 10.1111/jipb.12654.

Ramegowda V, Senthil-Kumar M, Ishiga Y, Kaundal A, Udayakumar M & Mysore KS. 2013. Drought stress acclimation imparts tolerance to *Sclerotinia sclerotiorum* and *Pseudomonas syringae* in *Nicotiana benthamiana. International Journal of Molecular Sciences,* **14(5):** 9497–9513, doi: 10.3390/ijms14059497.

Rivero RM, Mittler R, Blumwald E & Zandalinas SI. 2022. Developing climate-resilient crops: Improving plant tolerance to stress combination. *The Plant Journal,* **109(2):** 373–389, https://doi.org/10.1111/tpj.15483.

Robison MM, Shah S, Tamot B, Pauls KP, Moffatt BA & Glick BR. 2001. Reduced symptoms of *Verticillium* wilt in transgenic tomato expressing a bacterial ACC deaminase. *Molecular Plant Pathology,* **2(3):** 135–145, doi: 10.1046/j.1364-3703.2001.00060.x.

Samain E, Ernenwein C, Aussenac T & Selim S. 2022. Effective and durable systemic wheat-induced resistance by a plant-growth-promoting rhizobacteria consortium of *Paenibacillus* sp. Strain B2 and *Arthrobacter* spp. Strain AA against *Zymoseptoria tritici* and drought stress. *Physiological and Molecular Plant Pathology,* **119:** 101830, https://doi.org/10.1016/j.pmpp.2022.101830.

Sharma M & Pande S. 2013. Unravelling effects of temperature and soil moisture stress response on development of dry root rot (*Rhizoctonia bataticola* (Taub.)) butler in chickpea. *American Journal of Plant Sciences,* **4(3):** 584–589, doi:10.4236/ajps.2013.43076.

Sinha R, Gupta A & Senthil-Kumar M. 2016. Understanding the impact of drought on foliar and xylem invading bacterial pathogen stress in chickpea. *Frontiers in Plant Science,* **7:** 902, https://doi.org/10.3389/fpls.2016.00902.

Sinha R, Gupta A & Senthil-Kumar M. 2017. Concurrent drought stress and vascular pathogen infection induce common and distinct transcriptomic responses in chickpea. *Frontiers in Plant Science,* **8:** 333, https://doi.org/10.3389/fpls.2017.00333.

Sinha R, Irulappan V, Mohan-Raju B, Suganthi A & Senthil-Kumar M. 2019. Impact of drought stress on simultaneously occurring pathogen infection in field-grown chickpea. *Scientific Reports,* **9:** 5577, https://doi.org/10.1038/s41598-019-41463-z.

Sunarti S, Kissoudis C, Van Der Hoek Y, Van Der Schoot H, Visser RG, Van Der Linden CG, et al. 2022. Drought stress interacts with powdery mildew infection in tomato. *Frontiers in Plant Science,* **13:** 845379, https://doi.org/10.3389/fpls.2022.845379.

Taketani RG, Lanconi MD, Kavamura VN, Durrer A, Andreote FD & Melo IS. 2017. Dry season constrains bacterial phylogenetic diversity in a semi-arid rhizosphere system. *Microb. Ecol,* **73:** 153–161.

Tarafdar A, Rani TS, Chandran US, Ghosh R, Chobe DR & Sharma M. 2018. Exploring combined effect of abiotic (soil moisture) and biotic (*Sclerotium rolfsii* Sacc.) stress on collar rot development in chickpea. *Frontiers in Plant Science,* **9:** 1154, https://doi.org/10.3389/fpls.2018.01154.

Vemanna RS, Bakade R, Bharti P, Kumar M, Sreeman SM, Senthil-Kumar M, et al. 2019. Cross-talk signaling in rice during combined drought and bacterial blight stress. *Frontiers in Plant Science,* **10:** 193, https://doi.org/10.3389/fpls.2019.00193.

Were JO, Ochuodho JO, Rop NK & Gyawali S. 2022a. Antagonistic effect of aluminium toxicity and water deficiency on the virulence of *Pyrenophora teres* in barley. *Plant Pathology Journal,* **21(1):** 1–11, doi: 10.3923/ppj.2022.1.11.

Were JO, Ochuodho JO, Rop NK & Gyawali S. 2022b. Biotic and abiotic stress tolerance induction by salicylic and abscisic acid phytohormones in barley. *American Journal of Plant Physiology,* **17(1):** 1–10, doi: 10.3923/ajpp.2022.1.10.

Xin XF, Nomura K, Aung K, Velasquez AC, Yao J, Boutrot F, et al. 2016. Bacteria establish an aqueous living space in plants crucial for virulence. *Nature,* **539:** 524–529.

Yang J, Duan G, Li C, Liu L, Han G, Zhang Y, et al. 2019. The crosstalks between jasmonic acid and other plant hormone signaling highlight the involvement of jasmonic acid as a core component in plant response to biotic and abiotic stresses. *Frontiers in Plant Science,* **10:** 01349, https://doi.org/10.3389/fpls.2019.01349.

Ye YX, Wen ZR, Yang H, Lu WP & Lu DL. 2020. Effects of post-silking water deficit on the leaf photosynthesis and senescence of waxy maize. *Journal of Integrated Agriculture,* **19(9):** 2216–2228, doi: 10.1016/S2095-3119(20)63158-6.

Zhang Y, Cai P, Cheng G & Zhang Y. 2022. A brief review of phenolic compounds identified from plants: Their extraction, analysis, and biological activity. *Natural Product Communications,* **17(1):** 1–14, doi: 10.1177/1934578X211069721.

10 | Molecular Mechanism of Water Stress Tolerance in Wheat

Anukriti Srivastava and Girish Chandra Pandey*

Department of Bioscience and Biotechnology, Banasthali Vidyapith, India

Climatic changes are major threats to crops globally. Water stress is also known as 'Drought'. It is one of the consequences of climatic changes that adversely affect crop development, growth and yield. Water scarcity, soil salinity and high temperature are reasons behind water stress. Crops show different morphological, physiological, molecular and biochemical ways to confront water stress. Wheat (*Tritium aestivum* L.), commonly called as bread wheat, is the staple food with 36% demand globally. Water stress negatively influenced the wheat crops, especially during the flowering and grain-filling stages. Water stress adversely affects growth, photosynthesis, nutrient assimilation and crop yield. Water stress can affect wheat in many different ways by changing proteins, by producing antioxidants, adjusting osmolytes, by changing hormone constitution, adjusting depth and extension of root, adjusting stomatal conductance, lower chlorophyll content, lowering transpiration, and inhibiting of crop growth. The complex trait of drought tolerance is governed by polygenes, whose expression is affected by a variety of environmental factors. Therefore, we now have some innovative biotechnological approaches that support a better grasp of the mechanisms behind molecular responses of plants to water stress. Numerous drought-induced genes have been discovered, some of which have even been cloned. The production of drought-tolerant germplasm is made possible through plant genetic engineering and molecular marker techniques. Molecular markers are being utilized to locate drought-related quantitative trait loci (QTL) and ensure their successful transfer into wheat crop varieties that are grown for commercial purposes. To manage water stress, transgenic plants expressing abiotic stress tolerance genes are being produced. Redox proteins, detoxifying enzymes, stress-induced LEA proteins, structural genes (such as proline, glycine betaine, mannitol, and trehalose), and

*Corresponding author: girishchandrapandey@banasthali.in

regulatory genes (such as dehydration-responsive, element-binding (DREB) factors, zinc finger proteins, and NAC transcription factor genes) are being used. Plant breeding has been revolutionized by the development of CRISPR (the clustered regularly interspaced short palindromic repeat technology), as a new frontier in crop improvement. In this chapter, we will focus on the limitation of water, its impact on wheat morphology, physiology, biochemical constitution, potential losses from water stress different biotechnological techniques for water stress management and developing transgenic wheat cultivars.

1. Introduction

Extreme drought hit at least 3% of the world's geographical area in 2017 (Fao, 2018). Nearly a billion people go hungry each year, which equates to around a fifth of the world's population not having enough food to live normally (Fao, 2018; Martignago et al., 2020). Many areas are currently experiencing drought, which is predicted to significantly dry up more than 50% of all arable land by 2050 (Vinocur and Altman, 2005). To meet the needs of the world's population, which is expected to exceed 9 billion people by 2050 (Virmani and Ilyas-Ahmed, 2007), food grain production must be doubled by then (Tilman et al., 2002). Wheat is a significant cereal crop grown worldwide. According to Dhanda et al. (2004), it is a staple meal in many nations, providing more than half of the calories and half of the protein in the diet of more than one-third of the world's population. The world's population is always growing, which has raised global demand for wheat. Rana et al. (2013) stated that abiotic factors can result in a 71% decrease in wheat production. According to Reynolds et al. (2001), developing countries cultivate 50% of their wheat using the rainfed method. Wheat production is reduced as a result of the drought stress that it experiences in many parts of the world. The vast majority of wheat is produced in rainy environments where changes in rainfall patterns have created water limitations, a significant contributor to the diminishing production of crops, particularly when they occur during the flowering and grain-filling stages (Bassi et al., 2017).

Due to climate change and a decrease in the amount of sub-surface water resources available for agriculture, there is a significant likelihood that there will be drought stress in the upcoming days. Abiotic stressors have a significant impact on wheat productivity which has been demonstrated in numerous studies. According to a study, there is a yield loss of between 4.1% to 6% for every 1°C rise in temperature (Liu et al., 2016). Unpredictable rainfall, which frequently reduces average yield by more than 50%, is the main issue restricting wheat production in many places. Although wheat can be grown in a wide range of agro-climatic conditions, the majority of these conditions are severely hampered by drought stress, which also reduces productivity. The frequency of droughts will rise due to the anticipated global warming and climatic changes, decreasing wheat yields. Agriculture has been impacted by the rise in the annual average temperature, variations in rainfall patterns, and the threat of drought in many locations. According to the IPCC (2013), the availability of sub-surface water will continue to decline as temperatures rise. According to Ashraf and Foolad (2007), up to 50% of agricultural land is vulnerable to regular drought, making it the main environmental stress reducing the production of cereal crops worldwide. In order to comprehend how plants might adapt significantly to lessen the impact of drought stress, it is necessary to grasp the physiological effects of drought from a context-dependent perspective (Vinocur & Altman, 2005). Plants that can withstand drought attempt to reduce the loss of water content, membrane stability, and photosynthetic activity. Due to their relationship to adaptation to future climatic scenarios, plant physio-morphological features are crucial for selection in a breeding programme to improve drought resistance (Bowne, 2012). Additionally, pinpointing the genes in charge of these physiological alterations may result in a rapid genetic advancement in a plant's capacity to withstand drought.

Whole-genome sequencing is now achievable for every genotype because of recent developments in genomics. Genotyping by sequencing, which produces a significant number of single nucleotide polymorphism (SNP) markers that span the wheat genome, is one of the most popular and commonly used techniques. Additionally, wheat reference genomes are accessible for imputation. Each SNP produced by genotyping-by-sequencing (GBS) can be accurately positioned and located on the chromosome using these genomic references.

This enormous collection of SNPs is used for quantitative trait loci (QTL) mapping and genome-wide association studies (GWAS) to analyze the genetics of complex characteristics by pinpointing genomic areas or genes that may be responsible for the target trait, in this instance, drought tolerance. The total number of genes discovered is determined by the number of traits linked to drought tolerance that have been measured, as well as by the size and proximity of the genes. Numerous genes affecting drought tolerance are discovered as more features are graded favorably (Ovenden et al., 2017; Qaseem et al., 2018). The number of genes regulating drought tolerance in each chosen genotype must be determined as well. For significant drought genes like the Dreb and Fehw3 genes, there are numerous particular molecular markers (Rasheed et al., 2016). As a result, any germplasm may be checked to determine if these two genes are present or absent. By finding new genes influencing drought using GWAS or QTL mapping, genetic improvement for drought tolerance can be accomplished (Zeng et al., 2014; Sukumaran et al., 2018).

2. Effect of Water-stress on Wheat

Water stress can have negative effects on a plant's entire physiological performance from morphological to molecular levels. The impacts of different types of drought stress are described below, along with their severity.

2.1 Morphological Changes

The wheat crop experiences a variety of morphological changes in response to drought, which are visible at various phases of plant development. The morphological response of wheat is divided into root and shoots parts. The shoot portion includes modifications to the leaf's form, expansion, area, size, pubescence, waxiness, cuticle tolerance, and reduction in shoot length. Grain yield and leaf area were shown to be positively correlated with plant height.

2.1.1 Variations in Plant Height

Wheat cell elongation can be changed by cutting off water passage from the xylem to the cells around it when there is severe water scarcity (Nonami, 1998). However, the length of the drought, its severity, and the stage at which the wheat plant is growing all affect how the drought stress is felt. At several stages of the wheat plant's development, including stem elongation, booting, and grain filling, many tests have been conducted. The results indicate that the plants that experienced drought stress beginning at the stem elongation stage suffered more than the others. Throughout the grain-filling stage, the plant length only decreased by 7%; however, during the stem extension stage and the booting stage, it decreased by 35% and 23%, respectively (Caverzan et al., 2016). Many other researchers have also discovered that wheat grows slowly when exposed to dry conditions in the roots and shoots (Azooz and Youssef, 2010; Farooq et al., 2013). So, one of the main causes of a general decline in the growth of wheat plants is drought.

2.1.2 Leaf Senescence

According to a study, drought stress during the reproduction stage increases the rate of senescence, which results in a striking loss in grain yield (Nawaz et al., 2013). Chlorosis can be the crucial factor responsible for leaf senescence that leads to reduction in photosynthesis (Ali et al., 2020). Senescence under drought stress can affect every part of the wheat plant, but it also makes it simpler for stored carbohydrates to be mobilized during flowering from the stem and leaves to produce new grains, thereby reducing the yield loss brought on by senescence (Farooq et al., 2014). Rubisco (Ribulose biphosphate carboxylase), glutamine synthetase, and total protein concentration were employed to identify the onset and progression of senescence in wheat.

Senescence typically starts in the older leaves, then moves on to the younger leaves. However, in some delicate cultivars, senescence is disrupted by dehydration and first appears in flag leaves before moving on to elder leaves. Younger leaves were compared to plants growing in settings with enough water, and it was discovered that the glutamine synthetase iso-enzyme content dramatically dropped and the order of senescence was slightly thrown off (Nagy et al., 2013). Drought is associated with the withering of wheat leaves, and by monitoring the breakdown of nitrogen and carbon; we can advance the development of varieties of wheat that are both drought-resistant and drought-sensitive.

2.2 Physiological Changes

There are many physiological reactions to drought stress that have been identified. Numerous physiological characteristics help wheat crops resist the effects of drought stress. Lack of water also causes membrane instability, a drop in water content, and a drop in chlorophyll levels (Sallam et al., 2019). To date, numerous studies have been conducted in an effort to comprehend how wheat reacts physiologically to drought stress. Drought stress caused a considerable drop in transpiration,which led to a gradual loss of heat from the leaves and an increase in leaf temperature, according to a group of researchers (Rosenberg et al., 1990).

And as a result, CO_2 levels rise, photosynthesis is affected, which in turn influences plant growth, and water use efficiency can be increased. Similar to this, these qualities are also in charge of breeding programs to produce drought-tolerant types that can thrive in areas of the world where water supplies are scarce.

2.2.1 Variations in Patterns of Cell Growth

Wheat is extremely susceptible to water stress circumstances, and when exposed to these kinds of conditions, the growth of plants varies dramatically. The impact of drought stress is further influenced by the length, timing, severity, and stage of the wheat crop. At several stages of the wheat plant's development, including stem elongation, booting, and grain filling, many tests have been conducted. The results indicate that the plants exposed to drought stress beginning at the stem elongation stage suffered more than the others.

Additionally, the length, kind, and severity of the drought, as well as the stage of plant development, control all potential modifications. There is plenty of information available on the development stage and the amount of drought stress tolerance of wheat cultivars. The length and kind of drought can also have an impact on plant development. A two-factor experiment was carried out by Shamsi et al. (2011) using three different wheat cultivars and three distinct growth stages. Up until harvest, grain filling, stem elongation, and booting were all affected by drought stress. Results indicated that, compared to the other two stages of plant growth, plants experiencing water stress during the stem elongation stage suffered more.

2.2.2 Variations in Chlorophyll Content and Photosynthetic Rate

In addition to directly reducing carboxylation, CO_2 reduction also causes more electrons to be directed toward producing reactive O_2 species. Reduced tissue water concentrations also make Rubisco binding inhibitors more effective. The green pigment chlorophyll is in charge of the photosynthetic process. Wheat primarily contains two forms of chlorophyll: chlorophyll *a* and chlorophyll *b*.

Chlorophyll *b* is more susceptible to the effects of environmental stress, and its abundance has decreased more than that of chlorophyll *a*. According to Fang et al. (1998) this is supported by the notion that chlorophyll *a* transformation to chlorophyll *b* contributed to some of the fall in chlorophyll *a*. According to researchers, once wheat plants are exposed to light, during chlorophyll synthesis an enzyme is activated that leads to an increase in the chlorophyll content of young leaves. On the other hand, as a result of the activation of chlorophyllase and the inactivation of enzymes during drought stress, the chlorophyll concentration declines by 13–15% in older leaves (Nikolaeva et al., 2010).

2.2.3 Relative Water Content (RWC)

Significant factors that affect the relationship between plants and water include relative water content, leaf water potential, stomata tolerance, transpiration rate and leaf temperature. When wheat leaves developed, their initial relative water content was higher; as the leaf grew older, this level decreased. When there is a water shortage, leaf RWC is seen as the most significant metric among the various types of water potential. The wheat RWC during its developmental phases is significantly reduced as a result of drought stress.

After six weeks of emergence, when drought has an impact on relative water content, nutrient uptake, growth, and yield, its effects on wheat are increasingly noticeable (Nawaz et al., 2014).

3. Biochemical Changes

Under stressful circumstances, wheat crops have an internal defense system with antioxidant enzymes like superoxide dismutase (SOD), catalase (CAT), and peroxidase (POX) for scavenging ROS (Chen et al., 2012).

3.1 Proline Content

According to reports, wheat plants exposed to drought stress have an increase in their proline content (Vendruscolo et al., 2007). When wheat is experiencing water stress, the amount of proline grows to its highest level during the heading stage. Various wheat genotypes have various degrees of production of proline under drought, which affects how well they can tolerate drought stress. Due to the fact that different wheat genotypes have varied water stress thresholds, the genotypes that accumulate more proline during droughts are better equipped to withstand drought stress. So selecting a variety of wheat that's drought-tolerant can benefit from knowing how much proline is in the wheat.

4. Biotechnological Approaches to Develop Drought Resistance Plants

The complexity of coping with drought, the lack of efficient selection procedures, and the limited genetic diversity of yield component factors under drought conditions have all been significant problems. Our knowledge of the mechanisms underpinning molecular and whole-plant responses of plants to drought has advanced quickly with the introduction of novel biotechnological methods. Numerous genes that are activated by drought have been discovered, and some of them have even been cloned. It is now possible to better understand the mechanisms behind drought tolerance using a variety of methods, including transgenic plants and gene expression patterns. To develop drought-tolerant germplasm, a variety of biotechnological methods are being used.

4.1 Transgenic Plant Development and Plant Genetic Engineering

Transgenesis is a method of genetically modified organisms (GMOs) that involves the transfer of specific replicated genes without co-transferring undesirable genes from donors. This method allows for gene stacking or pyramiding, and is being used to create transgenic plants with genes for abiotic stress tolerance (James, 2007). Over the past 15 years, various crop plants have been successfully transformed and produced as GMOs using recombinant DNA methods, gene transfer procedures, and tissue culture techniques.

5. Role of Redox Proteins and Detoxifying Enzymes During Drought

Oxidative stress is caused by an imbalance between the production and elimination of ROS (Halliwell, 1997). Stresses from drought and salinity are caused by the increased generation of reactive oxygen species (ROS). Oxidative stress is brought on by the rapid accumulation of harmful compounds (Rathinasabapathi and Kaur, 2006).Understanding the functions of particular antioxidant defenses under stress situations is made easier due to antioxidant genes (Zhang et al., 2007). For the complex trait of abiotic stress, gene pyramiding, or multi gene transfer, is essential.

6. Stress-Induced LEA Proteins

Low molecular weight (10–30 kDa) proteins called late embryogenesis-abundant (LEA) proteins help higher plants resist damage from environmental stresses, particularly water stress (Hong et al., 2005). Xu et al. (1996) demonstrated that second-generation transgenic plants exhibit enhanced water deficit tolerance. An effort was made to boost biomass productivity and water scarcity tolerance in transgenic wheat that is constantly releasing the barley HVA1 gene (Sivamani et al., 2000).

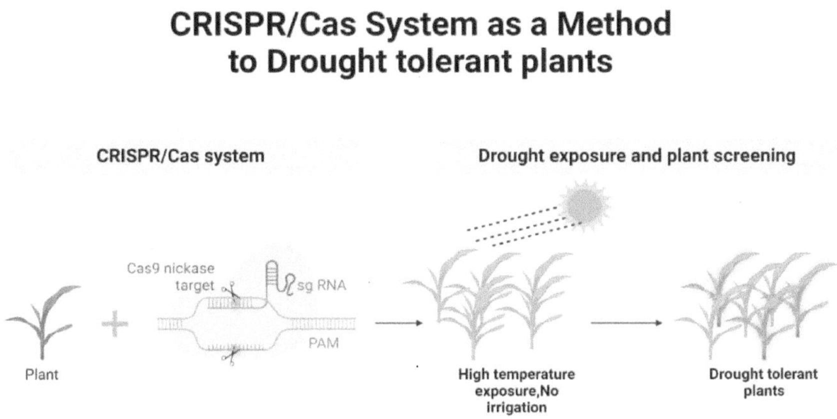

Figure 1: Effect of water stress on the severity of plant infections.

7. Application of the CRISPR–Cas on Drought Tolerance Mechanism

Plant breeding has advanced significantly as a result of the development of genome editing techniques. Sequence-definite nucleases are used in genome editing techniques to incorporate known changes into the genome (Haque, et al., 2018). The versatility and simplicity of the CRISPR-Cas genome editing technique has been widely praised.

In addition to morphological and physiological factors, which are crucial for plant growth, drought stress has an impact on a variety of biochemical aspects. Drought and soil water shortages can result in continuously reduced access to water for plants in a variety of climatic areas. Additionally, there are sporadic and unpredictable climate changes everywhere. The impacts of a water shortage may develop during the plant growth cycle.

8. Quantitative Trait Locus (QTL) of Drought Tolerance

The traditional QTL-mapping method has been used to identify dozens of significant genetic areas. This study clarified the genetic basis of this complicated feature in wheat and identified the loci causing variation in attributes related to drought tolerance. By analyzing attributes throughout the plant life cycle, it is possible to ascertain the impact of drought stress on plant performance, development, and yield. Such analysis aids in identifying the QTL of the trait or stage that is more susceptible to the effects of drought stress, as well as any common QTLs for drought tolerance between the developmental stages of traits and final yield.

Despite this, drought stress affects seed germination, seedling vigor, and growth (Samarah et al., 2011; Czyczyło-Mysza et al., 2014). Additionally, a limited amount of progress has been made in establishing the genetic causes of drought stress during the extremely delicate reproductive phase, which will determine the final grain yield (Campoli et al., 2014). The majority of QTL research in wheat has concentrated on the elements of the final grain yield under drought-stress conditions (Quarrie et al., 1994).

9. Molecular-marker Technology

Numerous applications for crop development can be made using molecular methods for identifying variations in individual plants' DNA. These variations are referred to as molecular markers because they are frequently linked to particular genes and serve as 'signposts' 'to those genes. The simplest type of marker-assisted selection can be used to select indirectly for the desired allele using such markers, which are closely connected to genes of interest. Marker-assisted selection (MAS) is currently being used to quicken backcross breeding and pyramid a number of desired alleles (Singh et al., 2001). Therefore, molecular markers can be employed to track any number of genes during the breeding programme. The existence (or disappearance) of a particular gene in the progeny can be selected by choosing a marker next to the desired gene (Paran and Michelmore, 1993). The identification of molecular markers has made it possible to separate quantitative features into their individual genetic components (Tanksley, 1993; Pelleschi et al., 2006; Bernier et al., 2007; Kato et al., 2008) and assisted in the selection, pyramiding, and MAS of QTL alleles (Ribaut et al., 2004; Neeraja et al., 2007; Ribaut & Ragot, 2007). In this light drought tolerance can be improved in crops including wheat.

10. Conclusions

The growth and output of the wheat crop are significantly reduced when there is drought stress. Wheat under drought stress experiences a range of morphological, physiological, and biochemical impacts. It typically inhibits seed germination and hinders the establishment of the stand. It results in the lowering of leaf area, overall plant growth, and the buildup of total dry mass in wheat. It has a detrimental impact on the photosynthetic process and the manufacture of chlorophyll, which is involved in gas exchange.

ROS are produced more frequently under drought stress, which disrupts different biological processes and results in DNA and RNA being oxidatively damaged. Transgenic plants that are tolerant to drought and stress have recently been successfully produced in a lab; this success must be taken advantage of in the future. Transgenic plants that can withstand drought and stress, as well as viruses, herbicides, and insects, are produced in commercial settings. Transgenic plants are still being tested in the field or in pot studies. Drought-related QTLs are being efficiently transferred into wheat crop varieties that are grown for commercial purposes using molecular markers.

References

Ali M, Gul A, Hasan H, Gul S, Fareed A, Nadeem M, et al. 2020. Cellular mechanisms of drought tolerance in wheat. *Climate Change and Food Security with Emphasis on Wheat*, **1:** 155–167.

Ashraf MF & Foolad MR. 2007. Roles of glycine betaine and proline in improving plant abiotic stress resistance. *Environmental and Experimental Botany,* **59(2):** 206–216.

Azooz MM & Youssef MM. 2010. Evaluation of heat shock and salicylic acid treatments as inducers of drought stress tolerance in Hassawi wheat. *American Journal of Plant Physiology,* **5(2):** 56–70.

Bassi F & Sanchez-Garcia M. 2017. Adaptation and stability analysis of ICARDA durum wheat elites across 18 countries. *Crop Science,* **57:** 2419–2430.

Bowne JB, Erwin TA, Juttner J, Schnurbusch T, Langridge P, Bacic A, et al. 2012. Drought responses of leaf tissues from wheat cultivars of differing drought tolerance at the metabolite level. *Molecular Plant,* **5(2):** 418–429.

Bernier J, Kumar A, Ramaiah V, Spaner D & Atlin G. 2007. A large-effect QTL for grain yield under reproductive-stage drought stress in upland rice. *Crop Science,* **47(2):** 507–516.

Campoli C & von Korff M. Genetic control of reproductive development in temperate cereals. 2014. *Advances in Botanical Research,* **72:** 131–158. Academic Press.

Caverzan A, Casassola A & Brammer SP. 2016. Antioxidant responses of wheat plants under stress. *Genetics and Molecular Biology,* **39(1):** 1–6.

Chen J, Xu W, Velten J, Xin Z & Stout J. 2012. Characterization of maize inbred lines for drought and heat tolerance. *Journal of Soil and Water Conservation,* **67:** 354–364.

Czyczyło-Mysza I, Marcińska I, Skrzypek E, Cyganek K, Juzoń K & Karbarz M. 2014. QTL mapping for germination of seeds obtained from previous wheat generation under drought. *Open Life Sciences,* **9(4):** 374–382.

Dhanda SS, GS Sethi & RK Behl. 2004. Indices of drought tolerance in wheat genotypes at early stages of plant growth. *Journal of Agronomy and Crop Science,* **19(1):** 6–8.

Fang Z, Bouwkamp JC & Solomos T. 1998. Chlorophyllase activities and chlorophyll degradation during leaf senescence in non-yellowing mutant and wild type of *Phaseolus vulgaris* L. *Journal of Experimental Botany,* **49(320):** 503–510.

Fao F. 2018. The impact of disasters and crises on agriculture and food security. Report.

Farooq M, Hussain M & Siddique KHM. 2014. Drought stress in wheat during flowering and grain-filling periods. *Critical Reviews in Plant Sciences,* **33(4):** 331–349.

Farooq M, Irfan M, Aziz T, Ahmad I & Cheema SA. 2013. Seed priming with ascorbic acid improves drought resistance of wheat. *Journal of Agronomy and Crop Science,* **199(1):** 12–22.

Halliwell B. 1997. Free radicals and human disease—Trick or treat? pp. 1–14. *In:* CE Thomas and B Kalyanaraman (eds), Oxygen Radical Sand the Disease Process, Newark, NJ: Harwood Academic.

Haque E, Taniguchi H & Hassan MM. 2018. Application of CRISPR/Cas9 genome editing technology for the improvement of crops cultivated in tropical climates: Recent progress, prospects, and challenges. *Frontiers in Plant Science,* **9:** 617.

Hong-Bo S, Zong-Suo L & Ming-An S. 2005. LEA proteins in higher plants: Structure, function, gene expression and regulation. *Colloids and Surfaces B: Biointerfaces,* **45(3–4):** 131–135.

IPCC. 2013. Climate Change. The Physical Science Basis. Working group-I, contribution to the fifth assessment report of the inter-governmental panel on climate change. 1–1535 Cambridge University Press 2013.

James C. 2007. Global status of commercialized biotech/GM crops, 1–115. ISAAA Brief No. 37. Ithaca, New York.

Kato Y, Hirotsu S, Nemoto K & Yamagishi J. 2008. Identification of QTLs controlling rice drought tolerance at seedling stage in hydroponic culture. *Euphytica,* **160:** 423–430.

Liu B, Asseng S, Müller C, Ewert F, Elliott J, Lobell DB, et al. 2016. Similar estimates of temperature impacts on global wheat yield by three independent methods. *Nature Climate Change,* **6(12):** 1130–1136.

Liu Q, Kasuga M, Sakuma Y, Abe H, Miura S & Yamaguchi-Shinozaki K.1998. Two transcription factors, DREB1 and DREB2, with an EREBP/AP2 DNA binding domain separate two cellular signal transduction pathways in drought- and low-temperature-responsive gene expression, respectively, in *Arabidopsis*. *The Plant Cell,* **10(8):** 1391–1406.

Martignago D, Rico-Medina A & Blasco-Escámez D. 2020. Drought resistance by engineering plant tissue-specific responses. *Frontiers in Plant Science,* **10:** 1676.

Nagy Z, Németh E, Guóth A, Bona L, Wodala, B & Pécsváradi A. 2013. Metabolic indicators of drought stress tolerance in wheat: Glutamine synthetase isoenzymes and Rubisco. *Plant Physiology and Biochemistry,* **67:** 48–54.

Nawaz A, Farooq M, Cheema SA,Yasmeen A & Wahid A. 2013. Stay green character at grain filling ensures resistance against terminal drought in wheat. *International Journal of Agriculture and Biology,* **15(6):** 1272–1276.

Nawaz F, Ashraf M, Ahmad R, Waraich E & Shabbir N. 2014. Selenium (Se) regulates seedling growth in wheat under drought stress. *Advances in Chemistry,* **14:** 143567.

Neeraja CN, Maghirang-Rodriguez R, Pamplona A, Heuer S, Collard BC, Septiningsih EM, et al. 2007. A marker-assisted backcross approach for developing submergence-tolerant rice cultivars. *Theoretical and Applied Genetics,* **115:** 767–776.

Nikolaeva MK, Maevskaya SN, Shugaev AG & Bukhov NG. 2010. Effect of drought on chlorophyll content and antioxidant enzyme activities in leaves of three wheat cultivars varying in productivity. *Russian Journal of Plant Physiology,* **57(1):** 87–95.

Nonami H. 1998. Plant water relations and control of cell elongation at low water potentials. *Journal of Plant Research,* **111:** 373–382.

Ovenden B, Milgate A, Wade LJ, Rebetzke GJ & Holland JB. 2017. Genome-wide associations for water-soluble carbohydrate concentration and relative maturity in wheat using SNP and DArT marker arrays. *G3: Genes, Genomes, Genetics,* **7(8):** 2821–2830.

Paran I & Michelmore RW. 1993. Development of reliable PCR-based markers linked to downy mildew resistance genes in lettuce. *Theoretical and Applied Genetics,* **85:** 985–993.

Pelleschi S, Leonardi A, Rocher JP, Cornic G, De Vienne D, Thevenot C, et al. 2006. Analysis of the relationships between growth, photosynthesis and carbohydrate metabolism using quantitative trait loci (QTLs) in young maize plants subjected to water deprivation. *Molecular Breeding,* **17:** 21–39.

Qaseem MF, Qureshi R, Muqaddasi QH, Shaheen H, Kousar R & Röder MS. 2018. Genome-wide association mapping in bread wheat subjected to independent and combined high temperature and drought stress. *PLoS One,* **13(6):** e0199121.

Quarrie SA, Steed A, Calestani C, Semikhodskii A, Lebreton C, Chinoy C, et al. 2005. A high-density genetic map of hexaploid wheat (*Triticum aestivum* L.) from the cross Chinese Spring × SQ1 and its use to compare QTLs for grain yield across a range of environments. *Theoretical and Applied Genetics,* **110:** 865–880.

Rana RM, Rehman SU, Ahmed J & Bilal M. 2013. A comprehensive overview of recent advances in drought stress tolerance research in wheat (*Triticum aestivum* L.). *Asian Journal of Agriculture and Biology,* **1(1):** 29–37.

Rasheed A, Wen W, Gao F, Zhai S, Jin H, Liu J, et al. 2016. Development and validation of KASP assays for genes underpinning key economic traits in bread wheat. *Theoretical and Applied Genetics,* **129:** 1843–1860.

Rathinasabapathi, B & Kaur, R. 2006. Metabolic engineering for stress. *Molecular Biology of Stress Tolerance in Plants*, 255–299.

Reynolds MP, Nagarajan S, Razzaque MA & Ageeb, OAA. 2001. Heat tolerance. *Application of Physiology in Wheat Breeding*, 124–135.

Ribaut JM, Bänziger M, Setter T, Edmeades G & Hoisington D. 2004. Genetic dissection of drought tolerance in maize: A case study. *Physiology and Biotechnology Integration for Plant Breeding,* **15:** 571–609.

Ribaut JM & Ragot M. 2007. Marker-assisted selection to improve drought adaptation in maize: The backcross approach, perspectives, limitations, and alternatives. *Journal of Experimental Botany,* **58(2):** 351–360.

Rosenberg NJ, Kimball BA, Martín P, Cooper CF & Waggoner PE. 1990. From climate and CO_2 enrichment to evapotranspiration.

Sallam A, Alqudah AM, Dawood MF, Baenziger PS & Börner A. 2019. Drought stress tolerance in wheat and barley: Advances in physiology, breeding and genetics research. *International Journal of Molecular Sciences,* **20(13):** 3137.

Samarah N & Alqudah A. 2011. Effects of late-terminal drought stress on seed germination and vigor of barley (*Hordeum vulgare* L.). *Archives and Agronomy and Soil Science,* **57:** 27–32.

Shamsi K, Petrosyan M, Noor-Mohammadi G, Haghparast A, Kobraee S & Rasekhi B. 2011. Differential agronomic responses of bread wheat cultivars to drought stress in the west of Iran. *African Journal of Biotechnology,* **10(14):** 2708–2715.

Singh S, Sidhu JS, Huang N, Vikal Y, Li Z, Brar DS, et al. 2001. Pyramiding three bacterial blight resistance genes (xa5, xa13 and Xa21) using marker-assisted selection into indica rice cultivar PR106. *Theoretical and Applied Genetics,* **102:** 1011–1015.

Sivamani E, Bahieldin A, Wraith JM, Al-Niemi T, Dyer WE, Ho TH, et al. 2000. Improved biomass productivity and water use efficiency under water deficit conditions in transgenic wheat constitutively expressing the barley HVA1 gene. *Plant Science,* **155(1):** 1–9.

Sukumaran S, Reynolds MP & Sansaloni C. 2018. Genome-wide association analyses identify QTL hotspots for yield and component traits in durum wheat grown under yield potential, drought, and heat stress environments. *Frontiers in Plant Science,* **9:** 330307.

Tanksley SD. Mapping polygenes. *Annual Review of Genetics*, **27(1):** 205–233.

Tilman D, Cassman KG, Matson PA, Naylor R & Polasky S. 2002. Agricultural sustainability and intensive production practices. *Nature,* **418(6898):** 671–677.

Vendruscolo ECG, Schuster I, Pileggi M, Scapim CA, Molinari HBC, Marur CJ, et al. 2007. Stress-induced synthesis of proline confers tolerance to water deficit intransgenic wheat. *Journal of Plant Physiology,* **164(10):** 1367–1376.

Vinocur B & Altman A. 2005. Recent advances in engineering plant tolerance to abiotic stress: Achievements and limitations. *Current Opinion in Biotechnology,* **16(2):** 123–132.

Virmani SS & Ahmed MI. 2007. Rice breeding for sustainable production. *Breeding Major Food Staples,* **2:** 141–191.

Xu D, Duan X, Wang B, Hong B, Ho TH & Wu R. 1996. Expression of a late embryogenesis abundant protein gene, HVA1, from barley confers tolerance to water deficit and salt stress in transgenic rice. *Plant Physiology,* **110(1):** 249–257.

Zeng ZB, Teulat B, Merah O, Sirault X, Borries C, Waugh RD, et al. 2014. Detection and validation of novel QTL for shoot and root traits in barley (*Hordeum vulgare* L.) *Journal of Experimental Botany,* **9:** 171–180.

Zhang Y, Yang C, Li Y, Zheng N, Chen H, Zhao Q, et al. 2007. SDIR1 is a RING finger E3 ligase that positively regulates stress-responsive abscisic acid signaling in *Arabidopsis. The Plant Cell,* **19(6):** 1912–1929.

Zhu B, Su J, Chang M, Verma DP, Fan YL & Wu R. 1998. Overexpression of a Δ1-pyrroline-5-carboxylate synthetase gene and analysis of tolerance to water- and salt-stress in transgenic rice. *Plant Science,* **139(1):** 41–48.

11 | Water Conservation Vis-à-Vis Sustainable Agricultural Practices

Sanjay Kumar Pandey*

Department of Botany, D.A.V.P.G. College, Buxipur, Gorakhpur - 273001, India

Water is a finite and essential resource, critical for human survival and agricultural production. Water conservation and sustainable agricultural practices are two critical components in addressing the growing global challenges of water scarcity and food security. This chapter explores the intricate relationship between water conservation and sustainable agriculture, highlighting the need for a holistic approach to address agricultural challenges. Water scarcity has become a pressing issue worldwide, exacerbated by climate change and unsustainable water management practices. Agriculture accounts for a significant share of global water usage, making it imperative to adopt sustainable practices that optimize water resources. Sustainable agriculture seeks to strike a balance between crop production and environmental preservation. Eco-friendly techniques deliver water directly to plant roots, minimizing evaporation and runoff compared to traditional flooding methods. This not only conserves water but also enhances crop yield and quality. Precision farming employs technology like sensors and data analysis to optimize water and nutrient application, resulting in more efficient resource utilization. Furthermore, the conservation of water bodies such as rivers, lakes, and aquifers are crucial for long-term sustainability. Sustainable agricultural practices can mitigate these risks. Crop residues and organic matter left in the fields improve soil structure, allowing for better water retention and reducing the risk of soil erosion. It also encourages the development of healthier, more water-efficient soils that can better retain moisture, reducing the reliance on irrigation. This, in turn, contributes to improved groundwater recharge, bolstering the overall water cycle. This symbiotic relationship holds the potential to address the pressing challenges of water scarcity and food security, ensuring a more sustainable and resilient future for agriculture and our planet.

*Corresponding author: drskp27@gmail.com

1. Introduction

Water is a natural resource which underpins survival on the planet and is an indispensable part of agriculture. Agriculture, in turn, is the primary source of food, fibre and livelihoods for billions of people worldwide. However, the unsustainable use of water in agriculture poses a severe threat to water resources and the environment. Water is the life blood of agriculture. It is a fundamental resource required for crop production, livestock farming, and food processing. Agriculture, besides being the largest consumer of fresh water globally, is also a significant contributor to water pollution and depletion. With the soaring population growth, estimated to exceed the 9.7 billion mark by 2050 the demand for food and, consequently, water for agriculture is expected to increase substantially (Kumar, 2022). This chapter explores the intricate relationship between water conservation and sustainable agricultural practices, examining their significance, challenges, and potential solutions.

Sustainable agricultural practices are methods and approaches used in farming that aim to minimize environmental impact, preserve natural resources, support local communities, and ensure the long-term viability of agriculture (Fig. 1). These practices are essential for addressing the challenges of climate change, food security, and the conservation of biodiversity. The primary objective of this chapter is to provide an in-depth analysis of water conservation vis-à-vis sustainable agricultural practices. It aims to provide a comprehensive overview of the challenges posed by water scarcity, the principles of sustainable agriculture, the role of innovative technology and the importance of policy and governance in ensuring the responsible use of water in agriculture. It seeks to elucidate the importance of water in agriculture, the principles of sustainable farming, various strategies for water conservation in agriculture, challenges encountered, and

Figure 1: Various sustainable agricultural practices.

the role of innovations, policies, and education in addressing these issues. This essay focuses on the global perspective of water conservation and sustainable agriculture, providing a comprehensive overview. It also includes case studies and examples from different regions to highlight the diversity of challenges and solutions.

2. Water in Agriculture

Water is a critical resource in agricultural practices, as it is essential for crop growth, livestock maintenance, and overall farm operations. Its proper management is crucial for the sustainability of agriculture and food production. Agriculture consumes the largest share of the world's freshwater resources, accounting for approximately 70% of total water use (Ingraoa et al., 2023). Without an adequate and reliable water supply, agricultural systems would collapse, leading to food scarcity, economic instability, and social unrest. Water scarcity, a growing global concern, arises from the imbalance between water availability and demand. Approximately 1.9 billion people live in areas with physical water scarcity, while 2.1 billion people experience economic water scarcity (UNESCO, 2021). In agriculture, water scarcity leads to reduced crop yields, increased production costs, and competition for limited water resources. This exacerbates food insecurity and environmental degradation. The demand for water in agriculture is projected to increase due to population growth, changing diets, and urbanization. To meet this demand, sustainable water management in agriculture is imperative. It is a fundamental resource in agriculture, and its management plays a pivotal role in food production and the sustainability of farming practices. Balancing the needs of agriculture with the conservation of water resources and the environment is a complex and essential task for modern agriculture.

3. Sustainable Agricultural Practices

Sustainable agriculture is an approach that seeks to balance agricultural production with environmental conservation and social well-being. It is guided by several key principles, including environmental stewardship, economic viability, social equity, resource efficiency and biodiversity conservation. Sustainable agricultural practices are vital to ensuring the long-term health and productivity of our planet (Brodt et al., 2011). As the global population continues to grow and climate change threatens our ecosystems, it is imperative that we rethink our approach to farming. Sustainable agriculture provides a blueprint for meeting current and future food demands while safeguarding the environment, conserving resources, and promoting economic stability. Sustainable agriculture minimizes the negative impacts like soil erosion and different pollution along with efficient use of natural resources. Maintaining and enhancing the local biodiversity and ensuring profitability for farmers and rural economy is an important component of the above practices. Sustainable agriculture is a holistic slant to farming that aims to meet current agricultural needs without compromising the capacity of future cohorts to meet their individual needs. It seeks to balance the commercial, public and environmental magnitudes of farming. Sustainable agriculture recognizes the finite nature of resources, including water, and aims to optimize their use while minimizing negative impacts. Sustainable practices can lead to healthier and more nutritious diverse food products which are more economical too. Few techniques of sustainable farming are no-till farming, organic farming, agroforestry and integrated pest management (IPM). The benefits of sustainable agriculture are:

Environmental Conservation: Sustainable practices minimize soil erosion, reduce chemical runoff, and promote biodiversity. This not only protects ecosystems but also ensures fertile land for future generations.

Food Security: By focusing on long-term sustainability, sustainable agriculture helps ensure a consistent food supply, even in the face of environmental challenges.

Economic Stability: Sustainable farming can lead to more stable incomes for farmers. It reduces the risks associated with unpredictable weather and market fluctuations.

Energy Efficiency: Sustainable agriculture often incorporates energy-saving technologies and practices, lowering the overall carbon footprint of the food manufacturing process.

Healthy Food Production: Reducing the use of synthetic chemicals in agriculture results in healthier, more nutritious food products.

Sustainable agricultural practices are not merely an option; they are a necessity for the health of our planet and future generations (Fig 2). These

Figure 2: Impacts of sustainable agriculture on environment and society (after Zhang et al., 2021).

practices embody principles of environmental stewardship, economic viability, social responsibility, and climate resilience. The benefits are multifold, ranging from environmental conservation to food security and economic stability (Zang et al., 2021).

4. Water Conservation Strategies in Agriculture

Efficient irrigation is crucial for water conservation in agriculture. Regular flood irrigation methods often result in noteworthy wastage of water. Recent irrigation techniques, such as sprinkler and drip systems, deliver water directly to the root zone of plants, reducing losses due to evaporation and runoff. Applying mulch to the soil's surface helps reduce water evaporation and maintain soil moisture. Reduced or no-till farming practices minimize soil disturbance, which helps retain moisture and reduce runoff. This can improve water retention in the soil. Organic mulch, such as straw or wood chips, is particularly effective. Rainwater harvesting involves capturing and storing rainwater for agricultural use. This can be as simple as collecting rain in barrels or as sophisticated as constructing large reservoirs. Rainwater harvesting helps supplement irrigation needs, especially in regions with erratic rainfall patterns. Soil conservation practices, such as contour farming and terracing, help retain soil moisture and prevent erosion. Healthy soils can hold more water, reducing the need for additional irrigation. Crop selection and rotation choosing drought-resistant crops and implementing crop rotation strategies can optimize water use in agriculture. Rotating crops with different water requirements can help maintain soil fertility and reduce water demand. Agricultural timing is crucial. Watering during the cooler parts of the day reduces evaporation. Monitoring weather conditions and adjusting irrigation schedules accordingly is important. Organic farming, with its emphasis on natural fertilizers and soil health, can reduce water pollution and improve water quality. Healthy soils in organic systems also require less water for irrigation.

5. Challenges to Water Conservation in Agriculture

Many agricultural regions rely heavily on groundwater for irrigation. Over-pumping of groundwater resources, often driven by unsustainable agricultural practices, leads to aquifer depletion and land subsidence. This exacerbates water scarcity and poses long-term risks. Agricultural runoff, laden with pesticides, fertilizers, and sediment, contributes to water pollution. Contaminated water bodies harm aquatic ecosystems and pose health risks to humans. Sustainable agricultural practices are essential to mitigate this pollution. Climate change disrupts traditional rainfall patterns and increases the frequency and severity of droughts and floods. These climate-related challenges affect water availability for agriculture and require adaptive strategies. Farmers often face economic constraints that make it difficult to invest in sustainable practices. Additionally, inadequate policies and regulations may not incentivize water-efficient agriculture. Policymakers must align incentives with sustainable practices to drive change.

Case Studies

Israel: This is a country with limited freshwater resources, and has become a global leader in water-efficient agriculture. Through innovations like drip irrigation, desalination, and wastewater recycling, Israel has achieved remarkable water-use efficiency in agriculture, setting an example for water-scarce regions.

India: India faces a dual challenge of water scarcity and the need to feed a growing population. Initiatives such as watershed management, community-based water management, and the promotion of drought-resistant crop varieties are helping India address its water-agriculture conundrum.

United States: The United States, a major agricultural producer, faces sustainability challenges due to varied climates and farming practices. While some regions employ sustainable practices like no-till farming and precision agriculture, others struggle with issues like excessive water use and soil degradation.

Rice Farming in Thailand: Rice is a water-intensive crop, and Thailand is one of the world's leading rice producers. To address water conservation in rice cultivation, Thai farmers have adopted the following innovative practices.

a. *System of Rice Intensification (SRI):* SRI is a method that optimizes rice cultivation with reduced water use. Farmers plant young seedlings singly instead of in clumps, maintain a shallow water layer in the field rather than flooding it, and use organic matter to improve soil health. This method reduces water usage by up to 50% compared to traditional rice farming (Uphoff, 2007).
b. *Alternate Wetting and Drying (AWD):* AWD is another water-saving technique used in rice cultivation. It involves intermittently allowing the fields to dry out between irrigation cycles, rather than continuously flooding the paddies. This practice can reduce water use by 15-30% (Sriphirom et al., 2019).
c. *Laser Land Leveling:* Precision land leveling using laser technology ensures an even distribution of water in the fields, preventing waterlogging in some areas while ensuring adequate moisture in others.
d. *Efficient Water Management:* Many Thai farmers are educated in efficient water management practices, including using water meters to monitor usage, creating shared water storage and distribution systems, and adopting eco-friendly pest control methods to prevent pollution of water sources.

These case studies demonstrate that sustainable agriculture practices, such as efficient irrigation, soil health management, and innovative farming techniques, can significantly reduce water usage and contribute to water conservation in agriculture, even in regions facing water scarcity challenges.

6. Innovations, Technologies and Policy

Farming drip irrigation systems deliver water directly to plant roots, reducing water wastage. Precision farming utilizes data and technology to optimize planting, irrigation, and fertilization, minimizing resource use. Agricultural desalination technologies are being explored to convert seawater into freshwater for agricultural use. While energy-intensive, desalination can be a valuable water source in arid regions, AI (Artificial Intelligence) and data analytics enable farmers to make data-informed decisions about resource allocation, crop management, and irrigation, enhancing efficiency and sustainability.

Effective water management policies must balance the needs of agriculture with environmental preservation. This includes regulating groundwater extraction, promoting efficient irrigation, and addressing water pollution. Government subsidies and incentives can encourage farmers to adopt sustainable practices. Redirecting subsidies away from resource-intensive crops and practices can drive positive change. International agreements, such as the United Nations Sustainable Development Goals, emphasize the importance of sustainable agriculture and water conservation. Collaboration among nations is vital to address global water challenges.

7. The Role of Education and Awareness

Education and awareness campaigns can promote sustainable farming techniques among farmers and stakeholders. Training programs can teach the principles of sustainable agriculture and water conservation. Technical training for farmers in water-efficient practices and technologies can accelerate adoption. Farmers' knowledge and skills are pivotal in implementing sustainable practices. Raising public awareness about the importance of water conservation in agriculture can garner support for policy changes and sustainable farming initiatives. The public can play a role in advocating for sustainable practices. The role of education and awareness in sustainable agriculture is crucial for addressing the environmental, social, and economic challenges associated with modern farming practices. Education helps farmers and agricultural practitioners acquire the knowledge and skills necessary to adopt sustainable farming practices. Educational institutions and research organizations play a critical role in advancing sustainable agriculture through research and development. These institutions conduct experiments, collect data, and disseminate findings that can lead to the development of new, sustainable farming techniques and technologies. Education and awareness also empower individuals and organizations to advocate for policies that support sustainable agriculture. By understanding the environmental and social impacts of various farming practices, people can engage with policymakers to promote regulations that incentivize sustainability. With changing environmental conditions and climate variability, education can help farmers adapt to new challenges. Awareness of climate-smart agricultural practices and the ability to implement them can improve the resilience of farming systems. Education can highlight the economic benefits of sustainable agriculture. This includes reduced input costs, enhanced crop yields, and access to premium markets for sustainably produced goods. Farmers who understand these economic advantages are more likely to embrace sustainable practices. Encouraging young people to pursue careers in agriculture with a focus on sustainability is essential for the sector's long-term viability. Education can inspire the next generation to become environmentally conscious farmers and agricultural scientists (Fig. 3).

Figure 3: Essential elements for sustainable agriculture under education and awareness.

8. Future Prospects and Recommendations

Efforts should be made to integrate sustainable agricultural practices and water conservation strategies into a holistic approach. This can include designing farming systems that maximize resource use efficiency and minimize environmental impacts. Investments in research and development are essential to develop new technologies and practices that further enhance water conservation in agriculture. Collaboration between scientists, farmers, and policymakers is crucial in this regard. Strengthening water conservation and sustainable agriculture requires international cooperation. Countries should share knowledge, best practices, and technology to collectively tackle the global water crisis.

Water conservation in agriculture is of critical importance for the sustainable future of food production. As the world's population continues to grow and water resources become increasingly scarce, it's essential to develop and implement water conservation models and practices in agriculture. Use of advanced technologies like drones and satellite imagery to monitor the moisture level and real time crop health help farmers optimize irrigation practices and reduce water waste. By adopting a combination of practices like crop rotation, training on water-efficient agricultural practices, government incentives for implementing water-saving technologies, regular monitoring can significantly reduce water usage while even increasing crop yields (Fig. 4). This will not only ensure a sustainable future for agriculture but also help address global water scarcity challenges.

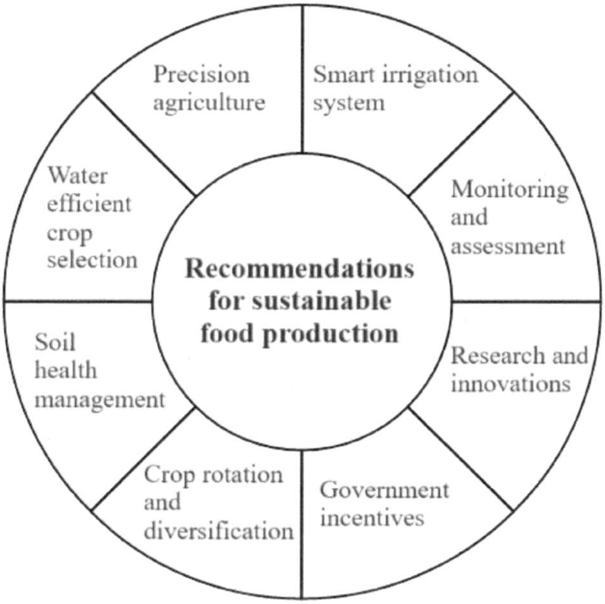

Figure 4: Recommendations for sustainable food production.

9. Conclusion

The aim of this essay is to shed light on the critical link between water conservation and sustainable agriculture, providing a foundation for informed decision-making and policy formulation in agriculture and water resource management. Reduced chemical usage, improved soil management, and the cultivation of drought-resistant crops help protect water quality and quantity. Organic farming avoids the use of synthetic chemicals, reducing water pollution and the subsequent need for water treatment. Incorporating agroforestry into farming systems can provide additional benefits in terms of water conservation. Policymakers, farmers, and researchers must collaborate to promote the adoption of sustainable agriculture as a key strategy in securing both water resources and food production for future generations.

References

Brodt S, Six J, Feenstra G, Ingels C & Campbell D. 2011. Sustainable agriculture. *Nature Education Knowledge,* **3(10):** 1.

Ingraoa C, Strippolia R, Lagioiaa G & Huisingh D. 2023. Water scarcity in agriculture: An overview of causes, impacts and approaches for reducing the risks. *Heliyon,* **9:** e18507, doi.org/10.1016/j.heliyon.2023.e18507.

Kumar L, Chhogyel N, Gopalakrishnan T, Hasan K, Jayasinghe SL, Kariyawasam CS, et al. 2022. Climate change and future of agri-food production. pp. 49–79. *In:* Bhat R (ed.), Future Foods: Global Trends, Opportunities, and Sustainability Challenges, Academic Press, Cambridge, USA.

Sriphirom P, Chidthaisong A & Towprayoon S. 2019. Effect of alternate wetting and drying water management on rice cultivation with low emissions and low water used during wet and dry season. *Journal of Cleaner Production,* **223:** 980–988.

UNESCO. 2021. The United Nations World Water Development Report. Published in 2021 by the United Nations Educational, Scientific and Cultural Organization, Paris, pp. 208. ISBN 978-92-3-100434-6.

Uphoff N. 2007. Reducing the vulnerability of rural households through agroecological practice: Considering the system of rice intensification (SRI). *Mondes en Développement n°,* **140(4):** 85–100. DOI:10.3917/med.140.0085.

Zhang X, Yao G, Vishwakarma S, Dalin C, Komarek MA, Kanter DR, et al. 2021. Quantitative assessment of agricultural sustainability reveals divergent priorities among nations. *One Earth,* **4:** 1262–1277.

12 | Updates on PGPR Potential on Plant Water Stress Management: A Global Perspective

Becky Nancy Aloo[1*], Javan Omondi Were[2], Vishal Tripathi[3], Lucy Wanjohi[4], Clabe Wekesa[5] and John Muoma[6]

[1] Department of Biological Sciences, University of Eldoret, Eldoret, Kenya
[2] Department of Seed, Crop & Horticultural Sciences, University of Eldoret, Eldoret, Kenya
[3] Department of Biotechnology, Graphic Era (Deemed to be University), Dehradun Uttarakhand, India
[4] Department of Environmental Sciences, University of Eldoret, Eldoret, Kenya
[5] Max Planck Institute for Chemical Ecology, Hans-Knöll-Straße 8, 07745 Jena, Germany
[6] Department of Biological Sciences, Masinde Muliro University of Science and Technology, Kakamega, Kenya

Water stress is among the many challenges caused by global climate change that brings forth a formidable problem to agriculture and global food security. Thus, there have been several attempts by researchers from all over the world to explore Plant Growth-Promoting Rhizobacteria (PGPR) as potential plant water stress alleviators. Nevertheless, there is a need to update our comprehension of PGPR potential for plant water stress management. This chapter explores the growing significance of PGPR for mitigating plant water stress from a global perspective. The chapter begins by elucidating the PGPR mechanisms for plant tolerance to water stress and delves into the diverse strategies employed by these beneficial rhizobacteria. Case studies from different regions and recent research findings presented in this chapter highlight the latest developments in PGPR-based interventions for plant water stress management. Furthermore, the challenges and prospects in the application of PGPR on a global scale are elucidated. Considerations such as regulatory frameworks, scalability, and the integration of PGPR into existing agricultural systems are examined. This chapter

*Corresponding author: baloo@uoeld.ac.ke

serves as a valuable resource for researchers, policymakers, and practitioners seeking insights into the dynamic field of PGPR and its pivotal role in addressing the global water stress crisis. It can thus contribute to the ongoing efforts to secure food production, conserve water resources, and promote sustainable agriculture.

1. Introduction

The global population is continually increasing and intensifying the need for increased agricultural production. Central to this challenge is water scarcity and water stress, which have detrimental consequences on crop productivity and food security (Figure 1). Plant water stress is the result of an insufficient supply of water relative to the atmospheric demand or plant needs (Kim et al., 2020). It manifests as a condition in which plants experience water deficits, affecting their physiological processes and overall health. Since almost 95% of plant biomass is made up of water (Seleiman et al., 2021), the element is indispensable for different processes (Rafique, 2023). Water deficiency affects several plant functions, including yield, morphology, and physiology (Amiri et al., 2018; Ghodke et al., 2018; Nemeskéri & Helyes, 2019; Sallam et al., 2019). Among the factors contributing to water stress are irregular rainfall patterns, prolonged droughts, limited irrigation, and suboptimal water management.

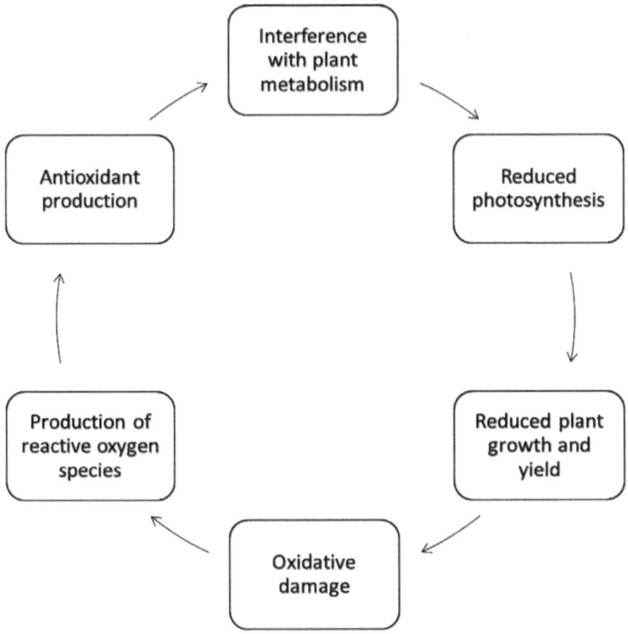

Figure 1: Impacts of water stress on crop productivity.

Nevertheless, plant water stress is now becoming a critical global issue fueled by climate change. The Food and Agriculture Organization of the United States (FAO) (2018) proposes that water stress causes agricultural losses to the tune of a billion dollars owing to detrimental effects on crop productivity. Thus, effective management of plant water stress is one of the ways through which agricultural productivity can be realized and sustained to feed the growing masses.

Plant growth-promoting rhizobacteria (PGPR) represent a subset of plant-root-inhabiting bacteria that have garnered increasing attention for their multifaceted roles in promoting plant development and controlling their invasion by pathogens (Agami et al., 2016; Ahad et al., 2014; Akintokun et al., 2019). The PGPR create symbiotic relationships with plants to enhance nutrient uptake (Ahmad et al., 2019; Kafle et al., 2018), root development (Patten & Glick, 2002; Shtark et al., 2021), and plant tolerance to several environmental stressors, like drought and water scarcity (Agami et al., 2016; Batool et al., 2020; Zhang et al., 2020). This remarkable coexistence between plants and PGPR has captured the interest of many researchers globally in search of sustainable avenues for improving crop resistance in the face of climatic change. Among their most sought-after functions is their ability to enhance plant resistance to water stress (Agami et al., 2016; Chandra et al., 2019; Mutumba et al., 2018).

An increasing amount of research suggests that PGPR can assist crops to withstand or lessen the effects of water stress in soil (see Section 2). The diverse mechanisms of PGPR, as detailed in Section 3, are an avenue for mitigating the detrimental impacts of water stress on crops. Nevertheless, there is a need to update our knowledge of PGPR potential on plant water stress management. In this chapter, we embark on a journey to unravel the latest advancements in PGPR research and application in managing plant water stress. The exploration goes beyond geographical boundaries, offering a comprehensive, global perspective on PGPR's efficacy in plant water stress management. The chapter delves into case studies and research findings that highlight the versatility and applicability of PGPR in plant water stress management in different countries while assessing their mechanisms and successful field trials and stories concerning their use in plant water stress management. The chapter also articulates the challenges and future directions concerning the limitations and constraints in PGPR utilization, strategies to overcome the existing barriers to their use, and areas for future research in this regard. The information contained in this chapter can reshape approaches to plant water stress management for sustainable agricultural production amid climate change.

2. Updates on PGPR Species with Water Stress Management Potential

Of late, the study of PGPR and their potential in mitigating plant water stress has gathered momentum. Researchers worldwide have attempted to identify and characterize specific PGPR species that exhibit remarkable abilities at

Table 1: Plant water stress management potential by PGPR

Rhizobacterial species	**Host/Test plant**	**Test conditions**	**Country of study**	**Reference**
Bacillus subtilis	Potato (*Solanum tuberosum*)	Controlled	Pakistan	(Batool et al., 2020)
Bacillus sp, *Pseudomonas* sp.	*Triticum aestivum*	Greenhouse	Chile	(Mutumba et al., 2018)
Bacillus sp.	Maize and common bean	Growth chamber	Brazil	(de Lima et al., 2019)
Cupriavidus necator sp. *P. fluorescens*	Maize (*Zea mays*)	Greenhouse Pot experiment	Portugal	(Banes et al., 2012)
B. subtilis	Wheat	Field test	India	(Sood et al., 2020)
Pseudomonas spp., *Bacillus* spp.	*Sorghum bicolor*	Greenhouse	South Africa	(Carlson et al., 2020)
Bacillus spp.	Sugarcane (*Saccharum* spp.)	Greenhouse	Not mentioned	(Chandra et al., 2018)
Pseudomonas sp, *Azotobacter* sp.	Lemongrass (*Cymbopogon citratus*)	Greenhouse	Iran	(Mirzaei et al., 2020)
P. lini, Serratia plymuthica	Jujube (*Ziziphus jujuba*)	Laboratory	China	(Zhang et al., 2020)
Rhizobium leguminosarum, P. putida	*Vicia faba*	Field experiment	Egypt	(Mansour et al., 2021)
Providencia rettgeri	Barley (*Hordeum vulgare* L.)	Not mentioned	Morocco	(Ferioun et al., 2023)
P. libanensis	Different cereals	Pot assays-lab and greenhouse	India	(Kour, Rana, Sheikh, et al., 2020)
Not mentioned	Rice (*Oryza sativa*)	Pot experiments	India	(Joshi et al., 2020)

enhancing plant water stress tolerance. This section provides an overview of the latest updates on PGPR species known for their water stress management potential, shedding light on their global applicability. Specific PGPR species with important roles in plant water stress alleviation are discussed in subsections 2.1 to 2.3. Similarly, studies that illustrate the potential of these species in plant water stress management are provided in Table 1.

2.1 *Bacillus subtilis*

Bacillus subtilis, a well-studied PGPR species, has emerged as a versatile ally in managing plant water stress (de Lima et al., 2019; Mutumba et al., 2018). The functions of this microbe in promoting plant development, nutrient absorption, and stress tolerance are well-documented (Table 1). Recent research has also illuminated the mechanisms through which *B. subtilis* enhances water stress resilience, such as by the synthesis of osmoprotectants, modulation of plant hormone levels, and improved root system development.

Bacillus subtilis has garnered significant recognition compared to other PGPR due to its versatile catabolic abilities and its capability to establish root colonization, and synthesis of various enzymes and metabolites which are important in enhancing plant growth, particularly in the face of abiotic and biotic stresses (Aloo et al., 2019; Gagné-Bourque et al., 2016). The adaptability of *B. subtilis* to various soil types and climatic conditions (Jayakumar et al., 2021) also underscores its global relevance in water stress management. Analysis of the microbe's complete genome sequences can bring new dimensions to the potential of this microbe in plant water stress management.

2.2 *Pseudomonas fluorescens*

Incidences of the potential of *P. fluorescens* in promoting plant growth under water stress tolerance are well-documented (Table 1). Like *B. subtilis*, *P. fluorescens* has earned recognition for its ability to enhance plant water stress tolerance through the production of stress-responsive compounds (Banes et al., 2012; Kour et al., 2020; Mutumba et al., 2018). A recent examination of the whole genome sequence of *P. fluorescens* from soybean (*Glycine max*) in Korea demonstrated the genomic and proteomic potential of the microbe at synthesizing stress-alleviating enzymes and metabolites (Nishu et al., 2020). The bacteria were identified as possessing 22 genes associated with antioxidant production as well as key gene clusters involved in efficient osmolyte biosynthesis, both of which are important in promoting the resistance of plants to drought stress.

Fluorescent Pseudomonads can also produce hormones like acetic acid and ACC deaminase, with important functions in plant water stress alleviation (Chandra et al., 2018). Moreover, this PGPR species exhibits an intriguing capacity to prompt plant systemic resistance to stressful conditions, hence enhancing their ability to withstand water stress in different countries. According to Pereira et al. (2020), *P. fluorescens* can promote efficient nutrient use and growth of maize under water deficit.

2.3 Other Bacterial Species

Apart from *P. fluorescens* and *B. subtilis*, lots of other rhizobacterial species, such as *Azospirillum* and *Enterobacter* spp., have been implicated in the alleviation of plant water stress. Although *Azospirillum* species have long been valued for their nitrogen-fixing capabilities, recent research has illuminated their role as valuable partners in water-limited environments (Marques et al., 2023). *Azospirillum* spp. promote water stress management by improving root structure, facilitating nutrient acquisition, and producing compounds that enhance plant stress tolerance. Their adaptability to arid and semi-arid regions positions them as promising candidates for sustainable agriculture in water-scarce areas.

Enterobacter species have emerged as promising players in the realm of PGPR with water stress management potential. Recent investigations have unveiled their role in enhancing drought resistance by modulating stress-responsive genes and producing exopolysaccharides that improve soil structure (Naveed et al., 2014). *Enterobacter* spp. offer exciting possibilities for harnessing the genetic diversity of these bacteria to tailor solutions for specific crops and environmental conditions, especially since they are associated with most plants as microbiomes.

Many rhizobacteria can assist plants in dealing with water stress. However, it is essential to note that different PGPR taxa have varying traits and, thus different degrees of drought-resistance implications on plants (Zhao et al., 2023).

3. Mechanisms of PGPR in Plant Water Stress Management

PGPR employs a range of intricate mechanisms to assist plants in managing water stress, thereby enhancing their resilience to drought and water deficit. A schematic illustration of the PGPR mechanisms in plant water stress resistance is provided in Figure 2. Understanding these mechanisms is important for harnessing the full potential of PGPR for agricultural sustainability. This section explores the mechanisms of PGPR that contribute to water stress management through different mechanisms.

Raheem et al. (2018) propose that some PGPR can facilitate plant water stress management by improving root development for water uptake from the soil, especially through the production of hormones. Studies have established that PGPR can lessen water stress in plants by enhancing root length and density and improving the capacity of plants to reach further into the soil for water. This phenomenon has recently been demonstrated in maize (*Zea mays*) inoculated with *Cupriavidus necator* and *P. fluorescens* under water deficit (Pereira et al., 2020). A very recent meta-analysis of the physiological, biochemical, and morphological PGPR-inoculated plant responses under water stress also reported increased length and dry weight of roots in inoculated plants as opposed to non-inoculated ones (Zhao et al., 2023).

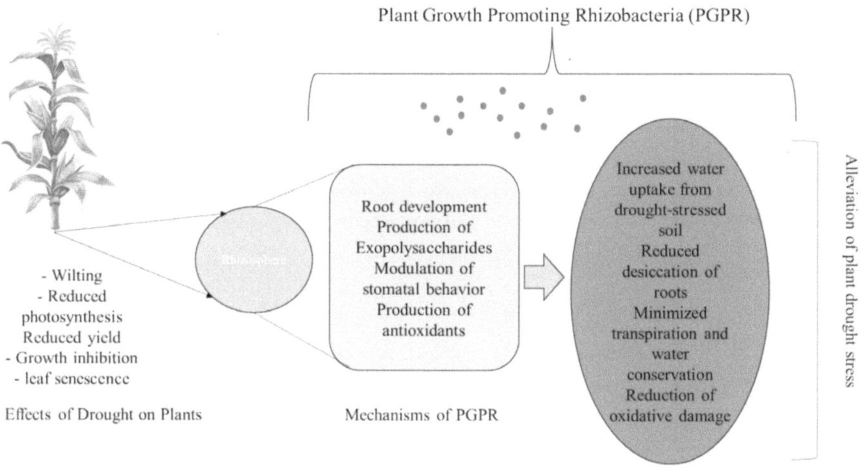

Figure 2: A schematic illustration of the PGPR mechanisms in plant water stress management.

Some PGPR can secret exopolysaccharides (EPS) that enhance soil structure and increase water retention in the root zone (Morcillo & Manzanera, 2021). Exopolysaccharides are natural microbial polysaccharides that assist plants with surviving unfavorable environments such as water scarcity by retaining water (Ahluwalia et al., 2021). The effects of bacterial EPS on plant abiotic stress mitigation are well articulated by Morcillo and Manzanera (2021). In general, EPS generated by PGPRs prevent desiccation through the hydrophilic biofilms they build on plant root surfaces that regulate water discharge. There have also been reports of EPS-producing bacteria helping plants to alleviate drought stress (Getahun et al., 2020; Ilyas et al., 2020; Latif et al., 2022; Shankar & Prasad, 2023).

Similarly, PGPR can modulate stomatal behavior to reduce water loss during water stress. Since stomata are small pores on leaves that control gas exchange and water loss in plants, through the production of certain compounds, PGPR can induce stomatal closure, minimizing transpiration rates and conserving water within the plant. While investigating the physiological and genetic modifications induced in tomatoes by PGPR under moderate water stress, Lucas et al. (2023) established that PGPR-produced ABA regulated stomatal closure. This strategic control of stomatal conductance helps plants maintain hydration under water-deficient conditions. Water stress often leads to osmotic imbalances within plant cells, causing cell shrinkage and impaired cellular functions; PGPR contributes to osmotic adjustment by producing osmoprotectants like proline and glycine betaine that sustain cellular turgor pressure and prevent cell desiccation. These osmoprotectants act as "osmotic buffers," enabling plants to withstand water stress better.

During water stress, plants experience oxidative stress due to the extreme formation of reactive oxygen species (ROS) like O_2^-, OH^-, H_2O_2, and O_2 (Hussain et al., 2019). Plant-associated bacteria can also alleviate the impacts of water deficit by decreasing ROS-induced oxidative damage by improving antioxidant enzyme activity. Regulation of the antioxidant defense system is important in detoxifying reactive ROS, thereby mitigating oxidative harm to plant cells. Consequently, plants remain healthier and more capable of coping with water-stress-induced oxidative stress. PGPR can also stimulate the synthesis of stress-responsive genes in plants, as recently established by Manjunatha et al. (2022) for *Cronobacter deblinensis* in pearl millet. This includes genes involved in stress signaling, transcription factors that regulate stress-related pathways, and genes responsible for synthesizing stress-related proteins. The upregulation of these genes improves plant adaptation to and survival of water stress conditions. Table 2 provides a summary of the different PGPR mechanisms in plant defense management.

Table 2: PGPR mechanisms in plant water stress management

No.	Mechanism	Reason	Reference
1	Enhanced root growth and water retention	Enhancing water uptake and absorption	(Carlson et al., 2020; Morcillo & Manzanera, 2021; Raheem et al., 2018)
2	Modulating stomatal behavior	Hormonal signaling controls stomatal movement	(Jing et al., 2023; Lucas et al., 2023)
3	Promoting Osmotic adjustment	Production of osmolytes	(Carlson et al., 2020)
4	Regulating antioxidant defense system	Production of antioxidants and destruction of ROS	(Batool et al., 2020; Carlson et al., 2020)
5	Formation of exopolysaccharides	Preventing desiccation on the plant root surface, hence, regulating water discharge	(Getahun et al., 2020; Ilyas et al., 2020; Latif et al., 2022; Niu et al., 2018; Shankar & Prasad, 2023)
6	Inducing stress-responsive gene expression	Transcription regulation of water stress	(Manjunatha et al., 2022; Yasmin et al., 2022)

ROS = Reactive Oxygen Species

4. A Global Perspective of Field Trials and Success Stories on PGPR-Mediated Plant Water Stress Management

The exploitation of PGPR can lessen the impacts of water and drought stress. Plant growth-promoting microorganisms like rhizobacteria are environmentally friendly and effective in the mitigation of drought stress in plants. Integration of PGPR as bio-inoculants has proved successful in reducing water stress (Khan et al., 2020). Various studies on rice, maize, guinea grass, peppermint, pulses, and foxtail millet have emphasized the importance of PGPR in minimizing the impacts of water stress (Danish et al., 2020; Fadiji et al., 2022; Saikia et al., 2018). The inoculation of plants with essential rhizobacteria improves their nutrient uptake. Inoculation of *B. thuringiensis* in water-deficient lavender plants increased both micronutrients (Mn^{2+}, Zn^{2+}, and Cu^{2+}) and macronutrients (K^{+}, Ca^{2+}, and Mg^{2+}) in the stem, leading to higher yields. Similarly, an indigenous bacterium increased K^{+} levels in Lavandula and Salvia plants causing reduced stomatal conductance that is necessary for supporting turgor pull in water stress conditions (Fadiji et al., 2022).

Some studies that have shown successful field trials concerning PGPR inoculation of plants for water stress management are provided in Table 3. According to Niu et al. (2018), millet inoculated with EPS-producing *P. fluorescens* increased the adhesion of soil aggregates to root tissue and soil moisture, which promoted plant growth by increasing water stress resistance. EPS-producing bacteria improve drought resistance by altering soil binding and structural characteristics (Ilyas et al., 2020). Kalman et al. (2023) also reported that *P. putida*-inoculated maize improved growth and enhanced water stress resistance. The seedling germination vigor under *P. putida* treatment was 55% higher, relative to the control. The *P. putida* treatment caused the highest ratio of fresh to dry weight relative to the control.

Kour et al. (2020) observed that the inoculation of *P. libanensis* increased plant growth and phosphorous uptake under water stress. In the same year, Jochum et al. (2019) observed that when maize and wheat were inoculated with *Enterobacter* sp., and *Bacillus* sp., drought symptoms were delayed. Coating the seeds with endophytic *B. subtilis* and *Corynebacterium hansenii* improved the dry and fresh biomass, and chlorophyll quantity under water stress (Bodhankar et al., 2019). Namwongsa et al. (2019) inoculated *Helianthus tuberosus* with *Micrococcus luteus* and *B. aquimaris* producing ACC deaminase and IAA in drought conditions and observed increased shoot weight and height, diameter, length, volume, and area of root culminating in an increase in harvest index. Zhang et al. (2020) also reported that ACC deaminase-producing *Serratia plymuthica* and *Pseudomonas lini* mitigated water stress in jujube plants by regulating IAA and ABA. *Enterobacter cloacae* and *Achromobacter xylosoxidans* produce ACC deaminase complexed with biochar and also enhance the yield and growth of maize under water stress (Danish et al., 2020).

Table 3: Field trials and success stories on PGPR-mediated plant water stress management

Microorganism	Country	PGP traits	Plant	Effect on plant parameter	Reference
Bacillus sp. *Enterobacter* sp.	Texas, USA	Indole 3 acetic acid (IAA), Salicylic acid,	Maize (*Zea mays*) and Wheat (*Triticum aestivum*)	Alleviated water stress	(Jochum et al., 2019)
Pseudomonas putida and *Bacillus pumilus*	Hungary	Hormones	Maize (*Zea mays*)	Alleviated water stress Improved germination Improved shoot, root biomass, and water content	(Kálmán et al., 2023)
Paenibacillus, Rhizobium, Bacillus, Azotobacter and *Pseudomonas*	Pakistan	Horomoes	Rice (*Oryza sativa* L.)	Increased harvest index	(Chen et al., 2023)
Bacillus aquimaris	Thailand	IAA, EPS, ACC deaminase	Jerusalem artichoke (*Helianthus tuberosus*)	Increase in leaf area, chlorophyll levels, photosynthesis, and harvest index	(Namwongsa et al., 2019)
Penicillium sp, *Acinetobacter calcoaceticus*	India	IAA, ACC deaminase,	Foxtail millet (*Setaria italica* L.)	Mitigation of effects of water stress by enhancing proline, sugars, and glycine betaine, and reducing lipid peroxidation	(Kour, Rana, Sheikh, et al., 2020)
Pseudomonas lin	China	ACC deaminase	Jujube	Promotes growth and development Reduces drought stress	(Zhang et al., 2020)

Many researchers have emphasized the benefits of plant yield and growth with most of these studies emphasizing plants above ground. In research carried out by Kalman et al. (2023) *Bacillus pumilus* and *P. putida* increased the yield of maize by 15%. These bacteria are recommended for enhancing yield potential and water stress tolerance in plants (Kálmán et al., 2023; Safari, 2019). Rocha et al. (2019) indicated that utilization of *P. putida* to inoculate chickpea with SiO_2 and starch increased yield as a result of increased biomass under drought stress.

5. Challenges and Constraints of PGPR Utilization for Plant Water Stress Management

Varying behavior of PGPRs and the inherent sensitivity to environmental and soil conditions contribute largely to inconsistent performance under different agricultural production systems. In comparison to the widely used synthetic pesticides, PGPRs have a narrow spectrum of biocontrol efficacies towards numerous pests and disease-causing agents which restricts utilization and preferences in medium and large scales of agricultural production. To realize the impacts of this bacteria as a biofertilizer for improved growth and quality in plants, several constraints need to be well understood. These include biosafety, carrier materials, quality control, field utilization, marketing, financial, infrastructural, regulatory, technical, and biological constraints (Basu et al., 2021).

5.1 Biosafety Constraints

Despite being considered as the best alternative for sustainable agriculture on the globe, PGPRs should not elicit negative impacts on the environment and living organisms including humans. However, adherence to the previously set biosafety levels which classify microorganisms into four risk groups according to their modes of transmission, pathogenicity levels to humans, and available treatments remains a major challenge to date (Meena et al., 2020).

5.2 Biofertilizer Carrier Constraints

Since bioinoculant agents possess a short shelf life they require a more suitable carrier material to ease their application when used as biofertilizers. Contrary to this requirement, the unavailability of the appropriate carrier and lack of desired quality proves a major challenge that limits their large-scale adoption in agricultural production systems (Mahajan & Gupta, 2009). Also, an ideal carrier should be cheap and readily available, have high water holding capacity, and organic matter. To realize these qualities, such carriers need to be sterile with zero moisture content, non-toxic, non-polluting, with almost neutral pH (Bashan et al., 2014). Getting such materials that meet the desired qualities and long-term compatibility with the bioinoculant is a major constraint.

5.3 Field Utilization Constraints

The slow response to biofertilizers by plants has been reported to be due to the time taken by the inoculum to accumulate its concentration in the soil and colonize the roots. In return, the acceptability level of farmers remains low. In addition to low preference, the effectiveness of the applied bioinoculants is significantly reduced by the unfavorable conditions caused by the abiotic conditions within the soils, the residual effects of artificial plant inputs, and unpredicted environmental stresses like drought in some parts of the world (Arora et al., 2010). Also, in most soils, the buildup of heavy metals like Hd, Cd, and Cr among others has been reported. As a result, these metals lessen the effectiveness of PGPR fertilizers especially when coupled with a deficit of critical nutrients such as P, Mo, Cu, and Co (Bhardwaj et al., 2014).

5.4 Quality Control, Regulatory, and Biological Constraints

The most important aspect considered by farmers when selecting plant nutrition alternatives is the quality and nutritional components, a trait which seems difficult to ascertain in biofertilizers which are natural products with a very short shelf-life (Wang et al., 2023). Currently, there is no quality check agency or organization for biofertilizers and to confirm their efficacies under multilocational environmental conditions. Hence there is an urgent need to set up quality control standards and institutionalize the organizations charged with the responsibility (Mahajan & Gupta, 2009; Meena et al., 2020). Moreover, the inconsistencies in the efficacy of these bioinoculants have made it very difficult when it comes to product registration and patenting around the globe due to varying effectiveness from one region to the other.

Also, the regulatory process in the registration of bioinoculants remains complex, expensive, and highly variable hence the lack of uniform regulatory policy (Timmusk et al., 2017). Further, the process of isolation, identification, and selection of an appropriate PGPR strain(s) is tedious in itself since the requirement of non-specificity to crops and broad host range is an important limiting factor. Nevertheless, the presence of other microbes that interact both compatibly and incompatibly with the PGPRs affects both efficacy and quality upon application, an interaction that can cause either synergistic or antagonistic effect, depending on the microbial species (Mahajan & Gupta, 2009; Meena et al., 2020).

5.5 Marketing and Financial Constraints

The commercial viability of PGPRs for plant water stress management is reliant on the performance and selection of robust strains, as well as the shelf life, viability of cells, cost-effectiveness, and convenient use, all of which are challenging factors. All over the world, there is a lack of proper transportation and storage facilities for biofertilizers due to the nature of their components and mode of action which largely depend on highly varying efficacy requirements by PGPR. This usually leads to delayed deliveries hence lowered quality and deterioration of biocontrol potentials. Other than these two primary marketing constraints, the insufficient financial need to support large-scale production is a major factor hindering the commercialization of these products for large-scale farm utilization in different agricultural production systems. As a result, low quantities are always produced and this further contributes to inaccessibility by the final consumers (Arora et al., 2010).

5.6 Technical and Infrastructural Constraints

Ensuring prolonged shelf life for the biofertilizers during processing and commercialization poses significant challenges to the industrialization of this product in most parts of the world. Therefore, the key players in the production of biofertilizers usually incur huge losses through recycling when a product is not sold before its expiry date. Since these products contain live microbial cells, the major technicality is experienced in storage and transportation to avoid deterioration. Moreover, the lack of soil-specific microbial strains per region further constrains the use of these products by farmers (Arora et al., 2010; Zandi & Basu, 2016). For quality biofertilizer products to be manufactured, very complex technologies alongside trained and qualified personnel are key and the lack of these cause major infrastructural constraints in this sector (Mahajan & Gupta, 2009).

6. Strategies to Overcome Barriers to Adoption of PGPRs

Overcoming major constraints in the use of PGPRs requires stringent measures to be put in place to maintain both excellent quality and efficacy when applied to plants. Presently, the major safety measures and guidelines required to overcome the barriers when using PGPRs as plant water stress alleviators are purely speculative (Mahajan & Gupta, 2009). Prior soil amendment procedures such as liming are needed especially when the soils are more acidic to create favorable conditions for the selected PGPR strain to colonize the roots and act efficiently on the plants.

For accuracy and precision in the application and to avoid wastage, biofertilizers should be applied as seed dressers where the majority of the recommendations suggest that 200 g can be applied to 10 kg of seeds before planting. For prolonged shelf life of these microbial products, it is recommended that storage before application in fields should be under shady conditions with a temperature ranging between 25 to 28°C. Due to the current specificity to crops, soils, and environmental conditions, there should be proper labeling of the biofertilizer products showing information such as PGPR strain, crop, soil, and environmental conditions at which it is more efficacious.

The inoculum should be applied in larger quantities during application in fields or the excess cultures should be put in furrows in them to enhance interaction with the microbiota in the plant root region (Xia et al., 2023), thus effecting pre-colonization of the plant root for increased efficacy. Field hygiene should be observed when using biofertilizers and for significant results to be realized, each gram of the biofertilizers should contain 10^7 viable cells as the inoculum obtained from a registered producer. Finally, due to the inherent specificity of the PGPR strains, their application and use should strictly adhere to the specified crop(s) indicated in the product label for good results to be realized.

7. Future Research Directions and Emerging Trends

The discovery of the potential of PGPRs in agricultural production and food security presents great opportunities for not only good agricultural practices but also environmental protection around the globe. To overcome the constraints surrounding their utilization as biofertilizers, further research should concentrate on enhancing the shelf life, quality, and alleviation of abiotic soil and environmental factors affecting the efficacy of products, and by-products. At the moment, the amount of research on PGPR gives more focus on enhancing efficacious activities through consortium exposure. Among the consortia in consideration include microbial consortium, rhizobacterial consortium, strategies for designing synthetic rhizobacterial, and selection strategies for consortium components (Singh et al., 2023). Also, with the growing concerns surrounding food safety, quality, and quantity, PGPRs need to be researched to improve the desired results.

Mycorrhizal fungi-PGPR synergy is emerging as a novel frontier. Beyond individual PGPR species, research has also delved into the synergistic interactions between PGPR and mycorrhizal fungi. This emerging frontier explores how the combined actions of these beneficial microorganisms can amplify the plant's ability to combat water stress. Recent studies have unveiled intriguing mechanisms through which mycorrhizal fungi and PGPR work to improve water uptake and nutrient assimilation. As our understanding of PGPR species and their contributions to water stress management continues to evolve, the potential for harnessing these beneficial microorganisms in sustainable agriculture becomes increasingly promising. The updates presented here offer a glimpse into the expanding horizon of PGPR research, and innovative approaches to address one of the most pressing challenges in global agricultural water stress. To overcome the challenges and barriers surrounding the utilization and improved efficacy of PGPRs, several genes identified to control distinct plant growth traits have been assembled for transfer into the rhizobacterial genome. This would minimize the limitations surrounding the requirements such as soil specificity by the PGPRs and improve efficiency towards traits such as N-fixation, phosphate solubilization, auxin biosynthesis, compatibility with other biocontrol agents and rhizoremediation (Haskett et al., 2021).

Some researchers have established that the simultaneous inoculation of PGPR and Arbuscular Mycorrhiza (AM) in plants can enable better alleviation of water stress. For example, the K content in *Trifolium* was enhanced by 2.17 after inoculation with AM and *B. thuringiensis* while it was enhanced by another 3.48 from inoculation with *P. putida* and AM (Ortiz et al., 2015). In addition, lettuce plants treated with *Klebsiella* and *Bacillus* had higher P^+ and K^+ concentrations and more biomass than plants treated with AM or bacteria only (Durán et al., 2016). According to Chen et al. (2023), a mixture of PGPR and AM resulted in better yields and nutritional concentrations. Certainly, such frontiers must be investigated to improve the overall plant water stress management.

8. Conclusions and Recommendations

The exploration of the potential of PGPR in mitigating plant water stress presents a promising avenue for addressing one of the most pressing challenges in agricutlure. This chapter reveals multifaceted mechanisms through which PGPR enhance plant resilience to water stress, and their potential in several parts of the world. The case studies and recent research findings discussed in this chapter underscore the dynamic nature of PGPR-based interventions for managing plant water stress. While the potential of PGPR is undeniable, several challenges have been identified as constraints of PGPR utilization in plant water stress management. Regulatory frameworks must be developed or adapted to ensure the safe and sustainable use of these beneficial rhizobacteria. Scalability and integration into existing agricultural systems are critical considerations for widespread adoption. As we foresee the complex landscape of water stress management due to the changing climatic conditions, PGPRs are emerging as a valuable tool for sustainable agriculture. It is evident that they hold great potential, and should be actively pursued for plant water stress management. This chapter presents the ever-evolving field of PGPR research and its potential to reshape the future of agriculture by ensuring food security, water conservation, and sustainable practices on a global scale. Moving forward, further research and explorations are critical to reveal the full potential of PGPR for plant water stress management.

References

Agami R, Medani R, Abd El-Mola I & Taha, R. 2016. Exogenous application with plant growth promoting rhizobacteria (PGPR) or proline induces stress tolerance in basil plants (*Ocimum basilicum* L.) exposed to water stress. *International Journal of Agricultural and Environmental Research,* **2(5):** 78–92.

Ahad A, Akhtar N, Vasmatkar P & Baral P. 2014. Biochemical characterization and molecular fingerprinting of plant growth promoting rhizobacteria. *International Journal of Agriculture, Environment and Biotechnology*, **7(3):** 491–498.

Ahluwalia O, Singh PC & Bhatia R. 2021. A review on drought stress in plants: Implications, mitigation and the role of plant growth promoting rhizobacteria. *Resources, Environment and Sustainability*, **5:** 100032.

Ahmad M, Adil Z, Hussain A, Mumtaz MZ, Nafees M, Ahmad I, et al. 2019. Potential of phosphate solubilizing Bacillus strains for improving growth and nutrient uptake in mungbean and maize crops. *Pakistan Journal of Agricultural Sciences,* **56(2):** 283–289.

Akintokun AK, Ezaka E, Akintokun PO, Shittu OB & Taiwo LB. 2019. Isolation, screening and response of maize to plant growth promoting. *Scientia Agriculturae Bohemica,* **50(3):** 181–190.

Aloo BN, Makumba BA & Mbega ER. 2019. The potential of Bacilli rhizobacteria for sustainable crop production and environmental sustainability. *Microbiological Research,* **219:** 26–39.

Amiri H, Dousty B & Hosseinzedeh SR. 2018. Water stress-induced changes of morphological, physiological and essential oil compounds in *Thymus eriocalyx* from Iran. *Journal of Essential Oil Bearing Plants,* **21(5):** 1210–1223.

Arora NK, Maheshwari DK & Khare E. 2010. Plant growth promoting rhizobacteria: Constraints in bioformulation, commercialization and future strategies. pp. 97–116. *In:* DK. Maheshwari (ed.), Bacteria and Plant Health, Springer.

Banes C, Huque R & Zipperer M. 2012. Towards the National Community Infrastructure Upgrading Program for Ghana. The World Bank.

Bashan Y, de-Bashan LE, Prabhu S & Hernandez J-P. 2014. Advances in plant growth-promoting bacterial inoculant technology: Formulations and practical perspectives (1998–2013) (A Marshner Review). *Plant and Soil,* **378(1):** 1–33.

Basu A, Prasad P, Das SN, Kalam S, Sayyed R, Reddy M, et al. 2021. Plant growth promoting rhizobacteria (PGPR) as green bioinoculants: Recent developments, constraints, and prospects. *Sustainability,* **13(3):** 1140.

Batool T, Ali S, Seleiman MF, Naveed NH, Ali, A, Ahmed, K, et al. 2020. Plant growth promoting rhizobacteria alleviates drought stress in potato in response to suppressive oxidative stress and antioxidant enzymes activities. *Scientific Reports,* **10(1):** 1–19.

Bhardwaj D, Ansari MW, Sahoo RK & Tuteja N. 2014. Biofertilizers function as key player in sustainable agriculture by improving soil fertility, plant tolerance and crop productivity. *Microbial Cell Factories,* **13(1):** 1–10.

Bodhankar S, Grover M & Reddy G. 2019. In planta screening of maize seed endophytic bacteria for potential applications under dryland conditions. *Indian Journal of Dryland Agricultural Research and Development,* **34(1):** 53–62.

Carlson R, Tugizimana F, Steenkamp PA, Dubery IA, Hassen AI & Labuschagne N. 2020. Rhizobacteria-induced systemic tolerance against drought stress in *Sorghum bicolor* (L.) Moench. *Microbiological Research,* **232:** 126388.

Chandra D, Srivastava R, Gupta VV, Franco CM, Paasricha N, Saifi SK, et al. 2019. Field performance of bacterial inoculants to alleviate water stress effects in wheat (*Triticum aestivum* L.). *Plant and Soil,* **441(1):** 261–281.

Chandra D, Srivastava R & Sharma AK. 2018. Influence of IAA and ACC deaminase producing fluorescent pseudomonads in alleviating drought stress in wheat (*Triticum aestivum*). *Agricultural Research,* **7(3):** 290–299. https://doi.org/10.1007/s40003-018-0305-y

Chandra P, Tripathi P & Chandra A. 2018. Isolation and molecular characterization of plant growth-promoting *Bacillus* spp. and their impact on sugarcane (*Saccharum* spp. hybrids) growth and tolerance towards drought stress. *Acta Physiologiae Plantarum,* **40:** 1–15.

Chen D, Saeed M, Ali MNHA, Raheel M, Ashraf W, Hassan Z, et al. 2023. Plant growth promoting rhizobacteria (PGPR) and arbuscular mycorrhizal fungi combined application reveals enhanced soil fertility and rice production. *Agronomy,* **13(2):** 550.

Danish S, Zafar-ul-Hye M, Mohsin F & Hussain M. 2020. ACC-deaminase producing plant growth promoting rhizobacteria and biochar mitigate adverse effects of drought stress on maize growth. *PLoS One,* **15(4):** e0230615.

de Lima BC, Moro AL, Santos ACP, Bonifacio A, Araujo ASF & de Araujo FF. 2019. *Bacillus subtilis* ameliorates water stress tolerance in maize and common bean. *Journal of Plant Interactions,* **14(1):** 432–439. https://doi.org/10.1080/17429145.2019.1645896

Durán P, Acuña JJ, Armada E, López-Castillo OM, Cornejo P, Mora MD, et al. 2016. Inoculation with selenobacteria and arbuscular mycorrhizal fungi to enhance selenium content in lettuce plants and improve tolerance against drought stress. *Journal of Soil Science and Plant Nutrition,* **16(1):** 211–225.

Fadiji AE, Orozco-Mosqueda M del C, Santos-Villalobos S de los, Santoyo G, et al. 2022. Recent developments in the application of plant growth-promoting drought adaptive rhizobacteria for drought mitigation. *Plants,* **11(22):** 3090.

FAO. 2018. Progress on Level of Water Stress 2018 Global baseline for SDG indicator 6.4.2. Food and Agriculture Organization of the United Nations.

Ferioun M, Srhiouar N, Tirry N, Belahcen D, Siang TC, Louahlia, S, et al. 2023. Optimized drought tolerance in barley (*Hordeum vulgare* L.) using plant growth-promoting rhizobacteria (PGPR). *Biocatalysis and Agricultural Biotechnology,* **50:** 102691.

Gagné-Bourque F, Bertrand A, Claessens A, Aliferis KA & Jabaji S. 2016. Alleviation of drought stress and metabolic changes in timothy (*Phleum pratense* L.) colonized with *Bacillus subtilis* B26. *Frontiers in Plant Science,* **7:** 584.

Getahun A, Muleta D, Assefa F & Kiros S. 2020. Plant growth-promoting rhizobacteria isolated from degraded habitat enhance drought tolerance of acacia (*Acacia abyssinica* Hochst. Ex Benth.) seedlings. *International Journal of Microbiology,* **2020:** 1–13.

Ghodke P, Andhale P, Gijare U, Thangasamy A, Khade Y, Mahajan V, et al. 2018. Physiological and biochemical responses in onion crop to drought stress. *International Journal of Current Microbiology and Applied Sciences,* **7(1):** 2054–2062.

Haskett TL, Tkacz A & Poole PS. 2021. Engineering rhizobacteria for sustainable agriculture. *The ISME Journal,* **15(4):** 949–964. https://doi.org/10.1038/s41396-020-00835-4

Hussain S, Rao MJ, Anjum MA, Ejaz S, Zakir I, Ali MA, et al. 2019. Oxidative stress and antioxidant defense in plants under drought conditions. pp. 207–219. *In:* M Hasanuzzaman, K Hakeem, K Nahar H. Alharby (eds), Plant Abiotic Stress Tolerance, Cham: Springer.

Ilyas N, Mumtaz K, Akhtar N, Yasmin H, Sayyed R, Khan W, et al . 2020. Exopolysaccharides producing bacteria for the amelioration of drought stress in wheat. *Sustainability,* **12(21):** 8876.

Jayakumar A, Nair IC & Radhakrishnan E. 2021. Environmental adaptations of an extremely plant beneficial *Bacillus subtilis* Dcl1 identified through the genomic and metabolomic analysis. *Microbial Ecology,* **81:** 687–702.

Jing D, Liu B, Ma H, Liu F, Liu X & Ren L. 2023. Effects of inoculation with different plant growth-promoting rhizobacteria on the eco-physiological and stomatal characteristics of walnut seedlings under drought stress. *Agronomy,* **13(6):** 1486.

Jochum MD, McWilliams KL, Borrego EJ, Kolomiets MV, Niu G, Pierson EA, et al. 2019. Bioprospecting plant growth-promoting rhizobacteria that mitigate drought stress in grasses. *Frontiers in Microbiology,* **10:** 2106. https://doi.org/10.3389/fmicb.2019.02106

Joshi B, Chaudhary A, Singh H & Kumar PA. 2020. Prospective evaluation of individual and consortia plant growth promoting rhizobacteria for drought stress amelioration in rice (*Oryza sativa* L.). *Plant and Soil,* **457:** 225–240.

Kafle A, Garcia K, Peta V, Yakha J, Soupir A & Bücking H. 2018. Beneficial plant microbe interactions and their effect on nutrient uptake, yield, and stress resistance of soybeans. *In:* Soybean – Biomass, Yield and Productivity. IntechOpen London, UK.

Kálmán CD, Nagy Z, Berényi A, Kiss E & Posta K. 2023. Investigating PGPR bacteria for their competence to protect hybrid maize from the factor drought stress. *Cereal Research Communications,* **52(1):** 129–150.

Khan N, Ali S, Tariq H, Latif S, Yasmin H, Mehmood A, et al. 2020. Water conservation and plant survival strategies of rhizobacteria under drought stress. *Agronomy,* **10(11):** 1683.

Kim Y, Chung YS, Lee E, Tripathi P, Heo S & Kim K-H. 2020. Root response to drought stress in rice (*Oryza sativa* L.). *International Journal of Molecular Sciences,* **21(4):** 1513.

Kour D, Rana KL, Sheikh I, Kumar V, Yadav AN, Dhaliwal HS, et al. 2020. Alleviation of drought stress and plant growth promotion by *Pseudomonas libanensis* EU-LWNA-33, a drought-adaptive phosphorus-solubilizing bacterium. *Proceedings of the National Academy of Sciences, India Section B: Biological Sciences,* **90(4):** 785–795. https://doi.org/10.1007/s40011-019-01151-4

Kour D, Rana KL, Yadav AN, Sheikh I, Kumar V, Dhaliwal HS, et al. 2020. Amelioration of drought stress in Foxtail millet (*Setaria italica* L.) by P-solubilizing drought-tolerant microbes with multifarious plant growth promoting attributes. *Environmental Sustainability,* **3(1):** 23–34. https://doi.org/10.1007/s42398-020-00094-1

Latif M, Bukhari SAH, Alrajhi AA, Alotaibi FS, Ahmad M, Shahzad AN, et al. 2022. Inducing drought tolerance in wheat through exopolysaccharide-producing rhizobacteria. *Agronomy,* **12(5):** 1140.

Lucas JA, Garcia-Villaraco A, Montero-Palmero MB, Montalban B, Ramos Solano B & Gutierrez-Mañero FJ. 2023. Physiological and genetic modifications induced by plant-growth-promoting rhizobacteria (PGPR) in tomato plants under moderate water stress. *Biology,* **12(7):** 901.

Mahajan A & Gupta RD. 2009. Integrated Nutrient Management (INM) in a Sustainable Rice—Wheat Cropping System. Dordrecht: Springer.

Manjunatha BS, Nivetha N, Krishna GK, Elangovan A, Pushkar S, Chandrashekar N, et al. 2022. Plant growth-promoting rhizobacteria *Shewanella putrefaciens* and *Cronobacter dublinensis* enhance drought tolerance of pearl millet by modulating hormones and stress-responsive genes. *Physiologia Plantarum,* **174(2):** e13676. https://doi.org/10.1111/ppl.13676

Mansour E, Mahgoub HA, Mahgoub SA, El-Sobky E-SE, Abdul-Hamid MI, Kamara MM, et al. 2021. Enhancement of drought tolerance in diverse *Vicia faba* cultivars by inoculation with plant growth-promoting rhizobacteria under newly reclaimed soil conditions. *Scientific Reports,* **11(1):** 24142.

Marques DM, Magalhães PC, Marriel IE, Gomes Júnior CC, da Silva AB, da Silva NTA, et al. 2023. *Azospirillum brasilense* reduces the effects of water stress and increases maize yield in irrigated areas with high soil nitrogen doses. *Journal of Plant Growth Regulation,* **42(7):** 4263–4274.

Meena M, Swapnil P, Divyanshu K, Kumar S, Harish, Tripathi YN, et al. 2020. PGPR-mediated induction of systemic resistance and physiochemical alterations in plants against the pathogens: Current perspectives. *Journal of Basic Microbiology,* **60(10):** 828–861.

Mirzaei M, Ladan Moghadam A, Hakimi L & Danaee E. 2020. Plant growth promoting rhizobacteria (PGPR) improve plant growth, antioxidant capacity, and essential oil properties of lemongrass (*Cymbopogon citratus*) under water stress. *Iranian Journal of Plant Physiology,* **10(2):** 3155–3166.

Morcillo RJ & Manzanera M. 2021. The effects of plant-associated bacterial exopolysaccharides on plant abiotic stress tolerance. *Metabolites,* **11(6):** 337.

Mutumba FA, Zagal E, Gerding M, Castillo-Rosales D, Paulino L & Schoebitz M. 2018. Plant growth promoting rhizobacteria for improved water stress tolerance in wheat genotypes. *Journal of Soil Science and Plant Nutrition,* **18(4):** 1080–1096.

Namwongsa J, Jogloy S, Vorasoot N, Boonlue S, Riddech N & Mongkolthanaruk W. 2019. Endophytic bacteria improve root traits, biomass and yield of *Helianthus tuberosus* L. under normal and deficit water conditions. *Environmental Microbiology and Biotechnology,* **29(11):** 1777–1789.

Naveed M, Mitter B, Reichenauer TG, Wieczorek K & Sessitsch A. 2014. Increased drought stress resilience of maize through endophytic colonization by *Burkholderia phytofirmans* PsJN and *Enterobacter* sp. FD17. *Environmental and Experimental Botany,* **97:** 30–39.

Nemeskéri E & Helyes L. 2019. Physiological responses of selected vegetable crop species to water stress. *Agronomy,* **9(8):** 447.

Nishu SD, Hyun HR & Lee TK. 2020. Complete genome sequence of Pseudomonas fluorescens DR397, a drought tolerant plant growth-promoting rhizobacterium. *The Microbiological Society of Korea,* **56(3):** 331–333.

Niu X, Song L, Xiao Y & Ge W. 2018. Drought-tolerant plant growth-promoting rhizobacteria associated with foxtail millet in a semi-arid agroecosystem and their potential in alleviating drought stress. *Frontiers in Microbiology,* **8:** 2580.

Ortiz N, Armada E, Duque E, Roldán A & Azcón R. 2015. Contribution of arbuscular mycorrhizal fungi and/or bacteria to enhancing plant drought tolerance under natural soil conditions: Effectiveness of autochthonous or allochthonous strains. *Journal of Plant Physiology,* **174:** 87–96. https://doi.org/10.1016/j.jplph.2014.08.019

Patten CL & Glick BR. 2002. Role of *Pseudomonas putida* indoleacetic acid in the development of the host plant root system. *Applied and Environmental Microbiology,* **68(8):** 3795–3801.

Pereira S, Abreu D, Moreira H, Vega A & Castro P. 2020. Plant growth-promoting rhizobacteria (PGPR) improve the growth and nutrient use efficiency in maize (*Zea mays* L.) under water deficit conditions. *Heliyon,* **6(10):** e05106.

Rafique S. 2023. Physiological and biochemical responses in maize under drought stress. pp. 117–136. *In:* S.H. Wani, Z.A. Dar and G.P. Singh (eds), Maize Improvement: Current Advances in Yield, Quality, and Stress Tolerance under Changing Climatic Scenarios, Cham: Springer International Publishing.

Raheem A, Shaposhnikov A, Belimov AA, Dodd IC & Ali B. 2018. Auxin production by rhizobacteria was associated with improved yield of wheat (*Triticum aestivum* L.) under drought stress. *Archives of Agronomy and Soil Science,* **64(4):** 574–587.

Rocha I, Ma Y, Vosátka M, Freitas H & Oliveira RS. 2019. Growth and nutrition of cowpea (*Vigna unguiculata*) under water deficit as influenced by microbial inoculation via seed coating. *Journal of Agronomy and Crop Science,* **205(5):** 447–459.

Safari D. 2019. Effect of seed priming with growth promoting bacteria on yield and yield components of corn under drought stress. *Agroecology Journal,* **14(4):** 57–65.

Saikia J, Sarma RK, Dhandia R, Yadav A, Bharali R, Gupta VK, et al. 2018. Alleviation of drought stress in pulse crops with ACC deaminase producing rhizobacteria isolated from acidic soil of Northeast India. *Scientific Reports,* **8(1):** 1–16.

Sallam A, Alqudah AM, Dawood MF, Baenziger PS & Börner A. 2019. Drought stress tolerance in wheat and barley: Advances in physiology, breeding and genetics research. *International Journal of Molecular Sciences,* **20(13):** 3137.

Seleiman MF, Al-Suhaibani N, Ali N, Akmal M, Alotaibi M, Refay Y, et al. 2021. Drought stress impacts on plants and different approaches to alleviate its adverse effects. *Plants,* **10(2):** 259.

Shankar A & Prasad V. 2023. Potential of desiccation-tolerant plant growth-promoting rhizobacteria in growth augmentation of wheat (*Triticum aestivum* L.) under drought stress. *Frontiers in Microbiology,* **14:** 1017167.

Shtark O, Puzanskiy R, Avdeeva G, Yemelyanov V, Shavarda A, Romanyuk D, et al. 2021. Metabolic alterations in Pisum sativum roots during plant growth and arbuscular mycorrhiza development. *Plants,* **10(6):** 1033.

Singh A, Yadav VK, Chundawat RS, Soltane R, Awwad NS, Ibrahium HA, et al. 2023. Enhancing plant growth promoting rhizobacterial activities through consortium exposure: A review. *Frontiers in Bioengineering and Biotechnology,* **11**. https://www.frontiersin.org/articles/10.3389/fbioe.2023.1099999

Sood G, Kaushal R & Sharma M. 2020. Significance of inoculation with Bacillus subtilis to alleviate drought stress in wheat (*Triticum aestivum* L.). *Vegetos*, **33:** 782–792.

Timmusk S, Behers L, Muthoni J, Muraya A & Aronsson A-C. 2017. Perspectives and challenges of microbial application for crop improvement. *Frontiers in Plant Science,* **8:** 49. https://doi.org/10.3389/fpls.2017.00049

Wang D, Wang C, Chen Y & Xie Z. 2023. Developing plant-growth-promoting rhizobacteria: A crucial approach for achieving sustainable agriculture. *Agronomy,* **13(7):** 1835.

Xia F, Hao H, Qi Y, Bai H, Li H, Shi Z, et al. 2023. Effect of salt stress on microbiome structure and diversity in Chamomile (*Matricaria chamomilla* L.) rhizosphere soil. *Agronomy,* **13(6)**. https://doi.org/10.3390/agronomy13061444

Yasmin H, Bano A, Wilson NL, Nosheen A, Naz R, Hassan MN, et al. 2022. Drought-tolerant *Pseudomonas* sp. showed differential expression of stress-responsive genes and induced drought tolerance in *Arabidopsis thaliana. Physiologia Plantarum,* **174(1):** e13497.

Zandi P & Basu SK. 2016. Role of plant growth-promoting rhizobacteria (PGPR) as biofertilizers in stabilizing agricultural ecosystems. pp. 71–87. *In:* D Nandwani (ed.), Organic Farming for Sustainable Agriculture, Cham: Springer International Publishing. https://doi.org/10.1007/978-3-319-26803-3_3

Zhang M, Yang L, Hao R, Bai X, Wang Y & Yu X. 2020. Drought-tolerant plant growth-promoting rhizobacteria isolated from jujube (*Ziziphus jujuba*) and their potential to enhance drought tolerance. *Plant and Soil,* **452:** 423–440.

Zhao X, Yuan X, Xing Y, Dao J, Zhao D, Li Y. et al. 2023. A meta-analysis on morphological, physiological and biochemical responses of plants with PGPR inoculation under drought stress. *Plant, Cell & Environment,* **46(1):** 199–214.

13 | Current Progress Towards Management of Drought Stress by Transgenic Plants

Neeraj Kumar Dubey[1]*, Satyendra Kumar Yadav[2], Ran Vijay Singh[2], Pradeep Kumar Yadav[3], Amarjeet Singh[4] and Jogeswar Panigrahi[5]

[1] Department of Botany, Rashtriya Snatkottar Mahavidyalaya, Jaunpur, India
[2] Department of Zoology, Rashtriya Snatkottar Mahavidyalaya, Jaunpur, India
[3] Department of Botany, Faculty of Science, University of Allahabad, Prayagraj, India
[4] Department of Botany, Post Graduate College, Ghazipur, India
[5] Department of Biosciences and Bioinformatics, Khallikote University, Konishi, India

Being sessile, plants have to face several stresses during their germination to seed setting. Among them, drought is an important stress and plays a significant role in the growth and development of the plant. The timing of drought stress is very important. If it happens during flowering, seed setting, and grain filling, the crop productivity is greatly hampered. Drought-mediated water stress disturbs the normal osmotic potential of the cytoplasm and this leads to the generation of reactive oxygen species (ROS) which damages the plant tissue. Further due to drought, the closing of stomata and entry of carbon dioxide are reduced thus the photosynthesis process is significantly reduced. During drought stress, a plant changes its metabolic, transcriptomic, and proteomics signature to manage the drought-derived loss. Plants manage the drought stress-derived osmotic disturbance by increasing proline, sugars, and antioxidant activities. Human population growth reduces both the agricultural crop area and allocation of water demand for irrigation and thus reduces crop performance in the form of productivity. Further due to anthropogenic activity, global warming-derived climate changes lead to a reduction in rain and the creation of drought conditions. These factors together reduce the crop yields in a synergetic way. To manage drought stress, drought-resistant plants are one of the important tools to enhance crop yield. A drought resistant line can be generated by introducing the desired trait through conventional breeding or transgenic approaches. The

*Corresponding author: neerajd19@gmail.com

present book chapter describes the current advances in drought stress and transgenic approaches to generate transgenic lines against drought stress. These conclusions can be used in crop improvement programs for developing a drought tolerant nature.

1. Introduction

Water is the most abundant material for the growth and normal metabolism of plant cells (Kramer, 1963; Pugnaire et al., 1999). Reduction of water content below a certain point leads to dehydration and damages the membrane integrity and leads to the death of plant cells (Kramer, 1963). However, every plant has to face some extent of water stress at certain points of their life cycle (Morgan, 1984). Different plant cultivars respond in different ways to water stress at different levels (Gerik et al., 1996; Kirda, 2002; Bukhov and Carpentier, 2004; Cherian et al., 2006). Water also plays an important role in plant metabolism and thus also affects plant productivity. The decrease in the availability of water has an immediate effect on plant health by affecting photosynthesis and mineral transport (Pugnaire et al., 1999). Due to the huge population and a rapid increase in the consumption of water to meet the municipal and industrial demand, the allocation of water for agriculture is decreasing (Kirda, 2002). Additionally, limited availability of surface or groundwater calls for innovations in irrigation technology on an urgent basis. This will allow an equal allocation of water for the irrigation of all crops (Kirda, 2002). Different crops have different critical points of drought sensitivity during the plant life cycle and this affects crop yield (Kirda, 2002). Plant growth is directly affected by the amount of water absorption which also impacts the soil-derived water stress (Kramer, 1963). Water stresses reduce the water content, chlorophyll, and dry matter which negates the growth rate and productivity of crop plants (Kaya et al., 2006). Different plants operate multiple biochemical pathways and protect chloroplast activity, maintain ion homeostasis, and detoxify radicals (Bohnert and Jensen, 1996). Water stresses also induce free radicals' formation and photorespiration (Bohnert and Jensen, 1996). Many drought stress-tolerant plant species accumulate different osmoprotectants like prolines, methylated metabolites, and radical scavengers (Bohnert and Jensen, 1996). Plants manage the upcoming water stress through these compounds (Barichello et al., 2017). Drought resistance is a big problem for plants. Plants try to escape or avoid drought stress by early flowering or by tolerating different inherent mechanisms (Levitt, 1980; Meshram et al., 2022). During drought plants increase their root length, reduce transpiration, and also reduce the number of leaves (Levitt, 1980; Abid et al., 2018; Meshram et al., 2022). They also change the cytoplasm content and increase the absorption power of root cells and the transportation of water to other parts of the plant (Abid et al., 2018; Meshram et al., 2022). Drought induces the closing of stomata, reduction in leaf number and stem length to manage the upcoming drought stress (Meshram et al., 2022). Drought-tolerant, wild species have some differences in their epidermal hair, leaves, and root structures (Abid et al., 2018). Due to drought stress plants reduce the metabolic activity related to carbon fixation in the form of net photosynthetic activity (Meshram et al., 2022). Drought stress management includes osmotic adjustment by increasing the concentration of amino acids like proline, and sugars, and also by increasing antioxidant activities to reduce ROS activity (Abid et al., 2018). To manage the upcoming water stress, several changes at the molecular level take place which induce several signal responses and phenotype changes by gene expression and regulation (Ni et al., 2009). Proteins like drought and ABA-Inducible AREB/ABF, stress and ABA-Inducible NAC (SNAC), ABA-Independent DREB/CBF, Zinc Finger Protein (ZFP) TFs, Nuclear Factor (NF-Y), WRKY TFs help in the management of drought stress at the transcriptional level (Yang et al., 2010). Regulatory proteins like chaperons (heat shock proteins), CBFs (C-repeat binding proteins), water channel proteins, plasma membrane proteins, late embryogenesis-abundant (LEA) proteins, osmolytes like fructans, myo-inositol, trehalose, mannitol, proline, polyamines, glycine, etaine, and ABA-responsive genes help to manage drought stress at the post-transcriptional level (Cherian et al., 2006; Cao et al., 2017). Several phytohormones like brassinosteroid (BR), ET (Ethylene), ABA, and JA also play an important role in the management of drought conditions (Martignago et al., 2020). Receptors of these phytohormones also play an important role in drought management (Martignago et al., 2020). An increase in the expression of BR receptors in a tissue specific manner confers resistance against drought without penalizing growth (Martignago et al., 2020). Improving the tolerance of crop plants to drought stress is essential for the food security of the growing population (Khan et al., 2019). Due to the growing population, an increase in the area of cultivation is not enough to provide food. Thus, to increase crop productivity, alterations in crop traits against drought become the main focus area for agriculture researchers (Khan et al., 2019). Conventional plant breeding by marker-assisted, mapping of quantitative trait locus and wild gene introgression become less attractive compared to a direct insertion of one or a group of genes through genetic engineering programs (Gosal et al., 2009; Mwadzingeni et al., 2016; Gupta et al., 2017; Bhatta et al., 2018; Przewieslik-Allen et al., 2019; Khan et al., 2019). Classical plant breeding methods also face the problem of a narrow genetic base, inability to use alien genes, difficulty in breaking gene linkages, and taking long in hybrid generations and their tests (Breseghello and Coelho, 2013). Transgenic approaches have more beauty than classical plant breeding. They are quick to generate huge plants of having traits of interest, and a huge diversity of significant genes for selection. This chapter reviews the current progress of transgenic approaches managing water stress.

2. Water Stress Mediated Yield Loss in Different Crops

Water is a limited natural resource and creates an important proportion of abiotic environmental stresses such as drought. Different plants respond to different levels of water stress (Bukhov and Carpentier, 2004). Only lower plants and cyanobacteria are capable of rapid restoration of the photosynthesis process after re-watering during drought (Bukhov and Carpentier, 2004). Droughts become a limiting factor for plant growth, development, and distribution (Ranjan et al., 2012; Dubey et al., 2013). Drought stress affects root proliferation, germination of seeds, biomass accumulation, and final yield of field crops. In higher plants, especially C_3 crops, stomata closure is one of the first responses to water stress. This results in more provision of oxygen to photosystem apparatuses to manage photorespiration (Bukhov and Carpentier, 2004). Potato and tomato are important vegetable crops and more than 60% yield loss has been reported in both crops during water stress (Krishna et al., 2019; Lahoz et al., 2016; Zaki et al., 2022). Drought along with other stresses damages plant growth and the development of different crops including rice (Zhang et al., 2018; Cheng et al., 2021; Panda et al., 2021; Salgotra and Chauhan, 2023). Ultimately, reductions in flower development, fruit setting, and grain yield have been reported in rice worldwide (Zhang et al., 2018; Cheng et al., 2021; Panda et al., 2021; Salgotra and Chauhan, 2023). Sometimes, as in hydrophyte crops, about 50-100% yield loss has been reported in rice cultivation (Zhang et al., 2018; Cheng et al., 2021; Panda et al., 2021; Salgotra and Chauhan, 2023). In another report, authors have shown that wheat and rice yields decreased by 27.5% and 25.4%, respectively (Zhang et al., 2018). About 40-70% yield losses have been reported due to drought stress in wheat and maize (Kamara et al., 2003). Similarly, drought also influences the productivity of protein crops like pigeon pea (Nam et al., 2001), chicken pea (Nayyar et al., 2006), and soya bean (Samarah et al., 2006). More than 40% of yield losses have been reported in all these crops (Hussain et al., 2019). The oil-producing crops are also damaged by drought stress and about 60% yield loss has been reported in sunflower, while 30 and 70% yield losses have been reported in canola and soya bean respectively (Sinaki et al., 2007; Samarah et al., 2006; Hussain et al., 2019). About 10–30% of yield losses have been reported in sugar-producing plants like sugarcane, sugarbeet, and sweet potato (Sanghera and Jamwal, 2019; Misra et al., 2020; Rajab and Taleghani, 2022; Ghaffari et al., 2022; Sapakhova et al., 2023). Among the fiber crops, cotton (*Gossypium* spp.) ranks as the world's most important fiber plant. It is grown in more than 80 countries globally (Wendel and Crown, 2003; Ranjan et al., 2012). Cotton is very sensitive to drought during various growth stages, especially during flowering and ball formation (Voloudakis et al., 2002; Wendel and Crown, 2003; Ranjan et al., 2012). As a result of drought the area of leaf, nodes, and ball number per plant reduced in cotton. Also a reduction of about 35% has been reported in cotton productivity by different authors (Voloudakis et al., 2002; Wendel and Crown, 2003; Ranjan et al., 2012; Rehman et al., 2022). Considering the dramatic crop yield loss in the above mentioed examples, it is well established that there is an urgent need for crops that can survive at different levels of drought stress so that farmers can grow them as drought-resistant or tolerant varieties.

3. Strategies of Transgenic Crop Preparation

Due to the growing population and fixed area of crop cultivation along with the negative impact of climate change on crops growing the desired crop with a bio-fortified and high productivity nature is necessary. This will fulfill the prevailing food requirement and result in improved health (Hefferon, 2015; Malik and Maqbool, 2020). Since crop plants are also the main food of primary consumers, plants' life cycles are affected by both biotic and abiotic factors (Dubey et al., 2017, 2019-2020, 2022; Baillo et al., 2019; Srivastava et al., 2022). To introduce the desired trait in crop plants, traditional breeding, and transgenic strategies are important methods (Breseghello and Coelho, 2013). Although great improvements have been achieved through traditional breeding methods plant breeders are facing different problems such as a narrow genetic base, inability to use alien genes, difficulty in breaking gene linkages, taking too much time in generating hybrids, and testing them for the desired modifications (Breseghello and Coelho, 2013). In comparison to conventional plant breeding, transgenic strategies have several advantages, like, less faster transgenic plant generation, adequate gene resources, and an ideal plant for gene testing before use in the desired crop (Breseghello and Coelho, 2013). Using transgenic strategies, several plants having resistant properties against biotic and abiotic stresses with bio-fortified nature have been produced for better crop yield (Hefferon, 2015; Malik and Maqbool, 2020; Dubey et al., 2020, 2022). Several examples of crop improvement by genetic engineering are available where one or a few traits have been selectively targeted for successful introduction into crop plants. The development and commercial release of transgenic plants is the most successful example of such applications. Several approaches, including Agrobacterium-mediated transformation, particle gun bombardment, and direct methods (like pollen transformation and shoot apex transformation) have been deployed for the development of transgenic plants (Herrera-Estrella et al., 2004). Agrobacterium-mediated transformation is considered as the most convenient and reproducible way of gene delivery (Herrera-Estrella et al., 2004). Besides transformation, protocols to regenerate fertile plants from transformed cells have to be developed. Other methods like direct DNA transfer to protoplasts of desired crops using electroporation, polyethylene glycol (PEG), and calcium phosphate, has also been shown for various crops (Herrera-Estrella et al., 2004). Transgenic crops have some disadvantages like development of resistant microbes, and allergenic and biodiversity losses. However, the desired timely expression of transgenes at the site, and marker-free transgenic plants can neutralize the above demerits (Dubey et al., 2018; He and Zhao, 2020).

4. Transgenic Approach to Manage Water Stress

Not only does drought signal water stress to plants, high concentrations of ions in soil and low temperature also produce similar water stress signals (Bohnert and Jensen, 1996). Each plant responds to water stress in different ways depending on their multigenic signaling behavior (Bohnert and Jensen, 1996). Transgenic technology is an emerging area of science to increase the plant's resistance to biotic and abiotic stresses including drought (Feng et al., 2017) (Table 1). Transgenic strategies have been employed to manage drought stress by focusing on engineering plants to enhance the content of osmolytes including proline, mannitol, glycine and betaine; an overexpression of enzymes for free radical scavenging and utilizing reactive oxygen species to manage drought stress (Bohnert and Jensen, 1996). Osmolyte-mediated enhancement in water retention and scavenging of free radicals and ROS to support plants to manage water stress in transgenic plants have been shown (Bohnert and Jensen, 1996). First of all, transgenic model plants were tried to cope with drought-like stress. Genes like BrLAS (GRAS Transcription Factor of *Brassica rapa*); and carrot DcDREB1A, are used to produce drought stress tolerance in transgenic *Arabidopsis* (Li et al., 2018; Li S et al., 2020). Such transgenic *Arabidopsis* plants showed increased peroxidase and superoxide dismutase activity along with decreased stomatal apertures with increased lignin amounts (Li et al., 2018; Li S et al., 2020) (Table 1). Similarly, transgenic tobacco expressing *Vigna aconitifolia* P5CS gene (P5CSF129A) and apple transcription factor (MdDREB76), exhibited drought tolerance and high proline levels with elevated osmoprotectants, lesser H_2O_2 content, lipid peroxidation, and electrolyte leakage for drought tolerance nature (Gubiš et al., 2007; Sharma et al., 2019). Since the growth in population and higher living standards the use of edible oil has increased tremendously (Zhu et al., 2016). Transgenic oil-producing plants were also made to manage drought tolerance. *Brassica* such as canola varieties, are one of the big sources of edible oil, and animal cakes, experiencing yield loss due to drought stress. Additionally, being a closely related species of *Arabidopsis thaliana*, the knowledge gained from the transcriptomic and proteomic of this model plant can be applied directly to generate transgenic *Brassica* for introducing drought tolerance (Zhu et al., 2016; Li et al., 2021). Among the fodder crops, different transgenic crops like alfalfa have also been made to enhance drought tolerance (Tang et al., 2013). The beauty of alfalfa is in its high yields with significant levels of minerals, proteins, vitamins, amino acids, and other nutritional content in higher concentrations required for lactating cattle (Zheng et al., 2017; Yin et al., 2022). Various genes like WXP1 (ethylene-responsive element-binding TF); GsZFP1 (*Glycine soja*'s C2H2-type zinc-finger motif), sweet potato Orange gene (IbOr); *Arabidopsis* enhanced drought tolerance 1 (AtEDT1) gene; MsTHI1 (*Medicago sativa*'s thiamine thiazole synthase), etc. have been used to generate transgenic alfalfa plants (Jiang et al., 2009; Tang et al., 2013; Wang et al., 2015; Zheng et al., 2017; Yin et al., 2022). These plants showed higher relative water content, enhanced salt tolerance resistant power, less Na^+ and K^+ accumulation; increased carotenoid content under drought conditions, and provision of drought tolerance (Jiang et al., 2009; Tang et al., 2013; Wang et al., 2015; Zheng et al., 2017; Yin et al., 2022). Tomato

Table 1: Transgenic plant approach to manage water stress

S. No.	Plant	Transgene	Promoter/level of expression	Changed Property in transgenic plants	Overall Effect	Authors
1	Arabidopsis	BRL3	Overexpression	Increased proline and sugars	Drought tolerance	Fàbregas et al, 2018
2	Rice	OsAPX2	CAMV35S and Ubi	Homeostasis (ROS)	Drought tolerance	Feng et al, 2017
3	Wheat	TaRAP2.1L; TaSHN1	pUbi and Dehydrin 8	Less stomata density	Drought tolerance	Amalraj et al, 2016; Bi et al, 2018
4	Maize	nced and rpk	Lea and SalT promoters	Changed ABA signaling	Drought tolerance	Muppala et al, 2021
5	Soybean	Soybean NF-Y gene GmNFYA5	GmNFYA5 promoter	Changed ABA signaling genes	Drought tolerance	Ma et al, 2020
6	Citrus	P5CS-F129A	CaMV35S promoter.	More proline and glutamate	Drought tolerance	Barichello et al, 2017
7	Sugarcane	RICK1 (BRK1	Over expression	Epidermal changes by actin polymerization.	Drought tolerance	Narayan et al, 2023
8	Brassica	At. TC, LEA protein, SR45a gene	CaMV35S promoter	Physiological changes	Drought tolerance	Sobhanverdi et al, 2021, Saha et al, 2016; Muthusamy et al, 2020
9	Tobacco	P5CSF129A; MdDREB76; DcPSY2		high proline, less H2O2 and less electrolyte leakage	Drought tolerance	Gubiš et al, 2007; Sharma et al, 2019; Acevedo et al, 2023
10	Cotton	GHSP26	CaMV 35S promoter		Drought tolerance	Shamim et al, 2013
11	Alfalfa	GsZFP1, bZIP28, WXP1,	pCaMV35S and CER6	Accumulation of less Na(+) and more K(+)	Drought tolerance	Tang et al, 2013; Kim et al, 2019; Jiang et al, 2009
12	Tomato	SlGATA17, BcZAT12	Bclea1a promoter	Enhanced phenylpropanoid pathway	Drought tolerance	Zhao et al 2021, Rai et al, 2013
13	Potato	TPS1, STANN1	pCaMV35S promoter		Drought tolerance	Yeo et al, 2000, Szalonek et al, 2015

(*Solanum lycopersicum* L.) is an important vegetable used throughout the world (Awais et al., 2018). Fulfilling tomato requirements is good for human health due to their nutritional value (Awais et al., 2018). Tomato crop yield losses upto 70% due to various abiotic stresses including drought (Krishna et al., 2019). To develop transgenic plants against drought stress, several efforts have been made by researchers (Redenbaugh, 1992; Foolad, 2007; Pineda et al., 2012; Conti et al., 2023) (Table 1). Several transgenic tomato lines to improve self life or nutritional value have been made and cultivated commercially worldwide, particularly in temperate areas (Redenbaugh, 1992; Foolad, 2007; Pineda et al., 2012). Due to lycopene and its anti-oxidative and other medicinal properties, the total uses and production of tomato is increasing (Gerszberg and Hnatuszko-Konka, 2017). Different transgenic tomato lines by expressing different genes viz., tomato ethylene responsive factors (LeERF1), glutathione S-transferases (LeGSTU2), C2H2 zinc finger protein (BcZAT12), *Arabidopsis thaliana* dehydration responsive element binding (AtDREB1), and GATA transcription factors (SlGATA17) have been made. These lines showed good drought tolerance (Lu et al., 2010, Xu et al., 2015; Rai et al., 2013; Krishna et al., 2021a; Zhao et al., 2021). Most of these lines are overexpressed transgenic lines utilizing CaMV35S promoter. To overcome the drastic effect of over-expression, stress-induced expression has also been tried by Rai et al (2013), through the utilization of stress-inducible, Bclea1a promoter. Krishna et al (2021b) tried to express two genes viz., *Arabidopsis thaliana* Dehydration Responsive Element Binding1A (AtDREB1) and *Brasica caranata* Zinc finger proteins (BcZAT12). The result of this tomato transgenic line was also interesting and the plant showed drought tolerance capacity (Krishna et al., 2021b). These transgenic lines generally induce and maintain higher relative water content (RWC), free proline and soluble sugar levels, increased activity of antioxidant enzymes, and accumulation of more antioxidants like glutathione and ascorbic acid (Lu et al., 2010; Xu et al., 2015; Rai et al., 2013; Krishna et al., 2021a-b; Zhao et al., 2021). Transgenic lines reduce the level of malondialdehyde (MDA) level and electrolyte leakage (Lu et al., 2010; Xu et al., 2015; Rai et al., 2013; Krishna et al., 2021a; Zhao et al., 2021). Further, transgenic ectopic expression also enhanced the drought tolerance like ectopically expressing aquaporin MdPIP1; increased size of fruit and also enhanced tolerance against drought (Wang et al., 2017). Similar to tomato, potato is another vegetable crop grown worldwide. It is very sensitive to heat, drought, and salt crops. These abiotic stresses directly reduce the yield and quality of potato tuber (Si et al., 2012). Several transgenic potato plants have been generated to provide better yield during drought stress (Si et al., 2012) (Table 1). Constitutive over expressive genes of other plant origins like trehalose-6-phosphate synthase (TPS1) of yeast, hexokinase 1, dehydration-responsive element binding factor 1 (DREB1B), and Cu/Zn superoxide dismutase of *Arabidopsis thaliana* (AtHXK1); potato orange gene (IbOr) of sweet potato have been made and these plants show better productivity compared to the control under drought stress conditions (Yeo et al., 2000; Lehretz et al., 2021;

Cho et al., 2016; Van der Mescht et al., 2007; Movahed et al., 2012). Similarly constitutive overexpression of potato genes occurs under transgenic conditions like Potato Annexin STANN1; StMYB1R-1 transcription factor (TF); proline dehydrogenase (ProDH); DREB1A showed similar drought-resistant nature (Szalonek et al., 2015; Gazendam et al., 2014; Li T et al., 2020; Bouaziz et al., 2013) (Table 1). Stress inducible expression of transcription factor (TF) StMYB1R-1, and sweet potato orange gene (IbOr), transcription factor (IbMYB1) through drought inducive (rd29A) and SWPA2 promoter respectively also enhance the drought tolerance in transgenic potato plants (Gazendam et al., 2014; Cho et al., 2016; Cheng et al., 2013). Overexpression of zinc finger proteins (bZIP28) and DREB1A/Bar also induces drought tolerance in transgenic potato plants (Jia et al., 2015; Kim et al., 2019). Transgenic coexpression of *Arabidopsis thaliana* hexokinase 1 (AtHXK1) in guard cells and self-pruning 6A (SP6A) by leaf/stem-specific StLS1 promoter increases tuberization during drought stress by enhancing water utilization (Lehretz et al., 2021). Further proline is another important component in managing drought stress. Due to the rate limit of proline dehydrogenase (ProDH) proline formation is limited hence changing their expression through a silencing mechanism which enhances the drought tolerance of transgenic potato (Li S et al., 2020). This transgenic enhanced drought resistance increases water utilization, enhancing chloroplast functions, decreasing stomatal conductance, increasing water in green tissues, and higher accumulation of xanthophyll (Lehretz et al., 2021; Li S et al., 2020; Li T et al., 2020). Droughts also affect the plants that are not primarily irrigated like lemon trees (Barichello et al., 2017). Drought damages crop productivity by flowering and fruit setting and also changes the quality of fruit juice (Barichello et al., 2017). Transgenic citrus rootstocks expressing P5CS-F129A (Δ1-pyrroline-5-carboxylate synthetase (P5CS)) gene of *Vigna aconitifolia* are developed to tolerate drought stress (for drought-sensitive genotypes) with higher productive lines (Barichello et al., 2017). These citrus plants produce more proline accumulation, higher stomatal conductance, and relative water content (RWC) without irrigation upto 26 to 34 days (Barichello et al., 2017).

5. Conclusion

Due to climate changes, drought-mediated loss in the agricultural sector, presents a major threat to food security in a growing population (Khan et al., 2019; Martignago et al., 2020). To keep pace with the increasing demand for food, drought tolerance plants exclusive to yield loss have become a major objective of plant genetic engineering (Khan et al., 2019; Martignago et al., 2020). So far most transgenic plants having a drought resistant nature show a reduction in crop yield in comparison to the same plant having irrigated conditions (Martignago et al., 2020). It shows that drought resistance is a multigenic and multievent process and there is still a need for the best approaches to deal with drought (Martignago et al., 2020). It has been reported that not only transgenic but other techniques like nitrate treatment reduce the effect of drought on rice (Cheng et al., 2021). The transgenic strategy uses a gene of a selectable marker to select transgenic plants on herbicides or antibiotics and such markers have reduced the public acceptance of these crops. Marker-free transgenic plants can be generated by particle bombardment strategies (Feng et al., 2017). Presently the marker-free transgenic crop is more accepted by society and in this area, several crops have been generated showing a drought-resistant property (Feng et al., 2017; Singh et al., 2022). Further to avoid transgene load on transgenic plants at the site the timely expression of drought inducible genes can provide effective approaches to deal with drought stress without expressing the transgene when there is no need (Dubey et al., 2018; Ta et al., 2020).

References

Abid M, Ali S, Qi LK, Zahoor R, Tian Z, Jiang D, et al. 2018. Physiological and biochemical changes during drought and recovery periods at tillering and jointing stages in wheat (*Triticum aestivum* L.). *Scientific Reports,* **8(1):** 4615.

Acevedo O, Contreras RA & Stange C. 2023. The carrot phytoene synthase 2 (DcPSY2) promotes salt stress tolerance through a positive regulation of abscisic acid and abiotic-related genes in *Nicotiana tabacum. Plants,* **12(10):** 1925.

Amalraj A, Luang S, Kumar MY, Sornaraj P, Eini O, Kovalchuk N, et al. 2016. Change of function of the wheat stress-responsive transcriptional repressor Ta RAP 2.1 L by repressor motif modification. *Plant Biotechnology Journal,* **14(2):** 820–832.

Awais M, Ahmad R, Khan N, Garapati P, Shahzad M, Afroz A, et al. 2018. Transformation of tomato variety Rio grande with drought resistant transcription factor gene ATAF1 and its molecular analysis. *Pak. J. Bot.,* **50(5):** 1811–1820.

Baillo EH, Kimotho RN, Zhang Z & Xu P. 2019. Transcription factors associated with abiotic and biotic stress tolerance and their potential for crops improvement. *Genes,* **10(10):** 771.

Barichello D, Farinacio R, dos Santos CC, Andrade GA, Nagashima GT, Garbúglio DD, et al. 2017. Drought-tolerant transgenic *Swingle Citrumelo* controls accumulation of proline modulating the expression of key genes of the proline metabolism. *Citrus Research & Technology,* **38(2):** 1–2.

Bhatta M, Morgounov A, Belamkar V & Baenziger PS. 2018. Genome-wide association study reveals novel genomic regions for grain yield and yield-related traits in drought-stressed synthetic hexaploid wheat. *Int. J. Mol. Sci.,* **19:** 3011.

Bi H, Shi J, Kovalchuk N, Luang S, Bazanova N, Chirkova L, Zhang D, et al. 2018. Overexpression of the TaSHN1 transcription factor in bread wheat leads to leaf surface modifications, improved drought tolerance, and no yield penalty under controlled growth conditions. *Plant Cell Environ,* **41:** 2549–2566.

Bohnert HJ & Jensen RG. 1996. Strategies for engineering water-stress tolerance in plants. *Trends in Biotechnology,* **14(3):** 89–97.

Bouaziz D, Pirrello J, Charfeddine M, Hammami A, Jbir R, Dhieb A, et al. 2013. Overexpression of StDREB1 transcription factor increases tolerance to salt in transgenic potato plants. *Molecular Biotechnology,* **54:** 803–817.

Breseghello F & Coelho AS. 2013. Traditional and modern plant breeding methods with examples in rice (*Oryza sativa* L.). *Journal of Agricultural and Food Chemistry,* **61(35):** 8277–8286.

Bukhov NG & Carpentier R. 2004. Effects of water stress on the photosynthetic efficiency of plants. pp. 623–635. *In:* Chlorophyll a Fluorescence: A Signature of Photosynthesis, Dordrecht: Springer Netherlands.

Cao MJ, Zhang YL, Liu X, Huang H, Zhou XE, Wang WL, et al. 2017. Combining chemical and genetic approaches to increase drought resistance in plants. *Nature Communications,* **8(1):** 1183.

Cheng B, Hu S, Cai M, Cao C & Jiang Y. 2021. Nitrate application induced a lower yield loss in rice under progressive drought stress. *Plant Growth Regulation,* **95:** 149–156.

Cheng Yu-Jie, Myoung-Duck Kim, Xi-Ping Deng, Sang-Soo Kwak & Wei Chen. 2013. Enhanced salt stress tolerance in transgenic potato plants expressing IbMYB1, a sweet potato transcription factor. *Journal of Microbiology and Biotechnology,* **23:** 1737–1746.

Cherian S, Reddy MP & Ferreira RB. 2006. Transgenic plants with improved dehydration-stress tolerance: Progress and future prospects. *Biologia Plantarum,* **50:** 481–495.

Cho KS, Han EH, Kwak SS, Cho JH, Im JS, Hong SY, et al. 2016. Expressing the sweet potato orange gene in transgenic potato improves drought tolerance and marketable tuber production. *Comptes Rendus Biologies,* **339(5–6):** 207–213.

Conti V, Parrotta L, Romi M, Del Duca S & Cai G. 2023. Tomato biodiversity and drought tolerance: A multilevel review. *International Journal of Molecular Sciences,* **24(12):** 10044.

Dubey NK, Eizenberg H, Leibman D, Wolf D, Edelstein M, Abu-Nassar J, et al. 2017. Enhanced host-parasite resistance based on down-regulation of *Phelipanche aegyptiaca* target genes is likely by mobile small RNA. *Frontiers in Plant Science,* **8:** 1574.

Dubey NK, Goel R, Ranjan A, Idris A, Singh SK, Bag SK, et al. 2013. Comparative transcriptome analysis of *Gossypium hirsutum* L. in response to sap sucking insects: Aphid and whitefly. *BMC Genomics,* **14(1):** 1–20.

Dubey NK, Gupta K, Singh SP, Panigrahi J & Singh SP. 2022. Role of ET and ROS in salt homeostasis and salinity stress tolerance and transgenic approaches to making salt-tolerant crops. *Ethylene in Plant Biology,* **23:** 259–276.

Dubey NK, Gupta K, Yadav P, Panigrahi J & Gupta AK. 2020. Transgenic strategies to develop resistant plant against the pathogen and pest. *Pesticides in Crop Production: Physiological and Biochemical Action,* **6:** 259–290.

Dubey NK, Mishra DK, Idris A, Nigam D, Singh PK & Sawant SV. 2018. Whitefly and aphid inducible promoters of Arabidopsis thaliana L. *Journal of Genetics,* **97:** 109–119.

Dubey, NK, Singh PK, Yadav SK & Singh KD. 2019. Overview of insecticidal genes used in crop improvement program. *OMICS-Based Approaches in Plant Biotechnology,* 19–53.

Fàbregas N, Lozano-Elena F, Blasco-Escámez D, Tohge T, Martínez-Andújar C, Albacete A, et al. 2018. Overexpression of the vascular brassinosteroid receptor BRL3 confers drought resistance without penalizing plant growth. *Nature Communications,* **9(1):** 4680.

Feng D, Wang Y, Wu J, Lu T & Zhang Z. 2017. Development and drought tolerance assay of marker-free transgenic rice with OsAPX2 using biolistic particle-mediated co-transformation. *The Crop Journal*, **5(4):** 271–281.

Foolad MR. 2007. Current status of breeding tomatoes for salt and drought tolerance. *Advances in Molecular Breeding Toward Drought and Salt Tolerant Crops*, 669–700.

Gazendam I, Greyling R, Laurie RN, Matsaunyane LB, Oelofse D & Rakuambo JA. 2014. Transgenic approach to improve the drought tolerance of potato. *XXIX International Horticultural Congress on Horticulture: Sustaining Lives, Livelihoods and Landscapes* (IHC2014): **1110:** 203–210.

Gerik TJ, Faver KL, Thaxton PM & El-Zik KM. 1996. Late season water stress in cotton: I. Plant growth, water use, and yield. *Crop Science,* **36(4):** 914–921.

Gerszberg A & Hnatuszko-Konka K. 2017. Tomato tolerance to abiotic stress: A review of most often engineered target sequences. *Plant Growth Regulation,* **83:** 175–198.

Ghaffari H, Tadayon MR, Bahador M & Razmjoo J. 2022. Biochemical and yield response of sugar beet to drought stress and foliar application of vermicompost tea. *Plant Stress,* **5:** 100087.

Gosal SS, Wani SS & Kang MS. 2009. Biotechnology and drought tolerance. *J. Crop Imp.,* **23:** 19–54.

Gubiš J, Vaňková R, Červená V, Dragúňová M, Hudcovicová M, Lichtnerová H, et al. 2007. Transformed tobacco plants with increased tolerance to drought. *South African Journal of Botany,* **73(4):** 505–511.

Gupta PP, Balyan SS & Gahlaut V. 2017. QTL analysis for drought tolerance in wheat: Present status and future possibilities. *Agronomy,* **7:** 5.

He Y & Zhao Y. 2020. Technological breakthroughs in generating transgene-free and genetically stable CRISPR-edited plants. *Abiotech,* **(1):** 88–96.

Hefferon KL. 2015. Nutritionally enhanced food crops: Progress and perspectives. *International Journal of Molecular Sciences,* **16(2):** 3895–3914.

Herrera-Estrella L, Simpson J & Martínez-Trujillo M. 2004. Transgenic plants: An historical perspective. *Transgenic Plants: Methods and Protocols,* 3–1.

Hussain S, Hussain S, Qadir T, Khaliq A, Ashraf U, et al. 2019. Drought stress in plants: An overview on implications, tolerance mechanisms and agronomic mitigation strategies. *Plant Science Today,* **6(4):** 389–402.

Jia X, Qi E, Ma S, Hu X, Wang Y, Wen G, et al. 2015. Analysis of drought tolerance and herbicide resistance in transgenic potato plants over-expressing DREB1A/Bar. *Acta Prataculturae Sinica,* **24(11):** 58–64.

Jiang Q, Zhang JY, Guo X, Monteros MJ & Wang ZY. 2009. Physiological characterization of transgenic alfalfa (*Medicago sativa*) plants for improved drought tolerance. *International Journal of Plant Sciences,* **170(8):** 969–978.

Kamara AY, Menkir A, Badu–Apraku B & Ibikunle O. 2003. The influence of drought stress on growth, yield and yield components of selected maize genotypes. *J Agr Sci.,* **141:** 43–50.

Kaya C, Tuna L & Higgs D. 2006. Effect of silicon on plant growth and mineral nutrition of maize grown under water-stress conditions. *Journal of Plant Nutrition,* **29(8):** 1469–1480.

Khan S, Anwar S, Yu S, Sun M, Yang Z & Gao ZQ. 2019. Development of drought-tolerant transgenic wheat: Achievements and limitations. *International Journal of Molecular Sciences,* **20(13):** 3350.

Kim CK, Oh JH, Na JK, Cho C, Kim KH, Yu GE, et al. 2019. The genes associated with drought tolerance by multi-layer approach in potato. *Plant Breeding and Biotechnology,* **7(4):** 405–414.

Kirda C. 2002. Deficit irrigation scheduling based on plant growth stages showing water stress tolerance. *Food and Agricultural Organization of the United Nations, Deficit Irrigation Practices, Water Reports,* **22(102)**.

Kramer PJ. 1963. Water stress and plant growth. *Agronomy Journal,* **55(1):** 31–35.

Krishna R, Ansari WA, Jaiswal DK, Singh AK, Prasad R, Verma JP, et al. 2021a. Overexpression of AtDREB1 and BcZAT12 genes confers drought tolerance by reducing oxidative stress in double transgenic tomato (*Solanum lycopersicum* L.). *Plant Cell Reports,* **40(11):** 2173–2190.

Krishna R, Ansari WA, Jaiswal DK, Singh AK, Verma JP & Singh M, et al. 2021b. Co-overexpression of AtDREB1A and BcZAT12 increases drought tolerance and fruit production in double transgenic tomato (*Solanum lycopersicum*) plants. *Environmental and Experimental Botany,* **184:** 104396.

Krishna R, Karkute SG, Ansari WA, Jaiswal DK, Verma JP & Singh M, et al. 2019. Transgenic tomatoes for abiotic stress tolerance: Status and way ahead. *3 Biotech,* **9:** 1–4.

Lahoz I, Pérez-de-Castro A, Valcárcel M, Macua JI, Beltran J, Roselló S, et al. 2016. Effect of water deficit on the agronomical performance and quality of processing tomato. *Scientia Horticulturae,* **200:** 55–65.

Lehretz GG, Sonnewald S, Lugassi N, Granot D & Sonnewald U. 2021. Future-proofing potato for drought and heat tolerance by overexpression of hexokinase and SP6A. *Frontiers in Plant Science,* **11:** 614534.

Levitt J. 1980. Responses of Plants to Environmental Stresses. Volume II. Water, Radiation, Salt, and Other Stresses. Academic Press.

Li J, Duan Y, Sun N, Wang L, Feng S & Fang Y 2021. The miR169n-NF-YA8 regulation module involved in drought resistance in *Brassica napus* L. *Plant Science,* **313:** 111062.

Li P, Zhang B, Su T, Li P, Xin X, Wang W, et al. 2018. BrLAS, a GRAS transcription factor from *Brassica rapa*, is involved in drought stress tolerance in transgenic Arabidopsis. *Frontiers in Plant Science,* **9:** 1792.

Li S, Zhang N, Zhu X, Ma R, Yang J, Tang X, et al. 2020. Enhanced drought tolerance with artificial microRNA-mediated StProDH1 gene silencing in potato. *Crop Science,* **60(3):** 1462–1471.

Li T, Huang Y, Khadr A, Wang YH, Xu ZS & Xiong AS. 2020. DcDREB1A, a DREB-binding transcription factor from *Daucus carota*, enhances drought tolerance in transgenic *Arabidopsis thaliana* and modulates lignin levels by regulating lignin-biosynthesis-related genes. *Environmental and Experimental Botany,* **169:** 103896.

Lu C, Li Y, Chen A, Li L, Zuo J, Tian H, et al. 2010. LeERF1 improves tolerance to drought stress in tomato (*Lycopersicon esculentum*) and activates downstream stress-responsive genes. *African Journal of Biotechnology,* **9(38):** 6294–6300.

Malik KA & Maqbool A. 2020. Transgenic crops for biofortification. *Frontiers in Sustainable Food Systems,* **4:** 571402.

Martignago D, Rico-Medina A, Blasco-Escámez D, Fontanet-Manzaneque JB & Caño-Delgado AI, et al. 2020. Drought resistance by engineering plant tissue-specific responses. *Frontiers in Plant Science,* **10:** 1676.

Ma XJ, Yu TF, Li XH, Cao XY, Ma J, Chen J, et al. 2020. Overexpression of GmNFYA5 confers drought tolerance to transgenic Arabidopsis and soybean plants. *BMC Plant Biology,* **20:** 1–8.

Meshram JH, Singh SB, Raghavendra KP & Waghmare VN. 2022. Drought stress tolerance in cotton: Progress and perspectives. *Climate Change and Crop Stress,* **1:** 135–169.

Misra V, Solomon S, Mall AK, Prajapati CP, Hashem A, Abd Allah EF, et al. 2020. Morphological assessment of water stressed sugarcane: A comparison of waterlogged and drought affected crop. *Saudi Journal of Biological Sciences,* **27(5):** 1228–1236.

Morgan JM. 1984. Osmoregulation and water stress in higher plants. *Annual Review of Plant Physiology,* **35(1):** 299–319.

Movahedi S, Sayed Tabatabaei BE, Alizade H, Ghobadi C, Yamchi A & Khaksar G. 2012. Constitutive expression of Arabidopsis DREB1B in transgenic potato enhances drought and freezing tolerance. *Biologia Plantarum,* **56:** 37–42.

Muppala S, Gudlavalleti PK, Malireddy KR, Puligundla SK & Dasari P. 2021. Development of stable transgenic maize plants tolerant for drought by manipulating ABA signaling through Agrobacterium-mediated transformation. *Journal of Genetic Engineering and Biotechnology,* **19(1):** 1–4.

Muthusamy M, Yoon EK, Kim JA, Jeong MJ & Lee SI. 2020. *Brassica Rapa* SR45a regulates drought tolerance via the alternative splicing of target genes. *Genes,* **11(2):** 182.

Mwadzingeni L, Shimelis H, Dube E, Laing MD & Tsilo TJ. 2016. Breeding wheat for drought tolerance: Progress and technologies. *J. Integr. Agric,* **15:** 935–943.

Nam NH, Chauhan YS & Johansen C. 2001. Effect of timing of drought stress on growth and grain yield of extra-short duration pigeon pea lines. *J Agr Sci,* **136:** 179–189.

Narayan JA, Manoj VM, Nerkar G, Chakravarthi M, Dharshini S, Subramonian N, et al. 2023. Transgenic sugarcane with higher levels of BRK1 showed improved drought tolerance. *Plant Cell Reports,* **14:** 1–8.

Nayyar H, Kaur S, Singh S & Upadhyaya HD. 2006. Differential sensitivity of Desi (small-seeded) and Kabuli (large seeded) chickpea genotypes to water stress during seed filling: Effects on accumulation of seed reserves and yield. *J Sci Food Agr,* **86:** 2076–2082

Ni FT, Chu LY, Shao HB & Liu ZH. 2009. Gene expression and regulation of higher plants under soil water stress. *Current Genomics,* **10(4):** 269–280.

Panda D, Mishra SS & Behera PK. 2021. Drought tolerance in rice: Focus on recent mechanisms and approaches. *Rice Science,* **28(2):** 119–132.

Pineda B, García-Abellán JO, Antón T, Pérez F, Moyano E, García Sogo B, et al. 2012. Tomato: Genomic approaches for salt and drought stress tolerance. *Improving Crop Resistance to Abiotic Stress,* **14:** 1085–1120.

Przewieslik-Allen AM, Burridge AA, Wilkinson PP, Winfield MM, Shaw DD, McAusland L, et al. 2019. Developing a high-throughput SNP-based marker system to facilitate the introgression of traits from Aegilops species into bread wheat (*Triticum aestivum*). *Front. Plant Sci,* **9:** 1993.

Pugnaire FI, Serrano LU & Pardos JO. 1999. Constraints by water stress on plant growth. *Handbook of Plant and Crop Stress,* **2:** 271–283.

Rai AC, Singh M & Shah K. 2013. Engineering drought tolerant tomato plants over-expressing BcZAT12 gene encoding a C2H2 zinc finger transcription factor. *Phytochemistry,* **85:** 44–50.

Rajabi A & Taleghani D. 2022. Drought stress management in sugar beet (*Beta vulgaris* L.) cultivation. pp. 429–440. *In:* Sugar Beet Cultivation, Management and Processing, Singapore: Springer Nature Singapore.

Ranjan A, Pandey N, Lakhwani D, Dubey NK, Pathre UV & Sawant SV, et al. 2012. Comparative transcriptomic analysis of roots of contrasting *Gossypium herbaceum* genotypes revealing adaptation to drought. *BMC Genomics,* **13:** 1–22.

Redenbaugh K. 1992. Safety Assessment of Genetically Engineered Fruits and Vegetables: A Case Study of the Flavr Savr Tomato. CRC Press. pp. 288.

Rehman T, Tabassum B, Yousaf S, Sarwar G & Qaisar U. 2022. Consequences of drought stress encountered during seedling stage on physiology and yield of cultivated cotton. *Frontiers in Plant Science,* **13:** 906444.

Saha B, Mishra S, Awasthi JP, Sahoo L & Panda SK. 2016. Enhanced drought and salinity tolerance in transgenic mustard (*Brassica juncea* (L.) Czern & Coss.) overexpressing Arabidopsis group 4 late embryogenesis abundant gene (AtLEA4-1). *Environmental and Experimental Botany,* **A128:** 99–111.

Salgotra RK & Chauhan BS. 2023. Ecophysiological responses of rice (*Oryza sativa* L.) to drought and high temperature. *Agronomy,* **13(7):** 1877.

Samarah NH, Mullen RE, Cianzio SR & Scott P. 2006. Dehydrinlike proteins in soybean seeds in response to drought stress during seed filling. *Crop Sci,* **46:** 2141–2150.

Sanghera GS & Jamwal NS. 2019. Perspective for genetic amelioration of sugarcane towards waterlogged conditions. *Int. J. Pure Appl. Biosci,* **7(3):** 484–502.

Sapakhova Z, Raissova N, Daurov D, Zhapar K, Daurova A, Zhigailov A, et al. 2023. Sweet potato as a key crop for food security under the conditions of global climate change: A review. *Plants,* **12(13):** 2516.

Shamim Z, Rashid B, Rahman S & Husnain T. 2013. Expression of drought tolerance in transgenic cotton. *Science Asia,* **39:** 1–1.

Sharma V, Goel P, Kumar S & Singh AK. 2019. An apple transcription factor, MdDREB76, confers salt and drought tolerance in transgenic tobacco by activating the expression of stress-responsive genes. *Plant Cell Reports,* **38:** 221–241.

Srivastava R, Bajpai R, Khan Z, Singh SP, Mehrotra R & Dubey NK et al. 2022. Insight into strigolactone hormone functions in plant parasitic weeds: A regulatory perspective. *Indian Journal of Experimental Biology* (*IJEB*), **60(09):** 659–666.

Si H, Zhang N & Wang D. 2012. Drought and salinity tolerance in transgenic potato. pp. 373–387. *In:* Sustainable Potato Production: Global Case Studies, Springer, Dordrecht.

Sinaki JM, Heravan EM, Rad AHS, Noormohammadi G & Zarei G. 2007. The effects of water deficit during growth stages of canola (*Brassica napus* L.). *Am Euras. J Agri Environ Sci,* **2:** 417–422.

Singh R, Kaur N, Praba UP, Kaur G, Tanin MJ, Kumar P, et al. 2022 Jun. A prospective review on selectable marker-free genome engineered rice: Past, present and future scientific realm. *Frontiers in Genetics,* **13:** 882836.

Sobhanverdi S, Majidi A, Abbasi A, Mollabashi ZA, Sabokdast M, Bayat H, et al. 2021. Constitutive overexpressing At. TC improves drought-mediated oxidative tolerance in transgenic *Brassica Napus* L. *Research Squar,* 1–13, DOI: https://doi.org/10.21203/rs.3.rs-751269/v1

Szalonek M, Sierpien B, Rymaszewski W, Gieczewska K, Garstka M, Lichocka M, et al. 2015. Potato annexin STANN1 promotes drought tolerance and mitigates light stress in transgenic *Solanum tuberosum* L. plants. *PloS One,* **10(7):** e0132683.

Ta DT, Ngo DM, Nguyen NH, Pham NB, Do PT & Chu HH, et al. 2020. Production of drought tolerant transgenic soybean expressing codA gene under regulation of a water stress inducible promoter. *Pak. J. Bot.,* **52(3):** 793–799.

Tang L, Cai H, Ji W, Luo X, Wang Z, Wu J, et al. 2013. Overexpression of GsZFP1 enhances salt and drought tolerance in transgenic alfalfa (*Medicago sativa* L.). *Plant Physiology and Biochemistry,* **71:** 22–30.

Van der Mescht A, De Ronde JA, Slabbert MM & Oelofse D. 2007. Enhanced drought tolerance in transgenic potato expressing the *Arabidopsis thaliana* Cu/Zn superoxide dismutase gene. *South African Journal of Science,* **103(3):** 169–173.

Voloudakis AE, Kosmas SA, Tsakas S, Eliopoulos E, Loukas M & Kosmidou K. 2002. Expression of selected drought-related genes and physiological response of Greek cotton varieties. *Functional Plant Biology,* **29(10):** 1237–1245.

Wang L, Li QT, Lei Q, Feng C, Zheng X, Zhou F, et al. 2017. Ectopically expressing MdPIP1; 3, an aquaporin gene, increased fruit size and enhanced drought tolerance of transgenic tomatoes. *BMC Plant Biology,* **17(1):** 1–3.

Wang Z, Ke Q, Kim MD, Kim SH, Ji CY, Jeong JC, et al. 2015. Transgenic alfalfa plants expressing the sweetpotato orange gene exhibit enhanced abiotic stress tolerance. *PLoS One,* **10(5):** e0126050.

Wendel JF & Crown RC. 2003. Polyploidy and the evolutionary history of cotton. *Advance in Agronomy,* **78:** 139–186.

Xu J, Xing XJ, Tian YS, Peng RH, Xue Y, Zhao W, et al. 2015. Transgenic Arabidopsis plants expressing tomato glutathione S-transferase showed enhanced resistance to salt and drought stress. *PloS One,* **10(9):** e0136960.

Yang S, Vanderbeld B, Wan J & Huang Y. 2010. Narrowing down the targets: Towards successful genetic engineering of drought-tolerant crops. *Molecular Plant,* **3(3):** 469–490.

Yeo ET, Kwon HB, Han SE, Lee JT, Ryu JC & Byu MO. 2000. Genetic engineering of drought resistant potato plants by introduction of the trehalose-6-phosphate synthase (TPS1) gene from *Saccharomyces cerevisiae. Molecules and Cells,* **10(3):** 263–268.

Yin H, Wang Z, Li H, Zhang Y, Yang M, Cui G, et al. 2022. MsTHI1 overexpression improves drought tolerance in transgenic alfalfa (*Medicago sativa* L.). *Frontiers in Plant Science,* **13:** 992024.

Zaki HE & Radwan KS. 2022. Response of potato (*Solanum tuberosum* L.) cultivars to drought stress under in vitro and field conditions. *Chemical and Biological Technologies in Agriculture,* **9:** 1–9.

Zhang J, Zhang S, Cheng M, Jiang H, Zhang X, Peng C, et al. 2018. Effect of drought on agronomic traits of rice and wheat: A meta-analysis. *International Journal of Environmental Research and Public Health,* **15(5):** 839.

Zhao T, Wu T, Pei T, Wang Z, Yang H, Jiang J, et al. 2021. Overexpression of SlGATA17 promotes drought tolerance in transgenic tomato plants by enhancing activation of the phenylpropanoid biosynthetic pathway. *Frontiers in Plant Science,* **12:** 634888.

Zheng G, Fan C, Di S, Wang X, Xiang C & Pang Y. 2017. Over-expression of Arabidopsis EDT1 gene confers drought tolerance in alfalfa (*Medicago sativa* L.). *Frontiers in Plant Science,* **8:** 2125.

Zhu M, Monroe JG, Suhail Y, Villiers F, Mullen J, Pater D, et al. 2016. Molecular and systems approaches towards drought-tolerant canola crops. *New Phytologist,* **210(4):** 1169–1189.

Index